Linear and Multilinear Algebra and Function Spaces

CONTEMPORARY MATHEMATICS

750

Centre de Recherches Mathématiques Proceedings

Linear and Multilinear Algebra and Function Spaces

International Conference
Algebra and Related Topics (ICART 2018)
July 2–5, 2018
Mohammed V University, Rabat, Morocco

A. Bourhim
J. Mashreghi
L. Oubbi
Z. Abdelali
Editors

Editorial Committee of Contemporary Mathematics

Dennis DeTurck, Managing Editor

Michael Loss   Kailash Misra   Catherine Yan

Editorial Committee of the CRM Proceedings and Lecture Notes

Vašek Chvatal   Lisa Jeffrey   Nicolai Reshetikhin
Hélène Esnault   Ram Murty   Christophe Reutenauer
Pengfei Guan   Robert Pego   Nicole Tomczak-Jaegermann
Véronique Hussin   Nancy Reid   Luc Vinet

2010 *Mathematics Subject Classification.* Primary 15-XX, 16-XX, 30-XX, 46-XX, 47-XX.

Library of Congress Cataloging-in-Publication Data

Names: International Conference on Algebra and Related Topics (2018 : Rabat, Morocco), author. | Bourhim, Abdellatif, 1972– editor. | Mashreghi, Javad, editor. | Oubbi, Lahbib, 1960– editor. | Abdelali, Zine El Abidine, 1974– editor.
Title: Linear and multilinear algebra and function spaces : International Conference on Algebra and Related Topics: Linear and Multilinear Algebra and Function Spaces, July 2–5, 2018, Université Mohammed V, Rabat, Marocco / A. Bourhim, J. Mashreghi, L. Oubbi, Z. Abdelai, editors.
Description: Providence, Rhode Island : American Mathematical Society, [2020] | Series: Contemporary mathematics, 0271-4132 ; volume 750 | Includes bibliographical references.
Identifiers: LCCN 2019051607 | ISBN 9781470446932 (paperback) | ISBN 9781470456078 (ebook)
Subjects: LCSH: Algebras, Linear–Congresses. | Multilinear algebra–Congresses. | Function spaces–Congresses. | AMS: Linear and multilinear algebra; matrix theory. | Associative rings and algebras For the commutative case, see 13-XX. | Functions of a complex variable For analysis on manifolds, see 58-XX. | Functional analysis For manifolds modeled on topological linear spaces, see 57Nxx, 58Bxx. | Operator theory.
Classification: LCC QA184.2 .I58 2020 | DDC 512/.5–dc23
LC record available at https://lccn.loc.gov/2019051607
Contemporary Mathematics ISSN: 0271-4132 (print); ISSN: 1098-3627 (online)
DOI: https://doi.org/10.1090/conm/750

Copying and reprinting. Individual readers of this publication, and nonprofit libraries acting for them, are permitted to make fair use of the material, such as to copy select pages for use in teaching or research. Permission is granted to quote brief passages from this publication in reviews, provided the customary acknowledgment of the source is given.

Republication, systematic copying, or multiple reproduction of any material in this publication is permitted only under license from the American Mathematical Society. Requests for permission to reuse portions of AMS publication content are handled by the Copyright Clearance Center. For more information, please visit www.ams.org/publications/pubpermissions.

Send requests for translation rights and licensed reprints to reprint-permission@ams.org.

© 2020 by the American Mathematical Society. All rights reserved.
The American Mathematical Society retains all rights
except those granted to the United States Government.
Printed in the United States of America.

∞ The paper used in this book is acid-free and falls within the guidelines
established to ensure permanence and durability.
Visit the AMS home page at https://www.ams.org/

10 9 8 7 6 5 4 3 2 1   25 24 23 22 21 20

Contents

Preface	vii
Biduality in weighted spaces of analytic functions Christopher Boyd and Pilar Rueda	1
Interpolation with functions in the analytic Wiener algebra O. El-Fallah and K. Kellay	9
Jordan isomorphisms as preservers Lajos Molnár	19
Multiplicatively pseudo spectrum-preserving maps Zine El Abidine Abdelali and Hamid Nkhaylia	43
On algebraic characterizations of advertibly complete algebras Martin Weigt and Ioannis Zarakas	71
Polar decomposition, Aluthge and mean transforms Fadil Chabbabi and Mostafa Mbekhta	89
Recent progress on local spectrum-preserving maps Abdellatif Bourhim and Javad Mashreghi	109
Sadovskii-type fixed point results for edge-preserving mappings M. R. Alfuraidan and N. Machrafi	153
The joint numerical radius on C^*-algebras Mohamed Mabrouk	163
Weighted composition operators on non locally convex weighted spaces with operator-valued weights Mohammed Klilou and Lahbib Oubbi	175
Zero product preserving maps on matrix rings over division rings Matej Brešar and Peter Šemrl	195

Preface

The International Conference on Algebra and Related Topics (ICART 2018) was held at Faculty of Sciences, Mohammed V University in Rabat, Morocco, from July 2 to 5, 2018. It covered various research areas presented in three parallel sessions. "Linear and Multilinear Algebra, and Function spaces (in short LMAFS)" was one of them. The topics of interest of this session included linear and nonlinear preserver problems, Banach algebras, topological algebras, operator theory, and weighted function spaces. Numerous international experts in these areas presented their ongoing research, and interacted with their colleagues and Ph.D students working in their fields.

Linear preserver problems demand the characterization of linear maps between algebras that leave invariant certain properties or subsets or relations. The earliest result on linear preserver problems was established by Frobenius in 1896. Frobenius characterized all bijective linear transformations on the algebra $M_n(\mathbb{C})$ of complex matrices that preserve the determinant of matrices. His result was generalized in 1925 by Schur for subdeterminants of a fixed order and in 1949 by Dieudonné to arbitrary fields and for linear maps preserving the set of singular matrices. Since then, various linear preserver problems have been considered and a number of techniques have been developed to treat them. One of the most intractable unsolved problems in this active research area is the famous Kaplansky's conjecture that asserts that every surjective unital invertibility preserving linear map between two semisimple Banach algebras is a Jordan homomorphism. This conjecture has not been fully solved yet and remains open even for general C^*-algebras, but it has been confirmed, in particular, for von Neumann algebras and for the algebra of all bounded linear operators on a Banach space.

More recently, there has been an upsurge of interest in nonlinear preservers, where the maps studied are no longer assumed linear but instead a weak algebraic condition is somehow involved through the preserving property. The well-known theorem of Gleason-Kahane-Żelazko in the theory of Banach algebras states that every unital invertibility preserving linear map from a Banach algebra to a semisimple commutative Banach algebra is multiplicative. This ressult has been generalized in many directions. In particular, a number of techniques have been developed to treat nonlinear preservers related to various fields such as Algebra, Analysis, Functional Analysis, Geometry, Linear Algebra, Mathematical Physics and Operator Theory

As to the weighted spaces, they have been the subject matter of a lot of work over the last few decades. The weighted spaces of scalar-valued continuous functions appeared first in the work of Nachbin, specially in connection with approximation theory. But, over the years, weighted spaces and algebras of holomorphic or harmonic functions have also been studied. Such spaces have provided a general setting for the study of several function spaces encountered in analysis (e.g., in distributions and measure theory). Therefore, they have been investigated in several directions, mainly, in the approximation theory, in the theory of composition and multiplication operators, in connection with duality problems, with embedding problems into sequence spaces, with dynamical systems, with evolution equations, and so on.

This proceedings volume is the outcome of LMAFS, which brought mathematicians from various areas who are either working on or interested in preserver problems and/or function spaces. The editors would like to thank the Faculty of Sciences of Rabat for hosting this event and for its generous financial support for the invited speakers. The last day and the closing day of LMAFS was hosted by l'Ecole Normale Suprieure de Rabat. The editors would also like to thank l'Ecole Normale Suprieure, the Centre Régional des Métiers de l'Enseignement et de Formation, IMU-CDC, the Clay Mathematics Institute, the Centre National de Recherche Scientifique et Technique, Acadmie Hassan 2, and the Rabat City Hall for their financial support.

Rabat, Morocco
Summer 2018

A. Bourhim
J. Mashreghi
L. Oubbi
Z. Abdelali

Biduality in weighted spaces of analytic functions

Christopher Boyd and Pilar Rueda

ABSTRACT. We study new conditions for non necessarily radial weights implying that the weighted Banach space $\mathcal{H}_v(U)$ of analytic functions f such that vf is bounded on U, is canonically isometrically isomorphic to the bidual of $\mathcal{H}_{v_o}(U)$, its closed subspace formed by those functions f such that vf converges to 0 on the boundary of U. We provide several examples of weights that satisfy these conditions. As an application, we show that whenever $\mathcal{H}_v(U) = \mathcal{H}_{v_o}(U)''$ the norm-attaining functions are dense in $\mathcal{H}_v(U)$.

Introduction

The paper deals with the Biduality Problem, that addresses the question as to when the weighted Banach space $\mathcal{H}_v(U)$ of analytic functions f such that vf is bounded on U, is canonically isometrically isomorphic to the bidual of the space $\mathcal{H}_{v_o}(U)$, formed by all analytic functions f such that vf converges to 0 on the boundary of U.

Let us start by giving the basic definitions and notation used throughout the paper. Let U be an open subset of $\mathbf{C}^n$. A function $v : U \to (0, \infty)$ is called a weight if it is bounded, strictly positive and continuous. By $\mathcal{H}_v(U)$ we denote the Banach space of all holomorphic functions f on U such that $\|f\|_v := \sup_{z \in U} v(z)|f(z)| < \infty$, endowed with the weighted norm $\|\cdot\|_v$. The closed subspace $\mathcal{H}_{v_o}(U)$ is formed by all f in $\mathcal{H}_v(U)$ with the property that $|f(z)|v(z)$ converges to 0 as z converges to the boundary of U, i.e. given $\epsilon > 0$ there is a compact subset K of U such that $v(z)|f(z)| < \epsilon$ for z in $U \setminus K$.

The functions in $\mathcal{H}_v(U)$ satisfy a growth rate of order $O(1/v(z))$ while the growth rate of all functions in $\mathcal{H}_{v_o}(U)$ is of order $o(1/v(z))$. The following question, known as the Biduality Problem, has been a subject of much research over the last 50 years: '*Is $\mathcal{H}_v(U)$ isometrically isomorphic, in a canonical way, to the bidual of $\mathcal{H}_{v_o}(U)$?*' To date no infinite dimensional counterexample to this question has been found. Moreover, most of the known positive examples involve radial weights. Let U be balanced, i.e. $U = \bigcup_{|\lambda| \leq 1} \lambda U$, a weight $v : U \to (0, \infty)$ is *radial* if $v(\lambda x) = v(x)$ for all $x \in U$ and all complex numbers λ of modulus 1. Let us recall what we mean by canonically isometrically isomorphic. Let $\mathcal{G}_v(U)$ denote the space

---

2010 *Mathematics Subject Classification.* Primary 46E15, 32A37; Secondary 47B38.

Key words and phrases. weighted Banach spaces, biduality problem, spaces of analytic functions.

The second author was supported by Ministerio de Economía, Industria y Competitividad and FEDER under project MTM2016-77054-C2-1-P.

of all linear functionals on $\mathcal{H}_v(U)$ whose restriction to the closed unit ball of $\mathcal{H}_v(U)$ is continuous for the compact open topology, τ_o. The space $\mathcal{G}_v(U)$ is endowed with the topology inherited from $\mathcal{H}_v(U)'$. Consider the restriction mapping $\tilde{R} \colon \mathcal{G}_v(U) \to \mathcal{H}_{v_o}(U)'$ given by $\tilde{R}(\phi) = \phi|_{\mathcal{H}_{v_o}(U)}$, and the evaluation mapping $\tilde{\Phi} \colon \mathcal{H}_v(U) \to \mathcal{G}_v(U)'$ given by $\tilde{\Phi}(f)(F) = F(f)$ for $f \in \mathcal{H}_v(U)$ and $F \in \mathcal{G}_v(U)$. It is known from [**BS**, Theorem 1.1] that $\tilde{\Phi}$ is always an isomorphism, while $\tilde{R}$ is an isometric isomorphism if and only if the closed unit ball of $\mathcal{H}_{v_o}(U)$ is dense in the closed unit ball of $\mathcal{H}_v(U)$ for the compact open topology. We shall say that $\mathcal{H}_{v_o}(U)''$ is canonically isometrically (topologically) isomorphic to $\mathcal{H}_v(U)$, if the mapping $\tilde{\Phi}^{-1} \circ \tilde{R}^t$ is an isometric (a topological) isomorphism from $\mathcal{H}_{v_o}(U)''$ onto $\mathcal{H}_v(U)$.

Anderson and Duncan [**AD**] provided some particular examples of non-radial weights where $\mathcal{H}_{v_o}(\mathbb{C})'' = \mathcal{H}_v(\mathbb{C})$ and they also gave a negative answer to the biduality problem for some finite dimensional weighted spaces of entire functions. In infinite dimensional weighted spaces of holomorphic functions, a negative example has not yet been found, while several examples with a positive solution have been provided. Here are some of these contributions for infinite dimensional weighted spaces of analytic functions.

• Rubel and Shields [**RS**] in 1970 provided a positive solution to the biduality problem whenever v is a decreasing radial weight on the open unit disc Δ of $\mathbb{C}$, which vanishes on the unit circle. Williams [**Wi**] proved a related result for entire functions.

• Anderson and Duncan [**AD**] also found positive solutions for radial weights v on the complex plane which satisfy the condition $\lim\limits_{r \to \infty} \dfrac{v(r)}{v(tr)} = 0$ for all t in $(0,1)$.

• Bierstedt and Summers [**BS**] showed that a necessary and sufficient condition for a positive solution to the biduality problem is that the closed unit ball of $\mathcal{H}_{v_o}(U)$ is dense in the closed unit ball of $\mathcal{H}_v(U)$ for the compact-open topology. In particular, when $U \subset \mathbb{C}^n$ is balanced and open and $v \colon U \to (0, \infty)$ is radial, then $\mathcal{H}_v(U) = \mathcal{H}_{v_o}(U)''$ whenever $\mathcal{H}_{v_o}(U)$ contains the polynomials. Taking U to be the open unit disc of $\mathbb{C}$, this result recovers the results of Rubel and Shields, [**RS**], along with those of Anderson and Duncan, [**AD**].

• Petunin and Plichko [**PP**] proved that, given a separable Banach space X, any separating closed subspace of X' of norm attaining functionals on X, is an isometric predual of X. This result was rediscovered by Godefroy [**G2**] and used in [**G3**] to prove that $\mathcal{H}_{v_o}(U)'' = \mathcal{H}_v(U)$ whenever $\mathcal{H}_{v_o}(U)$ separates $G_v(U)$.

• Bierstedt, Bonet and Taskinen [**BBT**] stated that the biduality problem has a positive solution whenever $v \colon U \subset \mathbb{C}^n \to (0, \infty)$ is an arbitrary weight whose associated weights $\tilde{v}$ and $\tilde{v}_o$ coincide. Unfortunately, their proof contained a gap that has not been fixed so far. The associated weights $\tilde{v}$ and $\tilde{v}_o$ are defined by

$$\tilde{v}(z) = \frac{1}{\|\delta_z\|_{\mathcal{H}_v(U)'}} \quad \text{and} \quad \tilde{v}_o(z) = \frac{1}{\|\delta_z\|_{\mathcal{H}_{v_o}(U)'}}$$

for $z \in U$.

• In [**BR5**], the authors uncovered a relationship between the biduality problem and the M-ideal structure of weighted spaces of analytic functions. Given a Banach space E let B_E denote its closed unit ball. A subspace J of a Banach space E is said to be an M-ideal in E if $E' = J' \oplus_1 J^\perp$ (see [**HWW**] for the basics on M-ideals). Indeed, the authors proved Bierstedt, Bonet and Taskinen's result in the M-ideal setting: when v is a weight on an open subset U of $\mathbf{C}^n$ then $\mathcal{H}_v(U)$ is canonically the bidual of $\mathcal{H}_{v_o}(U)$ if and only if $\mathcal{H}_{v_o}(U)$ is an M-ideal in $\mathcal{H}_v(U)$ and we have equality of the associated weights $\tilde{v}_o$ and $\tilde{v}$. The authors also proved that $\mathcal{H}_v(U)$ is canonically the bidual of $\mathcal{H}_{v_o}(U)$ if and only if $\mathcal{H}_{v_o}(U)$ is an M-ideal in $\mathcal{H}_v(U)$ and the v-boundary of U, defined as those $z \in U$ such that $v(z)\delta_z$ is an extreme point of the closed unit ball of $\mathcal{H}_{v_o}(U)'$ and denoted by $\mathcal{B}_v(U)$, is a determining set for $\mathcal{H}_v(U)$. The condition that $\mathcal{B}_v(U)$ is a determining set for $\mathcal{H}_v(U)$ is trivially satisfied when $\mathcal{B}_v(U) = U$. Weights with this property are said to be complete. See [**BR2**] for examples of complete weights. As a consequence, we get that if v is a complete weight then $\mathcal{H}_v(U)$ is canonically the bidual of $\mathcal{H}_{v_o}(U)$ if and only if $\mathcal{H}_{v_o}(U)$ is an M-ideal in $\mathcal{H}_v(U)$.

It is worth mentioning that Werner [**W**] proved that $\mathcal{H}_{v_o}(\Delta)$ is an M-ideal in $\mathcal{H}_v(\Delta)$ whenever v is a radial weight on the open unit disc Δ of $\mathbb{C}$ that converges to 0 on the unit circle. Actually, it is proved in [**BR5**] that $\mathcal{H}_{v_o}(U)$ is an M-ideal in its bidual, for arbitrary weights v on open subsets of $\mathbb{C}^n$. Therefore, Werner's example also follows from the fact that $\mathcal{H}_{v_o}(\Delta)'' = \mathcal{H}_v(\Delta)$ for a radial weight v converging to 0 on the unit circle.

While there is an extensive literature on weighted spaces for radial weights, that for non radial weights is more illusive. This was realised by Bierstedt and Summers, [**BS**], who refer to some partial results in [**AD**] and indicate some of the difficulties that can arise when dealing with non radial weights. The examples that appear in [**AD**] are however finite dimensional. Another reference to non radial weights can be found in [**B**]. There it says that not too much is known about non radial weights. Moreover the examples that do arise are usually images under the Riemann Mapping Theorem of natural weights on halfplanes, strips or more general simply connected open domains. In that paper there appears a result on the biduality of weighted spaces of holomorphic functions on halfplanes credited to Holtmanns. The duality criterion in [**BS**] and the one in [**G3**] are valid for general weights. However they have been only applied to radial weights. In [**BR5**] the authors established necessary and sufficient conditions for a positive solution to the Biduality Problem for complete weights. It is worth mentioning that radial weights and complete weights are different, non-disjoint classes of weights: there are radial non complete weights and there are complete non radial weights.

In this paper we provide further examples of weighted spaces of holomorphic functions where there is a positive solution to the Biduality Problem. Our approach in finding these new examples will be to use the density criterion given in [**BS**], and provide examples of non-radial weights where the closed unit ball of $\mathcal{H}_{v_o}(U)$ is dense in the closed unit ball of $\mathcal{H}_v(U)$ for the compact-open topology and so, where the biduality problem has a positive solution. Finally, we shall see that a positive solution to the Biduality Problem leads us to a positive solution to the problem of when the norm-attaining functions are dense in $\mathcal{H}_v(U)$.

We have briefly mentioned the v-boundary of an open set $U \subset \mathbb{C}^n$. Some more details are in order. Let v be a weight on U. It is shown in [**BR1, BR4**] that the extreme points of $B_{\mathcal{H}_{v_o}(U)'}$ are contained in the set $\{\lambda v(z)\delta_z : z \in U, |\lambda| = 1\}$. Further, $v(z)\delta_z$ is an extreme point of $B_{\mathcal{H}_{v_o}(U)'}$ if and only if $\lambda v(z)\delta_z$ is an extreme point of $B_{\mathcal{H}_{v_o}(U)'}$ for every λ in $\mathbf{C}$ of modulus 1. Thus there is a distinguished subset U, called the v-boundary of U and denoted by $\mathcal{B}_v(U)$, formed by those $z \in U$ such that $v(z)\delta_z$ is an extreme point of $B_{\mathcal{H}_{v_o}(U)'}$. In other words, the set of all extreme points of $B_{\mathcal{H}_{v_o}(U)'}$ is given by $\{\lambda v(z)\delta_z : z \in \mathcal{B}_v(U), |\lambda| = 1\}$. See [**BR1**], [**BR2**] and [**BR3**] for a more detailed discussion on the properties of the v-boundary.

1. New solutions to the Biduality Problem

For weights such as $v(z) = (1+|z|^n)^{-1}$ or $v(z) = (1+|z|^n)^{-1}|e^z|^{-1}$, $n > 0$, on the complex plane we observe that $\mathcal{H}_{v_o}(\mathbf{C})$ and $\mathcal{H}_v(\mathbf{C})$ are finite dimensional spaces with the dimension of $\mathcal{H}_{v_o}(\mathbf{C})$ strictly less than that of $\mathcal{H}_v(\mathbf{C})$. Hence $\mathcal{H}_v(\mathbf{C})$ cannot be the bidual of $\mathcal{H}_{v_o}(\mathbf{C})$. Thus the condition that $\mathcal{H}_{v_o}(U)$ contains all polynomials is necessary for a successful solution to the Biduality Problem. See [**AD**] and [**BS**] for more details. We know of no infinite dimensional weighted space of holomorphic functions $\mathcal{H}_v(U)$ for which we have a negative solution to the Biduality Problem. In this section we provide further examples where we have $\mathcal{H}_{v_o}(U)'' = \mathcal{H}_v(U)$. We shall consider the additional assumption that our weight, $v\colon U \to \mathbf{R}^+$, converges to 0 on the boundary of U, i.e. given $\epsilon > 0$ there is a compact subset K of U such that $v(z) < \epsilon$ for z in $U \setminus K$. The closure of a set A is denoted $\overline{A}$. To accomplish our purpose, we need to introduce some definitions.

DEFINITION 1.1. *Let $0 \leq t < 1$. Let U be an open subset of $\mathbf{C}^n$ and $\{K_k\}_{k\in(0,1)}$ be a family of compact subsets such that $\bigcup_{k\in[0,1)} K_k = U$. A family of holomorphic functions, $(g_r)_{r\in[t,1)}$, on U is said to be holomorphically subordinate to the family of sets $\{K_k\}_{k\in[t,1)}$ if*

(1) *each $\overline{g_r(U)}$ is a compact subset of U,*
(2) *$g_r(K_k) \subseteq K_k$ for all r in $[t,1)$ and all k in $[0,1)$,*
(3) *g_r converges to the identity I, $I(x) = x$, uniformly on compact subsets of U as r tends to 1.*

DEFINITION 1.2. *Let U be an open subset of $\mathbf{C}^n$ and $v\colon U \to \mathbf{R}$ be a weight which converges to 0 on the boundary of U. For $k \in [0,1)$ consider the compact set $K_k := \{z : v(z) \geq (1-k)\max_{x\in U} v(x)\}$. A family of functions $(g_r)_{r\in[t,1)}$ defined on U is said to be holomorphically subordinate to the weight v if it is holomorphically subordinate to the family of compact sets $(K_k)_{k\in[0,1)}$.*

We shall use $B_r(x)$ to denote the closed ball in $\mathbf{C}^n$ of radius r centred at x.

THEOREM 1.3. *Let U be an open subset of $\mathbf{C}^n$ and $v\colon U \to \mathbf{R}$ be a weight which converges to 0 on the boundary of U. Suppose that there is a family of functions, $(g_r)_{r\in[t,1)}$, on U which is holomorphically subordinate to the weight v then $\mathcal{H}_v(U)$ is isometrically isomorphic in a canonical way to the bidual of $\mathcal{H}_{v_o}(U)$.*

PROOF. By [**BS**, Theorem 1.1] it suffices to show that the closed unit ball of $\mathcal{H}_{v_o}(U)$ is dense in the closed unit ball of $\mathcal{H}_v(U)$ for the compact-open topology. Let f belong to the closed unit ball of $\mathcal{H}_v(U)$.

Let $r \in [t,1)$. By continuity f is bounded on $g_r(U)$ and since v converges to 0 on the boundary of U, we conclude that the function $f \circ g_r$ belongs to $\mathcal{H}_{v_o}(U)$.

Let us prove that $f \circ g_r$ converges uniformly to f on compact subsets of U as r tends to 1. Given a compact subset K of U, consider $\epsilon_o > 0$ such that $K + \epsilon_o B_1(0) \subseteq U$. Let $\epsilon > 0$. Since f is uniformly continuous on $K + \epsilon_o B_1(0)$ there is $0 < \delta < \epsilon_o$ such that $|f(z) - f(z')| < \epsilon$ for every z, z' in $K + \epsilon_o B_1(0)$ with $|z - z'| < \delta$. Since g_r converges to I on K, there exists r_o such that for any $r > r_o$ we have $|g_r(z) - z| < \delta$ for all $z \in K$. Hence

$$g_r(z) \in K + \delta B_1(0) \subset K + \epsilon_o B_1(0)$$

for all $z \in K$, all $r > r_o$ and thus $|f(g_r(z)) - f(z)| < \epsilon$ for all $z \in K$ all $r > r_o$.

Let z be an arbitrary point of U. Then $v(z) = (1-k)\max_{x \in U} v(x)$ for some k in $[0, 1)$ and so z belongs to K_k. For each $r \in [t, 1)$ we have that $g_r(z)$ also belongs to K_k. It follows from the Maximum Modulus Theorem that we can find w in the boundary of K_k with $|f \circ g_r(z)| \leq |f(w)|$. By the continuity of v we have that $v(w) = (1-k)\max_{x \in U} v(x)$. Therefore we have

$$v(z)|f \circ g_r(z)| \leq v(z)|f(w)| = v(w)|f(w)| \leq 1$$

and hence for each r in $[t, 1]$, $f \circ g_r$ belongs to the closed unit ball of $\mathcal{H}_{v_o}(U)$.

Since $f \circ g_r$ converges uniformly to f on compact subsets of U as r tends to 1 the result follows. □

Given an open subset A of $\mathbf{C}^n$ let $\text{eq}(A) := \bigcup_{0 \leq r \leq 1} rA$ denote the $\mathbb{R}$-balanced hull of A. We shall say that A is star-shaped with respect to the origin if $\text{eq}(A) = A$.

THEOREM 1.4. *Let U be a bounded open subset of $\mathbf{C}^n$ which is star-shaped with respect to the origin and $v \colon U \to \mathbf{R}$ be a weight which converges to 0 on the boundary of U. Set $K_k = \{z \in U : v(z) \geq (1-k)\max_{x \in U} v(x)\}$. If the boundary of $\text{eq}(K_k)$ is contained in the boundary of K_k for each $k \in [0, 1)$, then $\mathcal{H}_v(U)$ is isometrically isomorphic in a canonical way to the bidual of $\mathcal{H}_{v_o}(U)$.*

PROOF. For $r \in [t, 1)$, let $g_r(z) = rz$. Let z be an arbitrary point of U and k in $[0, 1)$ be such that $v(z) = (1-k)\max_{x \in U} v(x)$. Since the boundary of $\text{eq}(K_k)$ is contained in the boundary of K_k, it follows from the Maximum Modulus Theorem that for each f in the unit ball of $\mathcal{H}_v(U)$ and each r in $[t, 1)$ there is w in the boundary of K_k so that $|f \circ g_r(z)| \leq |f(w)|$. Thus we obtain that $f \circ g_r$ has norm at most 1. Since $g_r \to I$ on compact subsets of U we see that $f \circ g_r$ converges uniformly to f on compact subsets of U. Therefore by Theorem 1.3 $\mathcal{H}_v(U)$ is isometrically isomorphic to the bidual of $\mathcal{H}_{v_o}(U)$. □

DEFINITION 1.5. Let U be a bounded open subset of $\mathbf{C}^n$ which is star-shaped with respect to the origin and $v \colon U \to \mathbf{R}$ be a weight which maximizes at 0 and which converges to 0 on the boundary of U. We shall say that v is star-shaped if $K_k = \{z \in U : v(z) \geq (1-k)v(0)\}$ is star-shaped for every k in $[0, 1)$.

When v is star-shaped, $\text{eq}(K_k) = K_k$ for each k in $[0, 1)$ and hence we obtain the following corollary to Theorem 1.4.

COROLLARY 1.6. *Let U be a bounded open subset of $\mathbf{C}^n$ which is star-shaped with respect to the origin and $v \colon U \to \mathbf{R}$ be a star-shaped weight which converges to 0 on the boundary of U. Then $\mathcal{H}_v(U)$ is isometrically isomorphic in a canonical way to the bidual of $\mathcal{H}_{v_o}(U)$.*

In particular we have:

THEOREM 1.7. *Let U be a bounded open subset of $\mathbf{C}^n$ which is star-shaped with respect to the origin. Let $v \colon U \to \mathbf{R}$ be a weight which converges to 0 on the boundary of U such that v is decreasing on each ray of U centred at 0. Then $\mathcal{H}_v(U)$ is isometrically isomorphic in a canonical way to the bidual of $\mathcal{H}_{v_o}(U)$.*

Setting $z = x + iy$, Theorem 1.7 gives us that each of the following weights gives a positive solution to the Biduality Problem on Δ.

(a) $v(z) = (1 - |z|^2)(1 - xy)$,

(b) $v(z) = (1-|z|^2)(1-x^2y)$,
(c) $v(z) = (1-|z|^2)(1-e^x xy)$,
(d) $v(z) = (1-|z|^2)(1-e^x x^2 y)$,
(e) $v(z) = (1-|z|^2)(1-x^2 e^y y)$,
(f) $v(z) = (1-|z|^2)(1-x^{2n}-y^{2m})$, $n,m \in \mathbf{N}$,
(g) $v(z) = (1-|z|^2)(1-x^{2n})(1-y^{2m})$, $n,m \in \mathbf{N}$.

Theorem 1.7 also shows that the weights

(h) $v(z) = (1-x^{2n})(1-y^{2m})$, $n,m \in \mathbf{N}$,
(i) $v(z) = (1-x^2)(1-y^2)(1-xy)$,

on the square $(-1,1) \times (-1,1)$ will give us examples where we have $\mathcal{H}_v((-1,1) \times (-1,1)) = \mathcal{H}_{v_o}((-1,1) \times (-1,1))''$.

The weights

(j) $v(z) = (1-x^2)(1-y^2)(1-x^2 e^y)$,
(k) $v(z) = (1-x^2)(1-y^2)(1-x^2 e^y y)$,
(l) $v(z) = \text{dist}(z, \text{bdry}(U))$ on any star-shaped domain U

will also give spaces of holomorphic functions (taking U as the set where v is strictly positive in cases j and k) where $\mathcal{H}_{v_o}(U)''$ is isometrically isomorphic to $\mathcal{H}_v(U)$.

Appealing to Theorem 1.4 directly we get that for $\alpha, \beta > 0$, $0 < \gamma < 1/2$, the weights

(m) $v(z) = \frac{1}{1+(\alpha x^2+\beta y^2-1)^2} - \gamma$
(n) $v(z) = \frac{1}{1+(\sqrt{\alpha x^2+\beta y^2}-1)^2} - \gamma$
(o) $v(z) = 10 - (x^2+2y^2-0.5)(x^2+2y^2-1)(x^2+2y^2-1.5)(x^2+2y^2-2)$,

defined on the open set U where v is strictly positive, have the property that $\mathcal{H}_v(U)$ is isometrically isomorphic in a canonical way to the bidual of $\mathcal{H}_{v_o}(U)$.

2. Density of norm-attaining functions in $\mathcal{H}_v(U)$

A positive solution to the biduality problem leads to a positive solution to a number of other questions. The example of this which we will concern ourselves with in this section is the density of the norm-attaining functions in $\mathcal{H}_v(U)$. Given an open subset U of $\mathbf{C}^n$ and a weight $v \colon U \to \mathbf{C}$ we shall say that f in $\mathcal{H}_v(U)$ attains its norm if there is z in U with $v(z)|f(z)| = \|f\|_v$. When we refer to the norm-attaining functions being dense in $\mathcal{H}_v(U)$ we shall mean that they are dense with respect to the norm topology on $\mathcal{H}_v(U)$.

A subset D of a Banach space E is said to be dentable if for every $\epsilon > 0$ there is x in D which does not belong to the closed convex hull of $D \setminus B_\epsilon(x)$. A subset H of a Banach space E is said to be hereditarily dentable if every closed convex subset of H is dentable. A Banach space E has the Radon-Nikodýn property if and only if every bounded subset of E is dentable (see [**DU**, Theorem 7 p.136]). We shall need the following Theorem, due to Stegall, which is stated in [**S**, Theorem 18].

THEOREM 2.1. (Stegall)[**S**] *Let X be a complex Banach space and D a bounded, closed convex, circled and hereditarily dentable subset of X. Let C be a closed subset of D. Then the set*

$$\{x^* \in X' : x^* \text{ attains its supremum of modulus on } C\}$$

contains a dense G_δ subset of X^*.

THEOREM 2.2. *Let U be an open subset of $\mathbf{C}^n$ and $v\colon U \to \mathbf{R}$ be a weight. Suppose that $\mathcal{H}_v(U)$ is isometrically isomorphic to the bidual of $\mathcal{H}_{v_o}(U)$. Then the set of functions in $\mathcal{H}_v(U)$ which attain their norm is dense in $\mathcal{H}_v(U)$.*

PROOF. Since U is separable and the mapping $z \to v(z)\delta_z$ is continuous the set $\{v(z)\delta_z : z \in U\}$ is separable. Hence, its closed linear span, $\mathcal{H}_{v_o}(U)'$, is also separable. Thus $\mathcal{H}_{v_o}(U)'$ is a separable dual space and so has the Radon-Nikodým property (see Theorem 1 in page 79 of [**DU**]).

Let f belong to $\mathcal{H}_v(U)$. If $f = 0$ then clearly f attains its norm. Therefore we may assume that $f \neq 0$. Let $\epsilon < \|f\|_v/2$. Applying Theorem 2.1 with $X = \mathcal{H}_{v_o}(U)'$, $D = B_{\mathcal{H}_{v_o}(U)'}$ and $C = \{\lambda v(z)\delta_z : |\lambda| = 1, z \in U\} \cup \{0\}$ we obtain g in $\mathcal{H}_v(U)$ such that $\|f - g\|_v < \epsilon$ and g attains its norm at some point of
$$\{\lambda v(z)\delta_z : |\lambda| = 1, z \in U\} \cup \{0\}.$$
However, as $\|f\|_v > \epsilon > \|f - g\|_v$, g cannot attain its norm at the linear functional 0 in $\mathcal{H}_{v_o}(U)'$ and thus there is z_o in U where g attains it norm. □

In particular from Theorem 1.4 we get:

COROLLARY 2.3. *Let U be a bounded open subset of $\mathbf{C}^n$ which is star-shaped with respect to the origin and $v\colon U \to \mathbf{R}$ be a weight which converges to 0 on the boundary of U. If the boundary of $\mathrm{eq}(K_k)$ is contained in the boundary of K_k for each $k \in [0,1)$, then the set of all f in $\mathcal{H}_v(U)$ which attain their norm is dense in $\mathcal{H}_v(U)$.*

Hence all the weights (a) to (o) of Section 1 give spaces of weighted holomorphic functions where the norm-attaining functions are norm dense in $\mathcal{H}_v(U)$.

The converse to Theorem 2.2 fails in general. To see this we observe that when we consider the weights $v(z) = (1 + |z|^n)^{-1}$ or $v(z) = (1 + |z|^n)^{-1}|e^z|^{-1}$ on the complex plane $\mathcal{H}_v(\mathbf{C})$ is finite dimensional. It therefore is reflexive and hence has the Radon-Nikodým property. Taking $X = \mathcal{H}_v(\mathbf{C})$, $D = B_{\mathcal{H}_v(\mathbf{C})'}$ and $C = \{\lambda v(z)\delta_z : |\lambda| = 1, z \in \mathbf{C}\} \cup \{0\}$, repeating the argument of Theorem 2.2 gives us that those functions in $\mathcal{H}_v(\mathbf{C})$ which attain their norm are dense in $\mathcal{H}_v(\mathbf{C})$. However, it follows from the discussion at the beginning of Section 1 that the Biduality Problem has a negative solution in both these cases.

Acknowledgement

This article is part of the proceedings of the International Conference on Algebra and Related Topics, held in Rabat in July 2018. The second author thanks the organizers for the excellent organization of the conference, and in particular Professor Oubbi for his warm welcome.

References

[AD]   J. M. Anderson and J. Duncan, *Duals of Banach spaces of entire functions*, Glasgow Math. J. **32** (1990), no. 2, 215–220, DOI 10.1017/S0017089500009241. MR1058534

[B]   K. D. Bierstedt, *A survey of some results and open problems in weighted inductive limits and projective description for spaces of holomorphic functions*, Bull. Soc. Roy. Sci. Liège **70** (2001), no. 4-6, 167–182 (2002). Hommage à Pascal Laubin. MR1904052

[BBT]   K. D. Bierstedt, J. Bonet, and J. Taskinen, *Associated weights and spaces of holomorphic functions*, Studia Math. **127** (1998), no. 2, 137–168. MR1488148

[BS]   K. D. Bierstedt and W. H. Summers, *Biduals of weighted Banach spaces of analytic functions*, J. Austral. Math. Soc. Ser. A **54** (1993), no. 1, 70–79. MR1195659

[BR1] C. Boyd and P. Rueda, *The v-boundary of weighted spaces of holomorphic functions*, Ann. Acad. Sci. Fenn. Math. **30** (2005), no. 2, 337–352. MR2173368

[BR2] C. Boyd and P. Rueda, *Complete weights and v-peak points of spaces of weighted holomorphic functions*, Israel J. Math. **155** (2006), 57–80, DOI 10.1007/BF02773948. MR2269423

[BR3] C. Boyd and P. Rueda, *Isometries between spaces of weighted holomorphic functions*, Studia Math. **190** (2009), no. 3, 203–231, DOI 10.4064/sm190-3-1. MR2470377

[BR4] C. Boyd and P. Rueda, *Isometries of weighted spaces of holomorphic functions on unbounded domains*, Proc. Roy. Soc. Edinburgh Sect. A **139** (2009), no. 2, 253–271, DOI 10.1017/S0308210507001230. MR2496963

[BR5] C. Boyd and P. Rueda, *The biduality problem and M-ideals in weighted spaces of holomorphic functions*, J. Convex Anal. **18** (2011), no. 4, 1065–1074. MR2917867

[DU] J. Diestel and J. J. Uhl Jr., *Vector measures*, American Mathematical Society, Providence, R.I., 1977. With a foreword by B. J. Pettis; Mathematical Surveys, No. 15. MR0453964

[G1] G. Godefroy, *Espaces de Banach: existence et unicité de certains préduaux* (French, with English summary), Ann. Inst. Fourier (Grenoble) **28** (1978), no. 3, x, 87–105. MR511815

[G2] G. Godefroy, *Boundaries of a convex set and interpolation sets*, Math. Ann. **277** (1987), no. 2, 173–184, DOI 10.1007/BF01457357. MR886417

[G3] G. Godefroy, *The use of norm attainment*, Bull. Belg. Math. Soc. Simon Stevin **20** (2013), no. 3, 417–423. MR3129049

[HWW] P. Harmand, D. Werner, and W. Werner, *M-ideals in Banach spaces and Banach algebras*, Lecture Notes in Mathematics, vol. 1547, Springer-Verlag, Berlin, 1993. MR1238713

[PP] Ju. Ī. Petunīn and A. N. Pličko, *Some properties of the set of functionals that attain a supremum on the unit sphere* (Russian), Ukrain. Mat. Ž. **26** (1974), 102–106, 143. MR0336299

[RS] L. A. Rubel and A. L. Shields, *The second duals of certain spaces of analytic functions*, J. Austral. Math. Soc. **11** (1970), 276–280. MR0276744

[S] C. Stegall, *Optimization and differentiation in Banach spaces*, Proceedings of the symposium on operator theory (Athens, 1985), Linear Algebra Appl. **84** (1986), 191–211, DOI 10.1016/0024-3795(86)90314-9. MR872283

[W] D. Werner, *New classes of Banach spaces which are M-ideals in their biduals*, Math. Proc. Cambridge Philos. Soc. **111** (1992), no. 2, 337–354, DOI 10.1017/S0305004100075447. MR1142754

[Wi] D. L. Williams, *Some Banach spaces of entire functions*, ProQuest LLC, Ann Arbor, MI, 1967. Thesis (Ph.D.)–University of Michigan. MR2616490

School of Mathematical Sciences, University College Dublin, Belfield, Dublin 4, Ireland
Email address: christopher.boyd@ucd.ie

Departamento de Análisis Matemático, Facultad de Matemáticas, Universitat de València, 46100 Burjassot, Valencia, Spain
Email address: pilar.rueda@uv.es

Interpolation with functions in the analytic Wiener algebra

O. El-Fallah and K. Kellay

ABSTRACT. Let A^+ be the analytic Wiener algebra and let $I^+(E)$ be the closed ideal of A^+ consisting of all functions in A^+ which vanish on the closed subset E of the unit circle. In this paper, we give estimates for the norm of e^{-int} in the quotient algebra $A^+/I^+(E)$, when E satisfies suitable geometric conditions. These estimates allow to give some classes of functions which can be interpolated with functions in A^+.

1. Introduction

The Wiener algebra A is the Banach algebra of continuous functions on the unit circle $\mathbb{T}$ with absolutely convergent Fourier series, equipped with the ℓ^1 norm of the Fourier coefficients. Namely,

$$A = \Big\{ f \in \mathcal{C}(\mathbb{T}) : \|f\|_A := \sum_{n \in \mathbb{Z}} |\widehat{f}(n)| < \infty \Big\}.$$

Let A^+ denote the closed subalgebra of all functions in A whose negative Fourier coefficients vanish. A closed subset E of $\mathbb{T}$ is called a ZA^+ set if there exists a nonzero function in A vanishing on E. In [2], A. Atzmon proved that if E is a ZA^+ set and if $f \in \mathcal{C}(\mathbb{T})$ is such that $|\widehat{f}(n)| = O(e^{-\varepsilon\sqrt{|n|}})$ for all $\epsilon > 0$, then $f_{|E}$ can be interpolated by a function in A^+, that is there exists $g \in A^+$ such that $f(z) = g(z)$ for all $z \in E$. More precisely, if E is a ZA^+ set, we denote by $I^+(E)$ the closed ideal of all functions in A^+ which vanish on E and by $A^+(E)$ the quotient algebra $A^+/I^+(E)$ equipped with the canonical quotient norm. Then, $A^+(E)$ can be identified with the restriction algebra $\{f_{|E} : f \in A^+\}$. If E is a ZA^+ set, then $e^{-int} + I^+(E) \in A^+(E)$. Let

$$\omega_E(n) = \|e^{int}\|_{A^+(E)} \quad (n \in \mathbb{Z}).$$

If we denote by A_{ω_E} the Beurling algebra given by

$$A_{\omega_E}(\mathbb{T}) = \Big\{ f \in \mathcal{C}(\mathbb{T}) : \sum_{n \in \mathbb{Z}} |\widehat{f}(n)|\omega_E(n) < \infty \Big\},$$

2010 *Mathematics Subject Classification.* 42A28 (30B30, 46J15).
Key words and phrases. Wiener algebra, zero sets, Carleson set.
Research partially supported by "Hassan II Academy of Science and Technology" for the first author. The second author was supported by the joint French-Russian Research Project PRC CNRS/RFBR 2017-2019 and by ANR-REPKA.

©2020 American Mathematical Society

then for every function $f \in A_{\omega_E}(\mathbb{T})$ there exists $g \in A^+$ such that $f_{|E} = g_{|E}$. See Proposition 3.1.

Atzmon proved in [2] that if E is a ZA^+ set, we have

$$(1.1) \qquad \lim_{n \to +\infty} \frac{\log \|e^{-int}\|_{A^+(E)}}{\sqrt{n}} = 0.$$

Conversely, Kahane and Katznelson proved in [15] that for every $\beta > 1$ there exists a closed set $E \subset \mathbb{T}$ which satisfies the Carleson condition

$$(1.2) \qquad \int_0^\pi N_E(t) \mathrm{d}t < \infty,$$

where N_E denotes the smallest number of closed arcs of length $2t$ cover E, as well as

$$\liminf_{n \to +\infty} \frac{\log \|e^{-int}\|_{A^+(E)} (\log n)^\beta}{\sqrt{n}} > 0.$$

Note that the condition (1.2) is equivalent to

$$(1.3) \qquad \int_{\mathbb{T}} \log \frac{1}{\mathrm{dist}(\zeta, E)} |\mathrm{d}\zeta| < \infty,$$

where $d(\zeta, E)$ denotes the distance of ζ from E. See Lemma 3.2 for equivalent conditions of Carleson type sets. Carleson proved in [6] that sets which satisfy the condition (1.2) are ZA^+ sets. In fact, condition (1.2) is a necessary and sufficient condition for the existence of nonzero analytic Lipschitz function vanishing on E. It should be mentioned that Carleson showed in [6] that there exist sets of Lebesgue measure zero that are not zero set for A^+. Kahane and Katznelson improved this result by showing in [15, Théorème 6], that there exist sets of measure zero relative to any Hausdorff measure that are not zero sets for A^+. On the other hand, Kaufman proved in [16] that there exists a perfect ZA^+ set which is not a Carleson set. Until now there is no explicit characteriztion of ZA^+ sets.

Throughout the paper, we use the following notations: $A \lesssim B$ means that there is an absolute constant C such that $A \leq CB$. We write $A \asymp B$ if both $A \lesssim B$ and $B \lesssim A$ hold.

2. Main results

Our approach in this paper is to improve (1.1), when the closed set E satisfies some suitable geometric conditions.

Let Λ be an increasing positive function on $(0, +\infty)$.

A closed subset E of $\mathbb{T}$ is said to be a Λ-Carleson set if

$$\int_{\mathbb{T}} \Lambda\left(\log \frac{1}{\mathrm{dist}(\zeta, E)}\right) |\mathrm{d}\zeta| < \infty.$$

Note that if Λ is a convex function and $t = O(\Lambda(t))$ every Λ-Carleson set is a ZA^+. In this paper we prove the following result

THEOREM 2.1. *Let Λ be a positive convex function. If E is a Λ-Carleson then*

$$\|e^{-int}\|_{A^+(E)} = O(\omega_n)(n \to +\infty),$$

where $\omega_n = \inf_{r<1} r^{-n} \exp\left[3\Lambda^{-1}\left(\frac{\varepsilon(1-r)}{1-r}\right)\right]$ *and* $\varepsilon(1-r) \to 0$ *as* $r \to 1-$.

The following corollary was obtained by Zerouali in [21]

COROLLARY 2.2. *Let E be a Λ-Carleson with $\Lambda(t) = t^p$, $p > 1$. Then*
$$\lim_{n \to +\infty} \frac{\log \|e^{-int}\|_{A^+(E)}}{n^{\frac{1}{1+p}}} = 0.$$

Note that the exponent $\frac{1}{p+1}$ in Corollary 2.2 is sharp. Indeed, for each $\gamma \in (0, \frac{1}{p+1})$, one can construct, using some arguments similar to those given in [15], a Λ-Carleson set E with $\Lambda(t) = t^p$ such
$$\liminf_{n \to +\infty} \frac{\log \|e^{-int}\|_{A^+(E)}}{n^\gamma} > 0.$$
For the proof of this results we refer to [11].

COROLLARY 2.3. *Let E be a Λ-Carleson with $\Lambda(t) = e^{\alpha t}$, $\alpha \in (0, 1)$. Then*
$$\|e^{-int}\|_{A^+(E)} = O(n^{3/\alpha}), \quad (n \to +\infty).$$

This latter corollary implies that for all $f \in \mathcal{C}^\infty(\mathbb{T})$ there exists $g \in A^+$ such that $f_{|E} = g_{|E}$.

Let E be a closed subset of $\mathbb{T}$ and let $A(E)$ be the algebras of restriction to E of elements of the Wiener algebras A. The set E is said to be AA^+-set if $A^+(E) = A(E)$. Note that E is AA^+ if and only if $\|e^{-nt}\|_{A^+(E)} = O(1)$. As examples we have:

- All closed countable sets of $\mathbb{T}$ are AA^+ sets.
- If E is a closed set such that $N_E(t) = o(\log(1/t))$, $t \to 0^+$, then E is AA^+ set.
- For $\zeta \in (0, 1/2)$, the perfect symmetric set E_ζ of constant ratio ζ where $1/\zeta$ is a Pisot number, is AA^+ set (see [14]). Recall that

$$E_\zeta = \left\{ \exp\left(2i\pi \sum_{n \geq 1} \varepsilon_n \zeta^{n-1}(1-\zeta)\right) : \varepsilon_n = 0 \text{ or } 1 \right\}.$$

In fact E_ζ is a an interpolating set for Lipschitz algebras. Such sets are called (after Kotocigov) K-sets. More precisely, we say that a closed subset E of $\mathbb{T}$ is a K–set if there exists $c_E > 0$ such that for all arcs $I \subset \mathbb{T}$
$$\frac{1}{|I|} \int_\mathbb{T} \frac{|d\zeta|}{\text{dist}(\zeta, E)^\alpha} \leq \frac{c_E}{|I|^\alpha}$$
for some $\alpha \in (0, 1)$, where $|I|$ denotes the length of I. See Lemma 3.3 for other equivalent conditions of K-sets. For other examples of K–sets see [10, 12]. Note that if E is a K-set, then E is necessarily Λ–Carleson set where $\Lambda(t) = t^\alpha$ for some $\alpha \in (0, 1)$. Let $\mathcal{A}(\mathbb{D})$ be the disc algebras and let $s \in (0, 1)$. Put
$$\Lambda_s(\mathbb{T}) = \left\{ f \in \mathcal{C}(\mathbb{T}) : \|f\|_s := \|f\|_\infty + \sup_{h \neq 0, \theta \in \mathbb{R}} \frac{|f(e^{i(\theta+h)}) - f(e^{i\theta})|}{|h|^s} \right\}$$
and $\mathcal{A}_s = \Lambda_s(\mathbb{T}) \cap \mathcal{A}(\mathbb{D})$. Let us denote
$$I_s(E) = \{f \in \Lambda_s(\mathbb{T}) : f_{|E} = 0\} \quad \text{and} \quad I_s^+(E) = I_s(E) \cap \Lambda^s,$$
and
$$\Lambda_s(E) = \Lambda_s(\mathbb{T})/I_s(E) \quad \text{and} \quad \mathcal{A}_s(E) = \mathcal{A}_s/I_s^+(E).$$

E.M. Dynkin showed in [7] that E is a set of interpolation for $\mathcal{A}_s$, that is $\Lambda_s(E) = \mathcal{A}_s(E)$, if and only if E is a K–set.

Wa have the following Esterle's result [12].

COROLLARY 2.4. *If E is a K-set then*
$$\|e^{-int}\|_{A^+(E)} = O(n^{\frac{1}{2}+\varepsilon}) \ (n \to +\infty).$$

Let E be a closed subset of $\mathbb{T}$ and suppose that there exists that $c_E > 0$ such that for all arcs $I \subset \mathbb{T}$

$$(2.1) \qquad \frac{1}{|I|} \int_I \log \frac{1}{\mathrm{dist}(\zeta, E)} |d\zeta| \leq \frac{c_E}{|I|^\alpha}$$

for some $\alpha \in (0,1)$. If E is a Λ-Carleson with $\Lambda(t) = t^p$, $p > 1$, then E satisfies (2.1) with $\alpha = 1/p$. Indeed, let I be an arc of $\mathbb{T}$. By Hölder's inequality, we have

$$(2.2) \qquad \int_I \log \frac{1}{\mathrm{dist}(\zeta, E)} |d\zeta| \leq \Big(\int_I \Lambda\Big(\frac{1}{\mathrm{dist}(\zeta, E)}\Big) |d\zeta| \Big)^{1/p} |I|^{\frac{p-1}{p}}.$$

The following result shows that a analogue of Corollary 2.2 is still valid for sets satisfaying (2.1).

PROPOSITION 2.5. *Let E be a closed subset satisfaying* (2.1), *then*
$$\|e^{-int}\|_{A^+(E)} = O(e^{cn^{\frac{\alpha}{1+\alpha}}}) \ (n \to +\infty).$$

Futhermore, if E is Λ-Carleson with $\Lambda(t) = t^{1/\alpha}$, $\alpha \in (0,1)$, then for all $\varepsilon > 0$
$$\|e^{-int}\|_{A^+(E)} = O(e^{\varepsilon n^{\frac{\alpha}{1+\alpha}}}) \ (n \to +\infty).$$

3. Technical tools

Let $\omega : \mathbb{Z} \to [1, +\infty)$ be a weight such that $\omega(0) = 1$ and $\omega(n+m) \leq \omega(n)\omega(m)$. Consider the weighted Beurling algebras

$$A_\omega(\mathbb{T}) = \Big\{ f \in \mathcal{C}(\mathbb{T}) : \sum_{n \in \mathbb{Z}} |\widehat{f}(n)| \omega(n) < \infty \Big\}$$

Clearly, $A_\omega(\mathbb{T})$ is a Banach algebra with respect to the norm

$$\|f\|_\omega = \sum_{n \in \mathbb{Z}} |\widehat{f}(n)| \omega(n).$$

Recall that if $\displaystyle\lim_{|n| \to +\infty} \frac{\log \omega(n)}{|n|} = 0$, then $\|e^{int}\|_\omega^{\frac{1}{|n|}} \to 1$ as $|n| \to \infty$. Consequently the character space of $A_\omega(\mathbb{T})$ can be identified, by

$$\chi_z(f) =: f(z), \quad (f \in A_\omega(\mathbb{T}), \ z \in \mathbb{T}),$$

to the unit circle $\mathbb{T}$.

If E is a ZA^+ set, then the character space of $A^+(E) = A^+/I^+(E)$ can be identified with E. Let π be the canonical map from A^+ onto $A^+(E)$ and let $u : e^{i\theta} \to e^{i\theta}$ for $\theta \in [0, 2\pi]$. Note that

$$\|e^{int}\|_{A^+(E)} = \|\pi(u)^n\|_{A^+/I^+(E)}, \qquad n \in \mathbb{Z}.$$

Set $\omega_E(n) = \|e^{int}\|_{A^+(E)}$ for $n \in \mathbb{Z}$. Since the spectrum of $\pi(u)$, $\sigma(\pi(u))$, is equal to E, then
$$\omega_E(n) = 1 \ (n \geq 0) \quad \text{and} \quad \lim_{|n| \to +\infty} \frac{\log \omega_E(n)}{|n|} = 0.$$
Hence A^+ can be identified isometrically as a subspace of A_{ω_E}.
We have the following classical proposition giving a relation between the asymptotic behavior of the sequence $(\|e^{-int}\|)_{n \geq 0}$ and interpolation with functions in A^+.

PROPOSITION 3.1. *If E is a ZA^+ set, then for every $f \in A_{\omega_E}(\mathbb{T})$ there exists $g \in A^+$ such that $f_{|E} = g_{|E}$.*

PROOF. Let $\Theta : A_{\omega_E}(\mathbb{T}) \to A^+(E)$ be the homomorphism defined by
$$\Theta(f) = \sum_{n \in \mathbb{Z}} \widehat{f}(n) \pi(u)^n, \qquad f \in A_{\omega_E}(\mathbb{T}).$$
For $f \in A_{\omega_E}$, there exists $g \in A^+$ such that $\Theta(f) = \pi(g)$. This implies that $\chi_z(\Theta(f)) = \chi_z \pi(g)$, for all $z \in E$. This means that $f_{|E} = g_{|E}$. and the proof is complete. $\square$

We have the following two Lemmas which give various conditions on Λ-Carleson sets and on K-sets. For the proofs we refer to [5, 9].
Let $E \subset \mathbb{T}$ we denote by
$$E_t = \{\zeta \in \mathbb{T} : \text{dist}(\zeta, E) \leq t\}.$$

LEMMA 3.2. *Let E be a closed subset of $\mathbb{T}$. Let $\Lambda : [1, \infty) \to [1, \infty)$ be a continous inceasing function such that $\Lambda(t) = o(e^t)(t \to +\infty)$. Then the following are equivalent:*

(1) $\int_{\mathbb{T}} \Lambda\big(\log 1/\text{dist}(\zeta, E)\big)|d\zeta| < \infty;$

(2) $|E| = 0$ and $\sum_j \int_0^{|I_j|/2} \Lambda(\log 1/t) dt < \infty$, where $(I_j)_j$ are the components of $\mathbb{T} \setminus E$;

(3) $\int_0^\pi |E_t| d\Lambda(\log 1/t) > -\infty$, where ;

(4) $\int_0^\pi t N_E(t) d\Lambda(\log 1/t) > -\infty$.

LEMMA 3.3. *Let E be a closed subset of $\mathbb{T}$. Then for each arc I in $\mathbb{T}$, the following are equivalent:*

(1) $\sup_{\zeta \in I} \text{dist}(\zeta, E) \leq c_1 |I|;$

(2) $\frac{1}{|I|} \int_I \log \text{dist}(\zeta, E) |d\zeta| \geq \log |I| - c_2;$

(3) $\frac{1}{|I|} \int_I \text{dist}(\zeta, E)^{-\alpha} |d\zeta| \leq c_3 |I|^{-\alpha}$ *for some* $\alpha \in (0, 1);$

(4) $\frac{1}{|I|} \int_I \text{dist}(\zeta, E)^\alpha |d\zeta| \geq c_4 |I|^\alpha$ *for some* $\alpha \in (0, 1).$

Where c_1, c_2, c_3, c_4 are constants independent of I.

Given $\lambda \in \mathbb{D}$ and $f \in A^+$, we define

$$T_\lambda f(z) = \begin{cases} \dfrac{f(z) - f(\lambda)}{z - \lambda}, & z \in \mathbb{D} \setminus \{\lambda\}, \\ f'(\lambda), & z = \lambda. \end{cases}$$

The following simple lemma is the first step to prove our estimate. We include a proof for sake of completeness.

LEMMA 3.4. T_λ is a bounded linear operator $A^+ \to A^+$ with $\|T_\lambda f\|_{A^+} \leq \|f\|_{A^+}/(1-|\lambda|)$. In addition, if $f \in I^+(E)$ then

$$f(\lambda)(\pi(u) - \lambda)^{-1} = -\pi(T_\lambda(f))$$

and $f(\lambda) \neq 0$ then

$$\|(\pi(u) - \lambda)^{-1}\|_{A^+/I^+(E)} \leq \frac{\|f\|_{A^+}}{|f(\lambda)|(1-|\lambda|)}.$$

PROOF. Note that $T_0 : A^+ \to A^+$ is a bounded linear operator and $\|T_0\| = 1$. Since $T_\lambda f - T_0 f = \lambda T_\lambda T_0 f$ for $f \in A^+$, we get $T_\lambda f = T_0(I - \lambda T_0)^{-1} f$ and $\|T_\lambda\| \leq \|f\|_{A^+}/(1-|\lambda|)$.

We now prove the second assertion. Let $f \in I^+(E)$, since

$$(z - \lambda) T_\lambda f(z) = f(z) - f(\lambda),$$

$(\pi(u) - \lambda)\pi(T_\lambda f) = -f(\lambda)$. So $f(\lambda)(\pi(u) - \lambda)^{-1} = -\pi(T_\lambda(f))$.
So, if $f(\lambda) \neq 0$, then

$$\|(\pi(u) - \lambda)^{-1}\|_{A^+(E)} \leq \frac{\|\pi(T_\lambda(f))\|_{A^+(E)}}{|f(\lambda)|} \leq \frac{\|f\|_{A^+}}{|f(\lambda)|(1-|\lambda|)}.$$

□

4. Λ-Carleson sets

We start with an upper estime of some outer functions.

LEMMA 4.1. Let E is a Λ-Carleson set. Then there exists $f \in A^+$ such that

$$\log \frac{1}{|f(z)|} = 3\Lambda^{-1}\left(\frac{o(1)}{1-|z|}\right) + O(1), \qquad |z| \to 1^-.$$

PROOF. Let f be the outer function such that

$$|f(\zeta)| = \operatorname{dist}(\zeta, E)^3, \qquad \text{a.e. on } \mathbb{T}.$$

Then $f \in A^+$ see [6, Theorem 1]. Let $1/2 < |z| = r = 1 - a < 1$ and let $I_k = [e^{ia_k}, e^{ia_{k+1}}]$ with $a_0 = 0$, $a_k = 2^k a$ ($k \geq 1$). Let N be the integer such that $2^N a \leq \pi < 2^{N+1} a$, and let $\varpi(r, I_k, \mathbb{D})$ be the harmonic measure of I_k at $r = |z|$ in $\mathbb{D}$. We have

$$f(z) = \exp \int_\mathbb{T} \frac{z + \zeta}{z - \zeta} \log \frac{1}{\operatorname{dist}(\zeta, E)^3} \frac{|d\zeta|}{2\pi}, \qquad z \in \mathbb{D}.$$

Without loss of generality, we may suppose that $2^N a = \pi$.

$$(4.1) \quad \frac{1}{3} \log \frac{1}{|f(z)|} = \sum_{k=0}^{N-1} \underbrace{\int_{I_k} \frac{1-|z|^2}{|1-z\bar\zeta|^2} \log \frac{1}{\operatorname{dist}(\zeta, E)} \frac{|d\zeta|}{2\pi}}_{\mathcal{J}_k}, \qquad z \in \mathbb{D}.$$

Since Λ is convex, by Jensen inequality,

$$\begin{aligned}
\mathcal{J}_k &= \frac{\varpi(r,I_k,\mathbb{D})}{\varpi(r,I_k,\mathbb{D})} \int_{I_k} \frac{1-|z|^2}{|1-z\bar\zeta|^2} \log \frac{1}{\mathrm{dist}(\zeta,E)} \frac{|\mathrm{d}\zeta|}{2\pi} \\
&= \varpi(r,I_k,\mathbb{D})\Lambda^{-1}\Big[\Lambda\Big(\frac{1}{\varpi(r,I_k,\mathbb{D})} \int_{I_k} \frac{1-|z|^2}{|1-z\bar\zeta|^2} \log \frac{1}{\mathrm{dist}(\zeta,E)} \frac{|\mathrm{d}\zeta|}{2\pi}\Big)\Big] \\
&\leq \varpi(r,I_k,\mathbb{D})\Lambda^{-1}\Big[\frac{1}{\varpi(r,I_k,\mathbb{D})} \int_{I_k} \frac{1-|z|^2}{|1-z\bar\zeta|^2} \Lambda\Big(\log \frac{1}{\mathrm{dist}(\zeta,E)}\Big) \frac{|\mathrm{d}\zeta|}{2\pi}\Big]
\end{aligned}$$

(4.2)

We have
$$\frac{1}{|1-z\bar\zeta|} \asymp \frac{1}{2^k}\frac{1}{1-|z|}, \qquad \zeta \in I_k,$$

and so
$$\varpi(r,I_k,\mathbb{D}) \asymp \frac{|I_k|}{2^{2k}(1-|z|)} \asymp \frac{1}{2^k}.$$

Hence, by (4.2),

(4.3) $$\mathcal{J}_k \leq \varpi(r,I_k,\mathbb{D})\Lambda^{-1}\Big[\frac{c}{2^k(1-|z|)} \int_{I_k} \Lambda\Big(\log \frac{1}{\mathrm{dist}(\zeta,E)}\Big) \frac{|\mathrm{d}\zeta|}{2\pi}\Big]$$

for some constant c independent of k. Since E is a Λ-Carleson, for all k

(4.4) $$\int_{I_k} \Lambda\Big(\log \frac{1}{\mathrm{dist}(\zeta,E)}\Big)|\mathrm{d}\zeta| = o(1), \qquad r \to 1-.$$

By (4.1), (4.3) and (4.4), we get
$$\begin{aligned}
\frac{1}{3}\log \frac{1}{|f(z)|} &\leq \sum_{k=1}^{N-1} \varpi(r,I_k,\mathbb{D})\Lambda^{-1}\Big(\frac{o(1)}{(1-|z|)}\Big) + O(1) \\
&\leq \Lambda^{-1}\Big(\frac{o(1)}{1-|z|}\Big) + O(1).
\end{aligned}$$

The proof is complete. $\square$

Proof of Theorem 2.1. Let $r \in (0,1)$. By Cauchy formula, we have
$$\pi(u)^{-n} = \frac{1}{2i\pi} \int_{|\lambda|=r} \lambda^{-n}(\lambda-\pi(u))^{-1} d\lambda, \qquad n \geq 0.$$

Therefore by Lemma 3.4 and Lemma 4.1, we get for every $\varepsilon > 0$
$$\begin{aligned}
\|\pi(u)^{-n}\|_{A^+(E)} &\leq r^{-n} \max_{|\lambda|=r} \|(\lambda-\pi(u))^{-1}\|_{A^+(E)} \\
&\lesssim \frac{|\lambda|^{-n}}{|f(\lambda)|(1-|\lambda|)}.
\end{aligned}$$

So,
$$\|\pi(u)^{-n}\|_{A^+(E)} \lesssim \inf_{0<r<1} r^{-n} \frac{\exp\big[3\Lambda^{-1}\big(\frac{\epsilon}{1-r}\big)\big]}{1-r}.$$

Proof of Corollary 2.2. If $\Lambda(t) = t^p$, $p > 1$, then $\Lambda^{-1}(t) = t^{1/p}$ and for $r_n = 1 - n^{-\frac{p}{1+p}}$ we get
$$\|\pi(u)^{-n}\|_{A^+(E)} \lesssim \inf_{0<r<1} r^{-n} \frac{\exp(\varepsilon(1-r)^{-1/p})}{1-r} = O(\exp(\varepsilon n^{\frac{1}{1+p}})).$$

Proof of Corollary 2.3. If $\Lambda(t) = e^{\alpha t}$ then $\Lambda^{-1}(t) = \frac{1}{\alpha}\log t$ and for $r_n = 1 - 1/n$ we get

$$\|\pi(u)^{-n}\|_{A^+(E)} \lesssim \inf_{0<r<1} r^{-n}\frac{1}{(1-r)^{3/\alpha}} = O(n^{3/\alpha}).$$

5. Interpolating sets

The following propostion was proved in [8] and improved in [1], see also [20]. Here we give a direct proof adapted to our case.

PROPOSITION 5.1. *Let $p \geq 1$ and let T be an invertible contraction on a Banach space X with spectrum K-set and $\|T^{-n}\| = O(n^p)$ then $\|T^{-n}\| = O(n^{\frac{1}{2}+\varepsilon})$.*

PROOF. Since $\|T^n\| = O(|n|^p)$, $|n| \to \infty$, we have

$$\|(T - zI)^{-1}\| \lesssim \frac{1}{|1 - |z||^{p+1}} \qquad |z| \leq 2 \text{ and } |z| \neq 1.$$

Using the fact that $z \to (T - zI)^{-1}$ is analytic in $\mathbb{C} \setminus \sigma(T)$, where $\sigma(T)$ denotes the spectrum of T, we get

(5.1) $$\|(T - zI)^{-1}\| \lesssim \frac{1}{d(z,\sigma(T))^{p+1}}, \qquad |z| < 2.$$

Let $x \in X$ and $l \in X^*$, we set

$$\varphi(z) = \langle (T-zI)^{-1}x, l\rangle, \qquad z \in \mathbb{C} \setminus \sigma(T).$$

For $f \in \mathcal{C}^\infty(\mathbb{T})$, we set

$$f(T) = \sum_{n \in \mathbb{Z}} \widehat{f}(n)T^n.$$

An easy computation shows that

(5.2) $$\langle f(T)x, l\rangle = \lim_{r \to 1^-} \frac{1}{2i\pi} \int_\mathbb{T} f(\zeta)[\varphi(r\zeta) - \varphi(\zeta/r)]\mathrm{d}\zeta.$$

Let $\mathcal{A}^\infty(\mathbb{D})$ be the algebra of infinitely differentiable functions on the closed disc $\overline{\mathbb{D}}$ which are analytic on $\mathbb{D}$. Since $\sigma(T)$ is a Carleson set, there exists an outer function $f \in \mathcal{A}^\infty(\mathbb{D})$ such that $f^{(m)}|\sigma(T) = 0$ for all $m \geq 0$. In particular we have $|f(\zeta)| = O(d(\zeta, \sigma(T)^{p+1})$ on $\mathbb{T}$. By (5.1) and (5.2) we get $\langle f(T)x, l\rangle = 0$. Hence $f(T) = 0$.

Let $s > 1/2$, By Bernstein inequality [13, p.13] we have $\mathcal{A}_s \subset A^+$. Therefore the functional calculus $\phi : \mathcal{A}_s \to \mathcal{L}(X)$ given by

$$\phi(g) := g(T) = \sum_{n \geq 0} \widehat{g}(n)T^n$$

is a continous homomorphism. Since Ker ϕ contains an outer function f such that $\mathcal{Z}(f) = \sigma(T)$, it follows from Matheson characterization of closed ideals of $\mathcal{A}_s$ that $I_s^+(\sigma(T)) \subset \mathrm{Ker}\phi$ (see [4, 18]). This allows to consider the homomorphism $\widetilde{\phi} : \mathcal{A}_s/I_s^+(\sigma(T)) \to \mathcal{L}(X)$ such that $\widetilde{\phi} \circ \pi = \phi$, where $\pi : \mathcal{A}_s(\sigma(T)) =: \mathcal{A}_s \to \mathcal{A}_s/I_s^+(\sigma(T))$ is the canonical map. Since $\sigma(T)$ is K-set (interpolating set for $\mathcal{A}_s$),

$$\Lambda_s(\sigma(T)) = \mathcal{A}_s(\sigma(T)).$$

Then

$$\|T^{-n}\| = O(\|e^{-int}\|_{\Lambda_s(\mathbb{T})}) = O(n^s) \qquad n \to +\infty.$$

The proof is complete. □

Proof of Corollary 2.4. If E is a K-set, then E is a Λ-Carleson set for some $\Lambda(t) = e^{\alpha t}$. By Corollary 2.3 $\pi(u) : A^+ \to A^+(E)$ is an invertible contraction with K-set spectrum and $\|\pi(u)^{-n}\| = O(n^{3/\alpha})$. The result follows from Proposition 5.1.

Proof of Proposition 2.5. Let f be the outer function considered in the proof of Lemma 2.3. Suppose that (2.1) is satisfied. By (4.1) we have

$$
\begin{aligned}
(5.3) \qquad \frac{1}{3}\log\frac{1}{|f(z)|} &\leq \sum_{k=0}^{N-1} \frac{1}{2^{2k}(1-|z|)} \int_{I_k} \log\frac{1}{\operatorname{dist}(\zeta,E)} \frac{|d\zeta|}{2\pi} \\
&\leq \sum_{k=0}^{N-1} \frac{c_E |I_k|^{1-\alpha}}{2^{2k}(1-|z|)} \\
&= \sum_{k=0}^{N-1} \frac{c_E}{2^{(1+\alpha)k}(1-|z|)^\alpha} \\
&\lesssim \frac{1}{(1-|z|)^\alpha}.
\end{aligned}
$$

As before

$$
\begin{aligned}
\|\pi(u)^{-n}\|_{A^+(E)} &\leq r^{-n}\max_{|\lambda|=r}\|(\lambda - \pi(u))^{-1}\|_{A^+(E)} \\
&\lesssim \frac{|\lambda|^{-n}}{|f(\lambda)|(1-|\lambda|)} \\
&\leq |\lambda|^{-n} e^{\frac{c}{(1-|\lambda|)^\alpha}}.
\end{aligned}
$$

For $1 - r_n = n^{-\frac{1}{1+\alpha}}$ we get

$$\|\pi(u)^{-n}\|_{A^+(E)} = \inf_{0<r<1} r^{-n} e^{\frac{c}{(1-r)^\alpha}} = O(e^{c' n^{\frac{\alpha}{1+\alpha}}})).$$

if E is a Λ-Carleson set with $\Lambda(t) = t^{1/\alpha}$, $\alpha \in (0,1)$, by (2.2)

$$
\begin{aligned}
\int_{I_k} \log\frac{1}{\operatorname{dist}(\zeta,E)}|d\zeta| &\leq \Big(\int_{I_k} \Lambda\Big(\frac{1}{\operatorname{dist}(\zeta,E)}\Big)|d\zeta|\Big)^{1/p}|I_k|^{\frac{p-1}{p}} \\
&= o(1)|I_k|^{\frac{p-1}{p}}, \quad r \to 1-.
\end{aligned}
$$

So by (5.3)

$$|\log|f(z)|| = o((1-|z|)^{-\alpha}).$$

The proof now is complete.

Acknowledgments. The authors are grateful to the referee for his valuable remarks and suggestions.

References

[1] C. Agrafeuil and K. Kellay, *Tauberian type theorem for operators with interpolation spectrum for Hölder classes*, Proc. Amer. Math. Soc. **136** (2008), no. 7, 2477–2482, DOI 10.1090/S0002-9939-08-09273-3. MR2390516

[2] A. Atzmon, *Boundary values of absolutely convergent Taylor series*, Ann. of Math. (2) **111** (1980), no. 2, 231–237, DOI 10.2307/1971199. MR569071

[3] A. Atzmon, *Operators which are annihilated by analytic functions and invariant subspaces*, Acta Math. **144** (1980), no. 1-2, 27–63, DOI 10.1007/BF02392120. MR558090

[4] B. Bouya, *Closed ideals in analytic weighted Lipschitz algebras*, Adv. Math. **219** (2008), no. 5, 1446–1468, DOI 10.1016/j.aim.2008.06.022. MR2458143

[5] J. Bruna, *Muckenhoupt's weights in some boundary problems of a complex variable*, Harmonic analysis (Minneapolis, Minn., 1981), Lecture Notes in Math., vol. 908, Springer, Berlin-New York, 1982, pp. 74–85. MR654180

[6] L. Carleson, *Sets of uniqueness for functions regular in the unit circle*, Acta Math. **87** (1952), 325–345, DOI 10.1007/BF02392289. MR50011

[7] E. M. Dyn'kin, *Free interpolation sets for Hölder classes* (Russian), Mat. Sb. (N.S.) **109(151)** (1979), no. 1, 107–128, 166. MR538552

[8] O. El-Fallah and K. Kellay, *Sous-espaces biinvariants pour certains shifts pondérés* (French, with English and French summaries), Ann. Inst. Fourier (Grenoble) **48** (1998), no. 5, 1543–1558. MR1662275

[9] O. El-Fallah, K. Kellay, and T. Ransford, *Cyclicity in the Dirichlet space*, Ark. Mat. **44** (2006), no. 1, 61–86, DOI 10.1007/s11512-005-0008-z. MR2237211

[10] O. El-Fallah, K. Kellay, and T. Ransford, *Cantor sets and cyclicity in weighted Dirichlet spaces*, J. Math. Anal. Appl. **372** (2010), no. 2, 565–573, DOI 10.1016/j.jmaa.2010.07.047. MR2678884

[11] O. El-Fallah, K. Kellay, Interpolating sets in analytic Wiener algebra, preprint.

[12] J. Esterle, *Distributions on Kronecker sets, strong forms of uniqueness, and closed ideals of A^+*, J. Reine Angew. Math. **450** (1994), 43–82, DOI 10.1515/crll.1994.450.43. MR1273955

[13] J.-P. Kahane, *Séries de Fourier absolument convergentes* (French), Ergebnisse der Mathematik und ihrer Grenzgebiete, Band 50, Springer-Verlag, Berlin-New York, 1970. MR0275043

[14] J.-P. Kahane and R. Salem, *Ensembles parfaits et séries trigonométriques* (French), Actualités Sci. Indust., No. 1301, Hermann, Paris, 1963. MR0160065

[15] J.-P. Kahane and Y. Katznelson, *Sur les algèbres de restrictions des séries de Taylor absolument convergentes à un fermé du cercle* (French), J. Analyse Math. **23** (1970), 185–197, DOI 10.1007/BF02795499. MR273299

[16] R. Kaufman, *On transformations of exceptional sets*, Bull. Soc. Math. Grèce (N.S.) **18** (1977), no. 2, 176–185. MR528178

[17] Y. Katznelson, *An introduction to harmonic analysis*, John Wiley & Sons, Inc., New York-London-Sydney, 1968. MR0248482

[18] A. Matheson, *Cyclic vectors for invariant subspaces in some classes of analytic functions*, Illinois J. Math. **36** (1992), no. 1, 136–144. MR1133774

[19] B. A. Taylor and D. L. Williams, *Ideals in rings of analytic functions with smooth boundary values*, Canadian J. Math. **22** (1970), 1266–1283, DOI 10.4153/CJM-1970-143-x. MR273024

[20] M. Zarrabi, *On operators with spectrum in Cantor sets and spectral synthesis*, J. Math. Anal. Appl. **462** (2018), no. 1, 764–776, DOI 10.1016/j.jmaa.2018.01.079. MR3771274

[21] E. Zerouali, *Estimates for inverses of e^{int} in some quotient algebras of A^+*, Proc. Amer. Math. Soc. **112** (1991), no. 3, 789–793, DOI 10.2307/2048702. MR1037228

LABORATOIRE ANALYSE ET APPLICATIONS, UNIVERSITÉ MOHAMED V RABAT, RABAT, MOROCCO
Email address: elfallah@fsr.ac.ma

UNIV. BORDEAUX, IMB, CNRS UMR 5251, F-33400 TALENCE, FRANCE.
Email address: kkellay@math.u-bordeaux.fr

Jordan isomorphisms as preservers

Lajos Molnár

ABSTRACT. This article is an extended version of the talk delivered by the author at The International Conference on Algebra and Related Topics (ICART) 2018 in Rabat, Morocco. In the first part we discuss certain preserver properties of Jordan *-isomorphisms on C^*-algebras and present some classical results showing that, to some extents, those properties in fact characterize Jordan *-isomorphisms among bijective linear transformations. In the second part we survey some of our recent results which can be regarded as non-linear characterizations of Jordan *-isomorphisms on the cones of positive elements in operator algebras and on the spaces of self-adjoint elements. Besides, we give a few new results which extend or unify certain former statements that have appeared in the literature. Finally, we also pose some open problems for further research.

1. Introduction

We begin with the necessary definitions. Let $\mathcal{A}, \mathcal{B}$ be complex algebras. The linear map $J: \mathcal{A} \to \mathcal{B}$ is called a Jordan homomorphism if it satisfies

$$J(a^2) = J(a)^2, \quad a \in \mathcal{A},$$

or, equivalently, it satisfies

$$J(ab + ba) = J(a)J(b) + J(b)J(a), \quad a, b \in \mathcal{A}.$$

If $\mathcal{A}, \mathcal{B}$ are *-algebras, the Jordan homomorphism $J: \mathcal{A} \to \mathcal{B}$ is called Jordan *-homomorphism if $J(a^*) = J(a)^*$, $a \in \mathcal{A}$ also holds. A bijective Jordan (*-)homomorphism is called a Jordan (*-)isomorphism.

Jordan homomorphisms of associative algebras (also of rings) play a significant role in various areas of mathematics. In particular, we mean ring theory and the theory of operator algebras and also refer to their important applications in physics, e.g., in the algebraic approach to quantum theory initiated originally in [40]. From the algebraic point of view, one of the most fundamental problems concerning Jordan homomorphisms is to find conditions under which such a map is necessarily either a homomorphism or an antihomomorphism, or the direct sum of transformations of these two latter kinds. Here we recall I. Herstein's fundamental theorem [16] stating that any Jordan homomorphism from a ring onto a prime ring

---

2010 *Mathematics Subject Classification.* 47B49, 46L40, 47L07.

Ministry of Human Capacities, Hungary grant 20391-3/2018/FEKUSTRAT is acknowledged, the work was supported also by the National Research, Development and Innovation Office of Hungary, NKFIH, Grant No. K115383.

©2020 American Mathematical Society

of characteristic different from 2 and 3 is a homomorphism or an antihomomorphism. N. Jacobson and C. Rickart proved in [18] that a Jordan homomorphism from the ring of $n \times n$ matrices ($n \geq 2$) over an arbitrary unital ring into another ring is the direct sum of a homomorphism and an antihomomorphism. This result was used by Kadison in [19] for proving that a Jordan *-homomorphism from a von Neumann algebra onto a C^*-algebra is the direct sum of a *-homomorphism and a *-antihomomorphism.

We continue with further terminology and notation. In what follows, when we speak about C^*-algebras, we always mean unital C^*-algebras. If $\mathcal{A}$ is a C^*-algebra, then $\mathcal{A}_s$ denotes the space of all of its self-adjoint elements. The set of positive elements of $\mathcal{A}$ (self-adjoint elements with non-negative spectrum) is denoted by $\mathcal{A}^+$. It is a cone that we call the positive semidefinite cone of $\mathcal{A}$. The symbol $\mathcal{A}^{++}$ stands for the positive definite cone of $\mathcal{A}$, i.e., the set of invertible elements in $\mathcal{A}^+$. The usual order $\leq$ on $\mathcal{A}_s$ is defined as follows: for any $a, b \in \mathcal{A}_s$ we write $a \leq b$ if and only if $b - a \in \mathcal{A}^+$. By a projection in $\mathcal{A}$ we mean a self-adjoint idempotent.

Jordan (*-)isomorphisms have several important properties that concern the preservation of some numerical quantities attached to elements, relations among elements, properties of elements, etc. Below we collect some of the most important such features.

Let $\mathcal{A}, \mathcal{B}$ be C^*-algebras and $J : \mathcal{A} \to \mathcal{B}$ be a Jordan *-isomorphism. We have the following.

(P1) J is a surjective linear isometry, i.e., $\|J(x)\| = \|x\|$ holds for all $x \in \mathcal{A}$. See, e.g., the introduction in [23].

(P2) J is a linear order isomorphism, i.e., we have

$$a \leq b \Longleftrightarrow J(a) \leq J(b), \quad a, b \in \mathcal{A}_s.$$

This property is easy to see, it follows from the apparent positivity of J (meaning that it sends positive elements to positive elements).

(P3) J preserves commutativity in both directions, i.e., we have

$$ab = ba \Longleftrightarrow J(a)J(b) = J(b)J(a), \quad a, b \in \mathcal{A}.$$

The validity of this property is not trivial, for a proof see, e.g., 6.3.4 Theorem in [38].

(P4) J preserves projections in both directions, i.e., for any $p \in \mathcal{A}$, we have

$$p \in \mathcal{A} \text{ is a projection} \Longleftrightarrow J(p) \text{ is a projection}.$$

This property is apparent.

(P5) J preserves invertibility in both directions, i.e., for any $a \in \mathcal{A}$, we have

$$a \text{ is invertible} \Longleftrightarrow J(a) \text{ is invertible}, \quad a \in \mathcal{A}.$$

In fact, we have $J(a^{-1}) = J(a)^{-1}$ for any invertible $a \in \mathcal{A}$, see, e.g., Proposition 1.3 in [43]. It clearly follows that J preserves the spectrum, i.e., we have

$$\sigma(J(a)) = \sigma(a), \quad a \in \mathcal{A}.$$

(P6) J preserves the unitaries in both directions, i.e., for any $u \in \mathcal{A}$, we have

$$u \in \mathcal{A} \text{ is unitary} \Longleftrightarrow J(u) \text{ is unitary}$$

which follows easily from the previous properties given in (P5).

2. Linear preserver characterizations of Jordan *-isomorphisms between C^*-algebras

In this section we present some classical results showing that the previously mentioned preserver properties in fact characterize Jordan *-isomorphisms to certain extents. We emphasize that the following results are considered as the most fundamental ones in that direction, several extensions, improvements were obtained which we do not present here due to space limitations. The reader can easily trace such results in the well-known mathematical data bases.

We also point out that in all results in this section we assume the linearity of the transformations under consideration. This is important to notice since in the next sections the focus will be on non-linear preservers.

In (P1) we have seen that every Jordan *-isomorphism between C^*-algebras is a surjective linear isometry. In [19], R. Kadison described the structure of all surjective linear isometries between such structures. His important result, which is a non-commutative extension of the famous Banach-Stone theorem (about the structure of the surjective linear isometries between the algebras of all continuous scalar valued functions on compact Hausdorff spaces) tells that every such map is necessarily a Jordan *-isomorphism followed by a multiplication by a fixed unitary element. The precise formulation reads as follows.

THEOREM 1 (Kadison (1951)). *Let $\mathcal{A}, \mathcal{B}$ be C^*-algebras. If $\phi : \mathcal{A} \to \mathcal{B}$ is a surjective linear isometry, then it is of the form*

$$\phi(a) = uJ(a), \quad a \in \mathcal{A},$$

*where $u \in \mathcal{B}$ is a unitary element and $J : \mathcal{A} \to \mathcal{B}$ is a Jordan *-isomorphism.*

In (P2) we have seen that Jordan *-isomorphisms between C^*-algebras are linear order isomorphisms. It was also R. Kadison who described the structure of all linear order isomorphisms between C^*-algebras [20] and proved that any such transformation is a Jordan *-isomorphism followed by a congruence transformation implemented by an invertible element. More precisely, his result is as follows.

THEOREM 2 (Kadison (1952)). *Let $\mathcal{A}, \mathcal{B}$ be C^*-algebras and let $\phi : \mathcal{A} \to \mathcal{B}$ be a bijective linear map which preserves the order in both directions, i.e., satisfies*

$$a \leq b \iff \phi(a) \leq \phi(b), \quad a, b \in \mathcal{A}_s.$$

Then ϕ is of the form

$$\phi(a) = tJ(a)t^*, \quad a \in \mathcal{A},$$

*where $t \in \mathcal{B}$ is an invertible element and $J : \mathcal{A} \to \mathcal{B}$ is a Jordan *-isomorphism.*

The property (P3) concerns the commutativity preserving property of Jordan (*)-isomorphisms. As for the converse, the situation is more complicated than what we have seen above. The reasons are manifold and, as one can easily figure, basically algebraic in nature. The first general result concerning commutativity preserving linear maps between operator algebras is due to R. Miers [24] and formulated in the following way.

THEOREM 3 (Miers (1988)). *Let $\mathcal{A}, \mathcal{B}$ be von Neumann algebras, $\mathcal{A}$ be a factor not of type I_2. Let $\phi : \mathcal{A} \to \mathcal{B}$ be a bijective linear map which preserves commutativity in both directions. Then ϕ can be written of the form*

(1) $$\phi(a) = c\Phi(a) + \ell(a)1, \quad a \in \mathcal{A},$$

where c is a nonzero scalar, Φ is a Jordan isomorphism and ℓ is a linear functional on $\mathcal{A}$.

Observe that the converse of the statement is obviously true, any transformation ϕ of the form (1) preserves the commutativity in both directions. We mention that the statement fails to be true in the case of 2×2 matrices, this is the reason for the assumption in the theorem which excludes the case of factors of type I_2. Also observe that the Jordan isomorphism Φ above is necessarily either an algebra isomorphism or an algebra antiisomorphism. This follows from the fact that factor von Neumann algebras are prime algebras and hence I. Herstein's classical result mentioned in the introduction applies. We note that far reaching generalizations of R. Miers' theorem were obtained by M. Brešar etal., see Chapter 7 (especially Section 7.1) in the book [6], and also see Section 6.5 in [1].

B. Russo and H. Dye determined the structure of all unitary preserving linear maps between C^*-algebras [39]. Their result states the following.

THEOREM 4 (Russo and Dye (1966)). *Let $\mathcal{A}, \mathcal{B}$ be C^*-algebras, $\phi : \mathcal{A} \to \mathcal{B}$ be a linear map sending unitaries to unitaries. Then ϕ is of the form*

$$\phi(a) = uJ(a), \quad a \in \mathcal{A},$$

*where $u \in \mathcal{B}$ is a unitary and $J : \mathcal{A} \to \mathcal{B}$ is a unital Jordan *-homomorphism (unital means that the unit is mapped into the unit).*

Concerning linear maps preserving projections we have the following result, see Theorem A.4 in [29]. We emphasize that the original idea of the proof showed up in [5] and since then the result appeared in several versions and in several places in the literature.

THEOREM 5. *Let $\mathcal{A}$ be a von Neumann algebra, $\mathcal{B}$ be a C^*-algebra and $\phi : \mathcal{A} \to \mathcal{B}$ be a continuous linear map sending projections to projections. Then ϕ is a Jordan *-homomorphism.*

The possibly deepest still open problem in the area of preservers is the famous Kaplansky's problem concerning invertibility preserving map. It was originally vagualy raised in Section 9 [21]. In its present precise form it reads as follows.

PROBLEM 6 (Kaplansky's problem). *Is every surjective unital linear map between unital Banach algebras which maps invertible elements to invertible elements necessarily a Jordan homomorphism?*

As mentioned above, after 50 years the problem is still unsolved and researchers have recently been tending to believe that the final answer might be negative. Nevertheless, the probably strongest and most general result in this direction was obtained by B. Aupetit [2]. It says the following.

THEOREM 7 (Aupetit (2000)). *Every spectrum preserving surjective linear map between von Neumann algebras is a Jordan homomorphism.*

To see the relation to Kaplansky's problem, observe that a unital linear map preserves the invertible elements in both directions if and only if it preserves the spectrum. We point out that a statement analogous to Kaplansky's was conjectured by L. Harris and R. Kadison in [13] concerning C^*-algebras but no proof is known yet even for that more specific problem.

3. Non-linear characterizations of Jordan *-isomorphisms on the self-adjoint and positive definite parts of C^*-algebras

As promised in the abstract, in this section and the next one we present some non-linear characterizations of Jordan *-isomorphisms on the self-adjoint and positive definite parts of C^*-algebras ('non-linear' meaning that we do not assume that the transformations we are considering are linear in any sense). Actually, we briefly survey our recent results obtained in this direction and, in addition, complement some of them. Below we first present our key result which turns to be the background for those characterizations. This is the description of so-called Thompson isometries of the positive definite cones of C^*-algebras. After that we exhibit several applications (recent or new) for solving various preserver problems that provide the mentioned characterizations of Jordan *-isomorphisms. Beside these we also formulate some hopefully interesting open problems.

The Thompson metric (or Thompson part metric) can be defined in a rather general setting involving normed linear spaces and certain closed cones. In the case of a C^*-algebra $\mathcal{A}$, the Thompson metric d_T on the positive definite cone $\mathcal{A}^{++}$ is defined by the formula

$$d_T(a,b) = \log \max\{M(a/b), M(b/a)\}, \quad a, b \in \mathcal{A}^{++},$$

where $M(x/y) = \inf\{t > 0 : x \leq ty\}$ for any $x, y \in \mathcal{A}^{++}$. It is easy to see that d_T can also be written as

$$d_T(a,b) = \left\|\log\left(a^{-1/2}ba^{-1/2}\right)\right\|, \quad a, b \in \mathcal{A}^{++}$$

(see, e.g., [30]). This metric has many applications, it is of fundamental importance in the theory of operator means, of great interest in non-linear Perron–Frobenius theory in geometric optimization, etc. In the paper [14] we determined the structure of all surjective Thompson isometries. Theorem 9 in [14] reads as follows.

THEOREM 8. *Let $\mathcal{A}, \mathcal{B}$ be C^*-algebras. The surjective map $\phi : \mathcal{A}^{++} \to \mathcal{B}^{++}$ is a Thompson isometry (i.e., a map which preserves the Thompson distance) if and only if there is a central projection p in $\mathcal{B}$ and a Jordan *-isomorphism $J : \mathcal{A} \to \mathcal{B}$ such that*

$$(2) \qquad \phi(a) = \phi(1)^{1/2}\left(pJ(a) + (1-p)J(a^{-1})\right)\phi(1)^{1/2}, \quad a \in \mathcal{A}^{++}.$$

PROOF. We present the main ideas of a possible proof which in fact differs from the one given in [14] at some points. First of all, one can show that for any given element $y \in \mathcal{B}^{++}$, the map $b \mapsto yby$ is a Thompson isometry of the positive definite cone $\mathcal{B}^{++}$. Therefore, multiplying ϕ by $\phi(1)^{-1/2}$ from both sides, we can and do assume that ϕ is unital, i.e., $\phi(1) = 1$ holds.

The classical Mazur-Ulam theorem says that surjective isometries between real normed linear spaces are affine, they preserve the operation of the arithmetic mean. The argument to show that result can essentially be applied to show the interesting fact that Thompson isometries necessarily preserve the operation of the geometric mean which is defined as

$$(3) \qquad a \sharp b = a^{1/2}\left(a^{-1/2}ba^{-1/2}\right)^{1/2} a^{1/2}$$

for any elements a, b in a positive definite cone (see the reasoning in [30]). It is a classical result (sometimes referred to as Anderson-Trapp theorem) that $a \sharp b$ is the unique positive invertible solution of the simple equation $xa^{-1}x = b$. It

follows easily that ϕ preserves the operation appearing on the left hand side of this equation (which turns to be a sort of abstract reflection), i.e., we have $\phi(xa^{-1}x) = \phi(x)\phi(a)^{-1}\phi(x)$, $a, x \in \mathcal{A}^{++}$. Since ϕ sends the unit to the unit, we obtain that ϕ preserves the Jordan triple product $(a, b) \mapsto aba$ of positive invertible elements, i.e., satisfies
$$\phi(aba) = \phi(a)\phi(b)\phi(a), \quad a, b \in \mathcal{A}^{++}.$$
This implies that $\phi(a^r) = \phi(a)^r$ holds for all $a \in \mathcal{A}^{++}$ and rational numbers r.

One can easily see that the topology what the Thompson metric generates coincides with the topology what the C^*-norm generates on the positive definite cone. Therefore, ϕ is continuous with respect to the norm topology. It then follows that $\phi(a^r) = \phi(a)^r$ holds for all $a \in \mathcal{A}^{++}$ and real number r.

The so-called symmetrized Lie-Trotter formula asserts that
$$\left(e^{-(t/2)x}e^{ty}e^{-(t/2)x}\right)^{1/t} \xrightarrow{t \to 0} e^{y-x}$$
in norm which implies that
$$(4) \qquad \frac{d_T(e^{tx}, e^{ty})}{t} \xrightarrow{t \to 0} \|x - y\|$$
holds for any $x, y \in \mathcal{A}_s$.

Define $F : \mathcal{A}_s \to \mathcal{B}_s$ by $F(x) = \log \phi(e^x)$, $x \in \mathcal{A}_s$. Then we have
$$e^{tF(x)} = (e^{F(x)})^t = \phi(e^x)^t = \phi((e^x)^t) = \phi(e^{tx}), \quad t > 0, x \in \mathcal{A}_s.$$
Applying the formula (4) and using the fact that ϕ is a Thompson isometry, we deduce that $\|F(x) - F(y)\| = \|x - y\|$, $x, y \in \mathcal{A}_s$. Therefore, $F : \mathcal{A}_s \to \mathcal{B}_s$ is a surjective isometry which also satisfies $F(0) = 0$. By the classical Mazur-Ulam theorem it follows that F is a linear surjective isometry between the self-adjoint parts of C^*-algebras. Kadison's result Theorem 2 in [20] describes the structure of those transformations. Namely, it tells that there is a central self-adjoint unitary element s in $\mathcal{B}$ and a Jordan *-isomorphism $J : \mathcal{A} \to \mathcal{B}$ such that $F(x) = sJ(x)$, $x \in \mathcal{A}_s$. Clearly, $s = 2p - 1$ holds with some central projection $p \in \mathcal{B}$. The rest of the proof of the necessity part of the theorem is simple calculation and so is its sufficiency part. □

From the above theorem we immediately have the following

COROLLARY 9. *Let* $\phi : \mathcal{A}^{++} \to \mathcal{B}^{++}$ *be an order isomorphism, i.e., a surjective mapping with the property that for any* $a, b \in \mathcal{A}^{++}$ *we have*
$$a \leq b \iff \phi(a) \leq \phi(b).$$
(Observe that this property obviously implies the injectivity of ϕ.) If ϕ is also positive homogeneous (i.e., satisfies $\phi(\lambda a) = \lambda \phi(a)$ for all $a \in \mathcal{A}^{++}$ and positive number λ), then ϕ is necessarily of the form
$$\phi(a) = \phi(1)^{1/2} J(a) \phi(1)^{1/2}, \quad a \in \mathcal{A}^{++},$$
where $J : \mathcal{A} \to \mathcal{B}$ *is a Jordan *-isomorphism.*

The proof of this result is obvious, one needs to observe the trivial fact that, by the definition of the Thompson metric, the above map ϕ is a surjective Thompson isometry, hence Theorem 8 applies. Since ϕ is positive homogeneous, $1 - p$ in (2) must be zero.

This corollary is our key result that we have recently used to describe several different kinds of preservers between positive definite cones of C^*-algebras. We especially mean maps which preserve certain quantum information related numerical quantities. In what follows we survey some of the corresponding results, complement them and also present some new observations.

We first mention that in [33] we determined the structure of Bures-Wasserstein isometries on positive definite cones in C^*-algebras. The Bures-Wasserstein metric appears in several different areas of mathematics and physics. For example, when defined on the state space of a quantum system, it is a quantum generalization of the Fisher information metric and plays an important role in quantum information geometry. When defined on the cone of positive definite matrices, it appears in various optimization problems, in the theory of optimal transport, etc. We also mention that very recently, in [3], Bhatia, Jain and Lim introduced and studied in details a new concept of matrix means based on this metric.

Let $\mathbb{P}_n$ be the positive definite cone in the algebra of all complex $n \times n$ matrices equipped with the usual trace functional Tr. The Bures-Wasserstein distance between the elements $a, b \in \mathbb{P}_n$ is defined as

$$d_{BW}(a,b) = \left(\operatorname{Tr}(a) + \operatorname{Tr}(b) - 2\operatorname{Tr}((a^{1/2}ba^{1/2})^{1/2})\right)^{1/2}.$$

We point out that the quantity $\operatorname{Tr}((a^{1/2}ba^{1/2})^{1/2})$ is in itself very important in quantum information theory. It is called fidelity which is a natural extension of the notion of transition probability from pure states to mixed states. It is not trivial at all to see that d_{BW} is a true metric. One can verify it first showing that

$$d_{BW}(a,b) = \inf\{\|a^{1/2} - ub^{1/2}\|_2 : u \text{ is unitary}\}, \quad a, b \in \mathbb{P}_n,$$

where $\|.\|_2$ stands for the Frobenius (or Hilbert-Schmidt) norm.

Another important metric on the positive definite cone of matrices is the Hellinger distance which is defined by

$$d_H(a,b) = \|a^{1/2} - b^{1/2}\|_2 = \left(\operatorname{Tr}(a) + \operatorname{Tr}(b) - 2\operatorname{Tr} a^{1/2}b^{1/2}\right)^{1/2}, \quad a, b \in \mathbb{P}_n.$$

It clearly majorizes the Bures-Wasserstein metric.

In what follows we present a result which, as a particular case, includes Theorem 2 in [33] describing the Bures-Wasserstein isometries between positive definite cones in C^*-algebras. In addition to that, also as particular cases, it gives the complete characterization of Hellinger isometries as well as extends a recent result of Gaál [11] describing transformations between positive definite cones of finite von Neumann algebras that preserve the trace distance between the arithmetic mean and the geometric mean of elements.

By a trace on a C^*-algebra $\mathcal{A}$ we mean a positive linear functional τ on $\mathcal{A}$ which satisfies $\tau(xy) = \tau(yx)$ for all $x, y \in \mathcal{A}$. The trace τ is said to be faithful if $\tau(a) = 0$, $a \in \mathcal{A}^+$, implies $a = 0$. Fundamental examples for C^*-algebras having faithful traces include UHF-algebras, finite von Neumann factors, irrational rotation algebras.

In order to formulate the above mentioned new general result, we consider a sort of operation denoted by $\bullet$ that is defined on the positive definite cones in all C^*-algebras (like the geometric mean $\sharp$ in (3)) and assume that it has the following properties. For any C^*-algebra $\mathcal{A}$ and for any elements $a, b \in \mathcal{A}^{++}$, central element $c \in \mathcal{A}^{++}$, sequence (b_n) in $\mathcal{A}^{++}$, and positive real number $\lambda > 0$ we have (p1) $(\lambda a) \bullet b = a \bullet (\lambda b) = \sqrt{\lambda}(a \bullet b)$;

(p2) $(ca) \bullet (cb) = c(a \bullet b)$;

(p3) $a \bullet b_n \to 0$ is valid whenever $b_n \to 0$ (convergence is meant in the C^*-norm).

Furthermore,

(p4) the restriction of any Jordan *-isomorphism between C^*-algebras to the corresponding positive definite cones is an isomorphism with respect to the operation $\bullet$.

We also assume that

(p5) for any C^*-algebra $\mathcal{A}$ with a faithful trace τ we have $\tau((a+b)/2) \geq \tau(a \bullet b)$ for all $a, b \in \mathcal{A}^{++}$

and finally that

(p6) for some given positive real number r, for any C^*-algebra $\mathcal{A}$ with faithful trace τ we have that for arbitrary $a, b \in \mathcal{A}^{++}$ the inequality $\tau(a \bullet x) \leq \tau(b \bullet x)$ holds for all $x \in \mathcal{A}^{++}$ if and only if $a^r \leq b^r$

which means that the operation $\bullet$ determines the rth power order on positive definite cones.

To give examples for such an operation, we assert that

$$(a,b) \mapsto (a^{1/2}ba^{1/2})^{1/2}, \quad (a,b) \mapsto a^{1/4}b^{1/2}a^{1/4}, \quad (a,b) \mapsto a\sharp b$$

all satisfy (p1)-(p6) above. Indeed, the validity of (p1)-(p3) is just obvious. The validity of (p4) follows from the the following properties of Jordan *-isomorphisms: for any such map J we have $J(aba) = J(a)J(b)J(a)$ (see 6.3.2 Lemma in [38]) which then implies that $J(f(a)) = f(J(a))$ holds for any self-adjoint element a and continuous scalar valued function on its spectrum. As for (p5), we observe that, in any C^*-algebra $\mathcal{A}$ with faithful trace τ, for all $a, b \in \mathcal{A}^{++}$ we have the following inequalities

(5) $\qquad \tau((a+b)/2) \geq \tau((a^{1/2}ba^{1/2})^{1/2}) \geq \tau(a^{1/4}b^{1/2}a^{1/4}) \geq \tau(a\sharp b).$

Here the first inequality is the tracial arithmetic-geometric mean inequality (see the first paragraph of the proof of Theorem 2.4 in [8]), the second one follows from $\tau(|x|) \geq |\tau(x)|$, $x \in \mathcal{A}$ (see the second paragraph of the proof of Proposition 2.7 in [9]), while the last one can be proved following the argument given in the last paragraph of the proof of Theorem 6 in [22]. To see the validity of (p6) we cite the following result from [34] which appears as Lemma 8 there.

LEMMA 10. *Let $\alpha \in]0, \infty[$ be a real number, $\mathcal{A}$ be a C^*-algebra with a faithful trace τ, and select $a, b \in \mathcal{A}^{++}$. We have $a \leq b$ if and only if $\tau((xax)^\alpha) \leq \tau((xbx)^\alpha)$ holds for all $x \in \mathcal{A}^{++}$.*

The lemma implies that the operation $(a,b) \mapsto (a^{1/2}ba^{1/2})^{1/2}$ satisfies (p6) with $r = 1$ and the operation $(a,b) \mapsto a^{1/4}b^{1/2}a^{1/4}$ satisfies (p6) with $r = 1/2$. As a particular case of Lemma 11 in [34] (using also the well-known commutativity property of the operation of the geometric mean $\sharp$), we obtain the following

LEMMA 11. *Let $\mathcal{A}$ be a C^*-algebra with a faithful trace τ, and select $a, b \in \mathcal{A}^{++}$. We have $a \leq b$ if and only if $\tau(a\sharp x) \leq \tau(b\sharp x)$ holds for all $x \in \mathcal{A}^{++}$.*

This means that the operation $(a,b) \mapsto a\sharp b$ satisfies (p6) with $r=1$.

We now extend the notions of Bures-Wasserstein and Hellinger metrics to the setting of general C^*-algebras in a natural way. Let $\mathcal{A}$ be a C^*-algebra with a faithful trace τ. We define the corresponding Bures-Wasserstein metric by

$$(a,b) \mapsto \left(\tau(a+b-2(a^{1/2}ba^{1/2})^{1/2})\right)^{1/2}.$$

Indeed, Farenick etal. defined this in [8], [9] on the set (called the τ-density space of $\mathcal{A}$) of all positive elements a in $\mathcal{A}$ with $\tau(a)=1$ and showed it is a true metric. But the proof they gave can easily be modified to show that the previous formula gives a true metric also on the whole positive definite cone $\mathcal{A}^{++}$.

The apparent definition of the corresponding Hellinger distance is

$$(a,b) \mapsto \left(\tau(a+b-2(a^{1/4}b^{1/2}a^{1/4}))\right)^{1/2}.$$

Obviously, this is exactly the distance between $a^{1/2}$ and $b^{1/2}$ in the norm that comes from the inner product what the trace τ generates in the natural way. Therefore, it a true metric, too. Finally, let us consider the trace distance between the arithmetic and the geometric means, more precisely the square root of the double of this quantity, i.e., the distance measure

$$(6) \qquad (a,b) \mapsto (\tau(a+b-2(a\sharp b)))^{1/2}.$$

Clearly, by (5), the latter function majorizes the Hellinger distance which majorizes the Bures-Wasserstine distance. We point out that we do not know if the function in (6) satisfies the triangle inequality or not and pose the following

PROBLEM 1. *Is the function in (6) a true metric?*

We believe this is a difficult problem, and in fact do not know the answer even in the most simple case, for matrices. Nevertheless, our conjecture is that the answer to the question is affirmative. [1]

After this we are in a position to present the following general result.

THEOREM 12. *Let $\circ$ and $\diamond$ be operations having the properties (p1)-(p6) above, assume that $\circ$ determines the αth power order and $\diamond$ determines the βth power order for some given positive real numbers α, β (in the sense described in (p6)). Suppose that on the positive definite cones of commutative C^*-algebras the operations $\circ$ and $\diamond$ coincide.*

Let $\mathcal{A}, \mathcal{B}$ be two particular C^*-algebras with faithful traces τ and ω, respectively. Define

$$\rho(a,b) = (\tau(a+b-2(a \circ b)))^{1/2}, \quad a,b \in \mathcal{A}^{++}$$

and

$$\delta(a,b) = (\omega(a+b-2(a \diamond b)))^{1/2}, \quad a,b \in \mathcal{B}^{++}.$$

For any given surjective map $\phi : \mathcal{A}^{++} \to \mathcal{B}^{++}$ we have that

$$(7) \qquad \delta(\phi(a), \phi(b)) = \rho(a,b), \quad a,b \in \mathcal{B}^{++}$$

---

[1] To our surprise, this conjecture has recently turned out to be false, see p. 1787 in the paper R. Bhatia, S. Gaubert, T. Jain, Matrix versions of the Hellinger distance. Lett. Math. Phys. **109** (2019), no. 8, 1777–1804.

if and only if $\tau(a \circ b) = \tau(a \diamond b)$, $a, b \in \mathcal{A}^{++}$, there is a Jordan *-isomorphism $J : \mathcal{A} \to \mathcal{B}$ and a central element $c \in \mathcal{B}^{++}$ such that $\phi(a) = cJ(a)$, $a \in \mathcal{A}^{++}$, and $\omega(cJ(x)) = \tau(x)$, $x \in \mathcal{A}$.

In the proof of this result we will use the interesting fact that between the positive definite cones of C^*-algebras the existence of a non-isometric homothety with respect to the Thompson metric implies that the underlying algebras are necessarily commutative. The corresponding result is given in [34] as Theorem 18.

THEOREM 13. *If $\mathcal{A}$ and $\mathcal{B}$ are C^*-algebras and there is a surjective map $\phi : \mathcal{A}^{++} \to \mathcal{B}^{++}$ such that*

$$d_T(\phi(a), \phi(b)) = \gamma d_T(a, b), \quad a, b \in \mathcal{A}^{++}$$

holds with some positive real number γ different from 1, then the algebras $\mathcal{A}, \mathcal{B}$ are necessarily commutative.

We will also need the following statement on centrality of elements that appears as Lemma 14 in [34].

LEMMA 14. *Let $\mathcal{A}, \mathcal{B}$ be C^*-algebras with faithful traces τ and ω, respectively. Let $J : \mathcal{A} \to \mathcal{B}$ be a Jordan *-isomorphism, $c \in \mathcal{B}^{++}$ and γ be a positive real number such that $\omega((cJ(a^{1/\gamma})c)^\gamma) = \tau(a)$ holds for all $a \in \mathcal{A}^{++}$. Then c is necessarily a central element in $\mathcal{B}$.*

After this, the proof of Theorem 12 is as follows.

PROOF OF THEOREM 12. We follow the basic idea given in [33]. Assume that $\phi : \mathcal{A}^{++} \to \mathcal{B}^{++}$ is a surjective map which satisfies (7). We claim that, for any $a, b \in \mathcal{A}^{++}$, the inequality $a^\alpha \leq b^\alpha$ holds if and only if the set

(8) $$\{\rho(a, x)^2 - \rho(b, x)^2 : x \in \mathcal{A}^{++}\}$$

is bounded from below. Indeed, we compute

(9) $$\rho(a, x)^2 - \rho(b, x)^2 = \tau(a) - \tau(b) + 2(\tau(b \circ x) - \tau(a \circ x)).$$

By the property (p6), if $a^\alpha \leq b^\alpha$, then $\tau(b \circ x) - \tau(a \circ x) \geq 0$, $x \in \mathcal{A}^{++}$, hence the quantity in (9) is greater than or equal to $\tau(a) - \tau(b)$ implying that the set in (8) is really bounded from below. Conversely, if the set is bounded from below, then $\tau(b \circ x) - \tau(a \circ x)$ must be nonnegative for all $x \in \mathcal{A}^{++}$ since otherwise, by (p1), multiplying x with positive real numbers, we would obtain that the set in (8) contains arbitrarily small elements and hence it is not bounded from below. Using (p6), this gives us that $a^\alpha \leq b^\alpha$.

Obviously, similar characterization holds concerning the βth power order on $\mathcal{B}^{++}$ involving the quantity δ. It follows that ϕ has the following property: for any $a, b \in \mathcal{A}^{++}$, we have $a^\alpha \leq b^\alpha$ if and only if $\phi(a)^\beta \leq \phi(b)^\beta$.

Let a_n be a sequence in $\mathcal{A}^{++}$. Apparently, the convergence $a_n \to 0$ holds if and only if for every $x \in \mathcal{A}^{++}$ the inequality $a_n^\alpha \leq x^\alpha$ is valid for large enough n. Therefore, ϕ sends zero sequences (i.e., sequences converging to zero) in $\mathcal{A}^{++}$ to zero sequences in $\mathcal{B}^{++}$. For any zero sequence (x_n) in $\mathcal{A}^{++}$ we have $\rho(a, x_n)^2 \to \tau(a)$ and similar observation holds in $\mathcal{B}^{++}$. We conclude that ϕ preserves the trace in the sense that

(10) $$\omega(\phi(a)) = \tau(a), \quad a \in \mathcal{A}^{++}.$$

This implies that we necessarily have
$$\omega(\phi(a) \diamond \phi(b)) = \tau(a \circ b), \quad a, b \in \mathcal{A}^{++}.$$

We next prove that ϕ is homogeneous. This easily follows from the equalities
$$\omega(\phi(\lambda a) \diamond \phi(x)) = \tau((\lambda a) \circ x) = \sqrt{\lambda}\tau(a \circ x) = \sqrt{\lambda}\omega(\phi(a) \diamond \phi(x)) = \omega(\lambda\phi(a) \diamond \phi(x))$$
which hold for all $x \in \mathcal{A}^{++}$ and from the order determining property (p6) of the operations $\circ$ and $\diamond$. Now, consider the map $\psi : a \mapsto \phi(a^{1/\alpha})^\beta$ from $\mathcal{A}^{++}$ to $\mathcal{B}^{++}$. It follows that ψ is an order isomorphism which is homogeneous of order β/α. From
$$a \leq tb \iff \psi(a) \leq t^{\beta/\alpha}\psi(b), \quad a, b \in \mathcal{A}^{++}$$
we deduce that
$$d_T(\psi(a), \psi(b)) = (\beta/\alpha)d_T(a, b), \quad a, b \in \mathcal{A}^{++}.$$
This means that ψ is a homothety between the positive definite cones $\mathcal{A}^{++}$ and $\mathcal{B}^{++}$ with coefficient β/α.

If β/α is different from 1, then by Theorem 13, we obtain that $\mathcal{A}, \mathcal{B}$ are necessarily commutative. But then it follows that ϕ is a positive homogeneous order isomorphism. By Corollary 9, ϕ is necessarily of the form $\phi(a) = cJ(a)c, a \in \mathcal{A}^{++}$ with some Jordan *-isomorphism $J : \mathcal{A} \to \mathcal{B}$ and element $c \in \mathcal{B}^{++}$. By the trace preserving property (10) of ϕ, we have $\omega(c^2 J(a)) = \omega(\phi(a)) = \tau(a), a \in \mathcal{A}^{++}$. Since $\mathcal{A}^{++}$ linearly generates $\mathcal{A}$, the latter equality holds on the whole algebra $\mathcal{A}$, too. This finishes the proof of the necessity part of our statement in the case where $\alpha \neq \beta$.

Assume now that $\alpha = \beta$. Then ψ is a homogeneous order isomorphism, hence by Corollary 9 we have that $\psi(a) = cJ(a)c, a \in \mathcal{A}^{++}$ holds with some element $c \in \mathcal{B}^{++}$ and Jordan *-isomorphism $J : \mathcal{A} \to \mathcal{B}$. For ϕ, this means that $\phi(a) = (cJ(a^\alpha)c)^{1/\alpha}, a \in \mathcal{A}^{++}$. Again, by the trace preserving property (10) of ϕ, it follows that $\omega((cJ(a^\alpha)c)^{1/\alpha}) = \tau(a), a \in \mathcal{A}^{++}$. Lemma 14 tells us that c is necessarily central and hence we obtain that $\phi(a) = dJ(a), a \in \mathcal{A}^{++}$ with a central element $d \in \mathcal{B}^{++}$. Using (10), (p4), (p2), we compute
$$\tau(a \diamond b) = \omega(\phi(a \diamond b)) = \omega(dJ(a \diamond b)) = \omega(d(J(a) \diamond J(b)))$$
$$= \omega(dJ(a) \diamond dJ(b)) = \omega(\phi(a) \diamond \phi(b)) = \tau(a \circ b), \quad a, b \in \mathcal{A}^{++}.$$

The proof of the necessity part of the statement is complete. The sufficiency is easy to verify, it requires only simple computation. $\square$

Concerning the equality $\tau(a \circ b) = \tau(a \diamond b), a, b \in \mathcal{A}^{++}$ that appears in Theorem 12, we have the following

PROPOSITION 15. *Let $\mathcal{A}$ be a C^*-algebra with a faithful trace τ. Consider the operations $(a, b) \mapsto (a^{1/2}ba^{1/2})^{1/2}$, $(a, b) \mapsto a^{1/4}b^{1/2}a^{1/4}$ and $(a, b) \mapsto a \sharp b$. If $\circ, \diamond$ denote any two different ones from these three operations and $\tau(a \circ b) = \tau(a \diamond b)$ holds for all $a, b \in \mathcal{A}^{++}$, then the algebra $\mathcal{A}$ is necessarily commutative.*

PROOF. By Lemma 10, we have that $\tau((x^{1/2}ax^{1/2})^{1/2}) \leq \tau((x^{1/2}bx^{1/2})^{1/2})$ holds for all $x \in \mathcal{A}^{++}$ if and only if $a \leq b$. The same lemma tells also that $\tau(x^{1/4}a^{1/2}x^{1/4}) \leq (x^{1/4}b^{1/2}x^{1/4})$ is valid for all $x \in \mathcal{A}^{++}$ if and only if $a^{1/2} \leq b^{1/2}$. By Lemma 11 we have that $\tau(a \sharp x) \leq \tau(b \sharp x)$ holds for all $x \in \mathcal{A}^{++}$ if and only if $a \leq b$.

Assume now that $\tau((a^{1/2}ba^{1/2})^{1/2}) = \tau(a^{1/4}b^{1/2}a^{1/4})$, $a, b \in \mathcal{A}^{++}$. It follows that we have $b \leq b'$ if and only if $b^{1/2} \leq b'^{1/2}$. Theorem 2 in [36] tells that if a non-concave continuous scalar valued function on the set of positive real numbers is operator monotone in a C^*-algebra, then the algebra is necessarily commutative. In our case the square function is operator monotone, hence applying that result we deduce the commutativity of $\mathcal{A}$. We argue in a similar way if $\tau(a^{1/4}b^{1/2}a^{1/4}) = \tau(a\sharp b)$, $a, b \in \mathcal{A}^{++}$. Finally, if $\tau((a^{1/2}ba^{1/2})^{1/2}) = \tau(a\sharp b)$, $a, b \in \mathcal{A}^{++}$, then by (5) we have $\tau((a^{1/2}ba^{1/2})^{1/2}) = \tau(a^{1/4}b^{1/2}a^{1/4}) = \tau(a\sharp b)$, $a, b \in \mathcal{A}^{++}$ and can complete the proof by referring back to the previous part. □

As a corollary we obtain the results mentioned above on the structures of Bures-Wasserstein isometries, Hellinger isometries, and maps preserving the trace distance between the arithmetic and geometric means of elements in positive definite cones.

COROLLARY 16. *Let $\mathcal{A}, \mathcal{B}$ be C^*-algebras with faithful traces τ and ω, respectively.*

The surjective map $\phi : \mathcal{A}^{++} \to \mathcal{B}^{++}$ is a Bures-Wasserstein isometry if and only if it is a Hellinger isometry if and only if it preserves the trace distance between the arithmetic and geometric means of elements. Namely, those maps are exactly the transformations of the form $\phi(a) = cJ(a)$, $a \in \mathcal{A}^{++}$, where $J : \mathcal{A} \to \mathcal{B}$ is a Jordan $*$-isomorphism and $c \in \mathcal{B}^{++}$ is a central element such that $\omega(cJ(x)) = \tau(x)$, $x \in \mathcal{A}$.

If there is a surjective isometry between $\mathcal{A}^{++}$ and $\mathcal{B}^{++}$, where $\mathcal{A}^{++}$ is equipped with the Bures-Wasserstein metric and $\mathcal{B}^{++}$ is equipped with the Hellinger distance, then $\mathcal{A}, \mathcal{B}$ are necessarily commutative in which case those two metrics are trivially coincide. Similar statements hold for the remaining two different pairs of the Bures-Wasserstein metric, Hellinger distance and the square root of the double of the trace distance between the arithmetic and geometric means of elements.

We note that beside the above considered particular distance measures, there is another very natural collection of metrics, namely the pth trace distances. Therefore, in view of the above results, the next problem is very natural.

PROBLEM 2. *Let $\mathcal{A}, \mathcal{B}$ be C^*-algebras with faithful traces τ, ω, respectively. What are the surjective isometries $\phi : \mathcal{A}^{++} \to \mathcal{B}^{++}$ with respect to the norm $(\tau(|.|^p))^{1/p}, (\omega(|.|^p))^{1/p}$?* (To see that those functions are true norms, we refer to [37].)

The really interesting case here is when $p = 1$, the reason being that then the corresponding norms are certainly not strictly convex.

We finally point out that in the recent paper [34] we have obtained results of very similar kinds concerning various types of the important quantum information related distance measures called Rényi divergences. Our strategy there has been similar, based on the use of order isomorphisms. To make this idea work we have needed various characterizations of the order expressed by those divergences.

In what follows we consider some other numerical quantities on positive definite cones and their related preservers. The distance measures above are connected to traces. In the next part of the section we deal with abstract determinants.

In the paper [12] Gaál and Nayak considered positive definite cones in finite von Neumann algebras with normalized faithful traces and mappings between them what we may call additively determinant-preserving. These are maps with the property that the determinant of the sum of elements coincide with the determinant of the sum of their images. In what follows we apply our method based on the statement in Corollary 9 to extend the result Theorem 4.2 in [12] to the context of C^*-algebras. Let $\mathcal{A}$ be a C^*-algebra and consider a 'determinant like' function on the positive definite cone of $\mathcal{A}$ by what we mean a function $\Delta : \mathcal{A}^{++} \to]0, \infty[$ with the following properties:

(d1) Δ is (norm-)continuous;
(d2) $\Delta(aba) = \Delta(a)\Delta(b)\Delta(a)$, $a, b \in \mathcal{A}^{++}$;
(d3) $\Delta(\lambda a) = \lambda \Delta(a)$, for any real number $\lambda > 0$ and $a \in \mathcal{A}^{++}$;
(d4) $\Delta(a + b) \geq \Delta(a) + \Delta(b)$, $a, b \in \mathcal{A}^{++}$;
(d5) for any $a, b \in \mathcal{A}^{++}$, if $\Delta(a+b) = \Delta(a) + \Delta(b)$, then $b = \lambda a$ holds for some real number $\lambda > 0$.
(d6) for any $a, b \in \mathcal{A}^{++}$, if $a \leq b$ and $\Delta(a) = \Delta(b)$, then we have $a = b$.

Observe that the Fuglede-Kadison determinant what the authors considered in [12] has the above properties. In order to apply our method based on the structure of positive homogeneous order isomorphisms, we need the following characterization of the order $\leq$ by the help of the function Δ.

LEMMA 17. *Let $\mathcal{A}$ be a C^*-algebra. Let $\Delta : \mathcal{A}^{++} \to]0, \infty[$ be a function with the poperties (d1), (d2), (d4), (d6). For any $a, b \in \mathcal{A}^{++}$, we have $a \leq b$ if and only if $\Delta(a + x) \leq \Delta(b + x)$ holds for all $x \in \mathcal{A}^{++}$.*

PROOF. To the necessity, let $a, b \in \mathcal{A}^{++}$ be such that $a \leq b$. Clearly, for any positive real number $\epsilon > 0$, by (d4) we have that

$$\Delta(b + \epsilon 1) = \Delta(a + \epsilon 1 + (b - a)) \geq \Delta(a) + \Delta(\epsilon 1 + (b - a)) \geq \Delta(a).$$

Letting $\epsilon \to 0$ and using (d1), we obtain $\Delta(b) \geq \Delta(a)$. This apparently implies the necessity part of the statement.

Let now $a, b \in \mathcal{A}^{++}$ and assume that $\Delta(a+x) \leq \Delta(b+x)$ holds for all $x \in \mathcal{A}^{++}$. Multiplying this inequality by $\Delta(b^{-1/2})$ from both sides, for $c = b^{-1/2}ab^{-1/2}$ we have

(11) $$\Delta(c + x) \leq \Delta(1 + x), \quad x \in \mathcal{A}^{++}.$$

Let $g :]0, \infty[\to]0, \infty[$ be the function which equals 1 for $0 < t \leq 1$, and $g(t) = t$ for $t \geq 1$. For any $n \in \mathbb{N}$ define the function $f_n :]0, \infty[\to]0, \infty[$ in the following way: $f_n(t)$ equals n for $0 < t \leq 1$, f_n is linear on $[1, 1+1/n]$ connecting the points $(1, n)$ and $(1 + 1/n, 0)$ in the plane, and $f_n(t) = 0$ for $t \geq 1 + 1/n$. We easily deduce that

$$\left| \frac{t + f_n(t)}{1 + f_n(t)} - g(t) \right| \leq 1/n, \quad t > 0,$$

and hence obtain that

$$\frac{t + f_n(t)}{1 + f_n(t)} \xrightarrow{n \to \infty} g(t)$$

uniformly for $t > 0$. It follows that $(c + f_n(c))(1 + f_n(c))^{-1} \xrightarrow{n \to \infty} g(c)$ and, by (d1), we have

$$\Delta((c + f_n(c))(1 + f_n(c))^{-1}) \xrightarrow{n \to \infty} \Delta(g(c)).$$

On the other hand, by (11), we know that

$$\Delta((c + f_n(c))(1 + f_n(c))^{-1}) \leq 1$$

holds for every n. Therefore, we obtain that $\Delta(g(c)) \leq 1$. Since $g(c) \geq 1$ clearly holds, using the obvious equality $\Delta(1) = 1$ and property (d6), it follows that $g(c) = 1$. This implies that $c \leq 1$ and hence that $a \leq b$. The proof of the lemma is complete. □

We are now in a position to prove the following extension of the result Theorem 4.2 in [**12**].

THEOREM 18. *Let $\mathcal{A}, \mathcal{B}$ be C^*-algebras and let $\Delta : \mathcal{A}^{++} \to]0, \infty[$, $\Gamma : \mathcal{B}^{++} \to]0, \infty[$ be functions with the properties (d1)-(d6). Let $\phi, \psi : \mathcal{A}^{++} \to \mathcal{B}^{++}$ be surjective functions. We have*

(12) $$\Gamma(\phi(a) + \psi(b)) = \Delta(a+b), \quad a, b \in \mathcal{A}^{++}$$

*if and only if $\phi = \psi$ and there is an element $c \in \mathcal{B}^{++}$ with $\Gamma(c) = 1$, and a Jordan *-isomorphism $J : \mathcal{A} \to \mathcal{B}$ with $\Gamma(J(a)) = \Delta(a)$, $a \in \mathcal{A}^{++}$ such that $\phi(a) = cJ(a)c$, $a \in \mathcal{A}^{++}$.*

PROOF. We begin with the necessity part. Assume ϕ, ψ satisfies (12). By Lemma 17, we have $\phi(a) \leq \phi(b)$ if and only if $a \leq b$ and similar observation holds for ψ, too. It follows that ϕ, ψ are order isomorphisms between $\mathcal{A}^{++}$ and $\mathcal{B}^{++}$. As in the proof of Theorem 12, for any sequence $a_n \in \mathcal{A}^{++}$ we have that $a_n \to 0$ if and only if for every $x \in \mathcal{A}^{++}$ there is an index n_0 such that $a_n \leq x$ holds for all $n \geq n_0$. It follows that $a_n \to 0$ implies $\phi(a_n), \psi(a_n) \to 0$. Using this, choosing a sequence (b_n) in $\mathcal{A}^{++}$ converging to zero, from (12) we can deduce that $\Gamma(\phi(a)) = \Delta(a)$, $a \in \mathcal{A}^{++}$. Obviously, similar identity holds for ψ, too. Select $a \in \mathcal{A}^{++}$. We compute

$$\Gamma(\phi(a) + \psi(a)) = \Delta(2a) = 2\Delta(a) = \Delta(a) + \Delta(a) = \Gamma(\phi(a)) + \Gamma(\psi(a)).$$

By (d5), we infer that $\psi(a) = \lambda\phi(a)$ holds for some $\lambda > 0$. On the other hand, we have

$$\Delta(a) = \Gamma(\psi(a)) = \Gamma(\lambda\phi(a)) = \lambda\Gamma(\phi(a)) = \lambda\Delta(a)$$

implying that $\lambda = 1$. It follows that $\phi = \psi$. In a similar fashion, we prove that ϕ is positive homogeneous. Indeed, for an arbitrary $a \in \mathcal{A}^{++}$ and positive real number λ, we have

$$\Gamma(\phi(a) + \phi(\lambda a)) = \Delta((1+\lambda)a) = (1+\lambda)\Delta(a) = \Delta(a) + \Delta(\lambda a) = \Gamma(\phi(a)) + \Gamma(\phi(\lambda a)).$$

This implies that $\phi(\lambda a) = \mu\phi(a)$ holds for some $\mu > 0$ and then we obtain

$$\lambda\Delta(a) = \Delta(\lambda a) = \Gamma(\phi(\lambda a)) = \Gamma(\mu\phi(a)) = \mu\Delta(a).$$

We infer $\phi(\lambda a) = \lambda\phi(a)$ and hence obtain that ϕ is a positive homogeneous order isomorphism. Applying Corollary 9, we get that $\phi(a) = cJ(a)c$ holds for some $c \in \mathcal{B}^{++}$ and Jordan *-isomorphism $J : \mathcal{A} \to \mathcal{B}$. Since $\Gamma(c^2) = \Gamma(\phi(1)) = \Delta(1) = 1$, we deduce $\Gamma(c) = 1$ completing the proof of the necessity part. The sufficiency is just obvious. □

It is a natural question to study the connection between 'determinants' on the positive definite cones of C^*-algebras and traces. By Lemma 16 in [**32**], if $\mathcal{A}$ is any C^*-algebra and $\Delta : \mathcal{A}^{++} \to]0, \infty[$ is a function with properties (d1), (d2), then there is a continuous linear functional $\varphi : \mathcal{A}_s \to \mathbb{R}$ such that $\Delta = \exp(\varphi(\log a))$,

$a \in \mathcal{A}^{++}$. Moreover, Lemma 17 in [32] asserts that if $\mathcal{A}$ is a von Neumann algebra, then a continuous linear functional $\varphi : \mathcal{A} \to \mathbb{C}$ which has real values on $\mathcal{A}_s$ satisfies

$$\exp(\varphi(\log(aba))) = \exp(\varphi(\log a)) \exp(\varphi(\log b)) \exp(\varphi(\log a)), \quad a, b \in \mathcal{A}^{++}$$

if and only is it is tracial meaning that $\varphi(ab) = \varphi(ba)$ holds for all $a, b \in \mathcal{A}$. In fact, one can check the proof in [32] to see that the sufficiency part of the statement holds true in general C^*-algebras. Therefore, we pose the following natural problem for further study.

PROBLEM 3. Assume $\mathcal{A}$ is a C^*-algebra and $\varphi : \mathcal{A} \to \mathbb{C}$ is a continuous linear functional which has real values on $\mathcal{A}_s$. Suppose further that

$$\exp(\varphi(\log(aba))) = \exp(\varphi(\log a)) \exp(\varphi(\log b)) \exp(\varphi(\log a)), \quad a, b \in \mathcal{A}^{++}.$$

Does it follow that the functional φ is necessarily tracial?

Above we have considered additively determinant-preserving transformations. The apparent dual problem concerns multiplicatively trace-preserving functions. Indeed, in [12], applying derivation in a suitable way, the authors showed that the additively Fuglede-Kadison determinant-preserving maps between positive definite cones in finite von Neumann algebras are in fact multiplicatively trace-preserving.

We have the following easy statement.

PROPOSITION 19. Let $\mathcal{A}, \mathcal{B}$ be C^*-algebras with faithful traces τ, ω, respectively. The surjective map $\phi : \mathcal{A}^{++} \to \mathcal{B}^{++}$ satisfies

(13) $$\omega(\phi(a)\phi(b)) = \tau(ab), \quad a, b \in \mathcal{A}^{++}$$

if and only if there is a central element $c \in \mathcal{B}^{++}$ and a Jordan *-isomorphism $J : \mathcal{A} \to \mathcal{B}$ such that $\phi(a) = cJ(a)$, $a \in \mathcal{A}^{++}$ and $\omega(c^2 J(x)) = \tau(x)$, $x \in \mathcal{A}$.

PROOF. Suppose that $\phi : \mathcal{A}^{++} \to \mathcal{B}^{++}$ is a surjective map which satisfies (13). By Lemma 10 (applied for $\alpha = 1$), we easily deduce that ϕ is a positive homogeneous order isomorphism. Therefore, by Corollary 9, there is an element $d \in \mathcal{B}^{++}$ and a Jordan *-isomorphism $J : \mathcal{A} \to \mathcal{B}$ such that $\phi(a) = dJ(a)d$, $a \in \mathcal{A}^{++}$. We also have

$$\omega(d^4 J(a)) = \omega(dJ(a)ddJ(1)d) = \tau(a), \quad a \in \mathcal{A}^{++}.$$

Applying Lemma 14, we have that d^4 and hence also d are central elements. Choosing $c = d^2$, the sufficiency part of the proof is now obvious. The necessity is also easy to verify, we omit the details. □

We note that with more efforts one can actually prove the following analogue of Theorem 18. With the notation of the proposition above, the pair ϕ, ψ of surjective functions from $\mathcal{A}^{++}$ onto $\mathcal{B}^{++}$ satisfies the identity

$$\omega(\phi(a)\psi(b)) = \tau(ab), \quad a, b \in \mathcal{A}^{++}$$

if and only if there is a central element $c \in \mathcal{B}^{++}$, an element $d \in \mathcal{B}^{++}$ and a Jordan *-isomorphism $J : \mathcal{A} \to \mathcal{B}$ such that $\phi(a) = dJ(a)d$, $\psi(a) = cd^{-1}J(a)d^{-1}$, $a \in \mathcal{A}^{++}$, and $\omega(cJ(x)) = \tau(x)$, $x \in \mathcal{A}$.

We continue along this line and in what follows consider so-called additively and multiplicatively spectrum-preserving maps between the Jordan algebras of all self-adjoint elements in operator algebras. The motivation to consider those structures comes from the fact that they play an essential role in the algebraic approach to quantum theory (that we have referred to in the introduction). Our aim is to give simple non-linear characterizations of the corresponding isomorphisms. The results in the remaining part of this section have recently appeared in our paper [35].

As for the case of additively spectrum-preserving maps, we have the following

PROPOSITION 20. *Let $\mathcal{A}, \mathcal{B}$ be C^*-algebras, $\phi : \mathcal{A}_s \to \mathcal{B}_s$ be a surjective map. We have*
$$\sigma(\phi(a) + \phi(b)) = \sigma(a+b), \quad a, b \in \mathcal{A}_s$$
*if and only if ϕ extends to Jordan *-isomorphism from $\mathcal{A}$ to $\mathcal{B}$.*

The proof of this result is easy, it follows straightforwardly from the classical Mazur-Ulam theorem and Kadison's result, Theorem 2 in [20], on the structure of linear surjective isometries between $\mathcal{A}_s$ and $\mathcal{B}_s$ that we have already referred to. The situation is very much different with multiplicatively spectrum-preserving maps. Concerning them we have employed the method based on Corollary 9.

Before presenting our corresponding result, let us point out that the study of multiplicatively spectrum-preserving maps started with our paper [27] where we studied such maps on full operator algebras and on the algebras of all continuous functions over compact Hausdorff spaces. The paper motivated a number of further investigations; here we mention only the works [7, 15, 17, 25] relating function algebras, matrix algebras, full or standard operator algebras over Banach spaces. Furthermore, we in particular mention the recent paper [4] concerning the general non-commutative setting. There the authors consider general semisimple Banach algebras with the additional condition of having essential socle. Regarding that assumption from the direction of operator algebras, it appears to be quite restrictive in the sense that, for example, type II and III von Neumann algebras (which play so important role in the theory of operator algebras and their applications) do typically not satisfy them. The following result (Theorem 2.2 in [35]) is about general von Neumann algebras and reads as follows. By a symmetry we mean a self-adjoint unitary element.

THEOREM 21. *Let $\mathcal{A}, \mathcal{B}$ be von Neumann algebras, $\phi : \mathcal{A}_s \to \mathcal{B}_s$ be a surjective map. We have*
$$\sigma(\phi(a)\phi(b)) = \sigma(ab), \quad a, b \in \mathcal{A}_s$$
*if and only if there is a central symmetry $s \in \mathcal{B}$ and a Jordan *-isomorphism $J : \mathcal{A} \to \mathcal{B}$ such that*
$$\phi(a) = sJ(a), \quad a \in \mathcal{A}_s.$$

As for the proof which can be found in [35], it worth mentioning that beside employing Theorem 8 (or rather Corollary 9), we also used the following spectral characterizations of elements which might be of independent interest. The first such result is about characterizing central symmetries.

LEMMA 22. *Let $\mathcal{A}$ be a von Neumann algebra. Assume $s \in \mathcal{A}$ is a symmetry such that for every symmetry $t \in \mathcal{A}$, the spectrum $\sigma(st)$ contains only real numbers. Then s is necessarily a central symmetry in $\mathcal{A}$.*

The second characterization is a spectral identification of self-adjoint elements.

LEMMA 23. *Let $\mathcal{A}$ be a von Neumann algebra. Pick $a, b \in \mathcal{A}_s$. If $\sigma(at) = \sigma(bt)$ holds for all $t \in \mathcal{A}^{++}$, then we have $a = b$.*

Having proven these results, the proof of the necessity part of Theorem 21 is simple. We first show that $\phi(1)$ is a central symmetry which allows us to assume without loss of generality that $\phi(1) = 1$. It follows that ϕ is a bijective map between positive definite cones which is a positive homogeneous Thompson isometry and we apply Corollary 9. Having known the form of ϕ on the positive definite cone and employing Lemma 23, we obtain the structure of ϕ on the whole set of self-adjoint elements.

As can be seen, our result Theorem 21 is proved only for von Neumann algebras. In view of Proposition 20 concerning additively spectrum-preserving maps on general C^*-algebras the following question is very natural to pose.

PROBLEM 4. Is the statement in Theorem 21 true for general C^*-algebras, too?

We have positive answer for a rather particular class of C^*-algebras called standard C^*-algebras. By such an algebra we mean a C^*-algebra of operators acting on a Hilbert space which contains the ideal of all finite-rank operators (hence that of the compact ones, too) and the identity. Important examples of such algebras include, beside the full operator algebra, the Toeplitz algebra (generated by a unilateral shift), the Laurent algebra (generated by a bilateral shift), or more generally, any extension of the algebra of compact operators by a commutative C^*-algebra, which concept plays a central role in the famous Brown-Douglas-Fillmore (BDF) theory.

Our related result presented in [**35**] is the following. The proof is quite similar in spirit to that of Theorem 21.

THEOREM 24. *Let $\mathcal{A}, \mathcal{B}$ be standard C^*-algebras. Suppose $\mathcal{A}$ is acting on the Hilbert space H and $\mathcal{B}$ is acting on the Hilbert space K. Let $\phi : \mathcal{A}_s \to \mathcal{B}_s$ be a surjective map. We have that*

$$\sigma(\phi(a)\phi(b)) = \sigma(ab)$$

holds for all $a, b \in \mathcal{A}_s$ if and only if there is a constant $\lambda \in \{-1, 1\}$ and either a unitary or an antiunitary operator $u : K \to H$ such that

$$\phi(a) = \lambda u a u^*, \quad a \in \mathcal{A}_s.$$

In [**35**] we also considered additively norm-preserving and multiplicatively norm-preserving maps. Observe that for a transformation $\phi : \mathcal{A}_s \to \mathcal{B}_s$, the property

(14) $$\|\phi(a)\phi(b)\| = \|ab\|, \quad a, b \in \mathcal{A}_s,$$

does not characterize Jordan *-isomorphisms as the multiplicatively spectrum-preserving property does in Theorem 21. Indeed, if $J : \mathcal{A} \to \mathcal{B}$ is a Jordan *-isomorphism and we have central symmetries $s_a \in \mathcal{B}$ varying as $a \in \mathcal{A}_s$ varies, then the map $a \mapsto s_a J(a)$ automatically satisfies (14). The situation becomes different if we consider our maps not on the self-adjoint parts of operator algebras but on positive semidefinite cones. In [**35**] we have obtained the following results. Our arguments have again been built on the use of the structure of Thompson isometries, especially on Corollary 9. Theorem 2.6 in [**35**] states the following.

THEOREM 25. *Let $\mathcal{A}, \mathcal{B}$ be von Neumann algebras. The surjective function $\phi : \mathcal{A}^+ \to \mathcal{B}^+$ satisfies*

$$\|\phi(a)\phi(b)\| = \|ab\|, \quad a, b \in \mathcal{A}^+$$

*if and only if there is a Jordan *-isomorphism $J : \mathcal{A} \to \mathcal{B}$ which extends ϕ.*

We note here that this statement is valid also in the setting of C^*-algebras and for multiplicatively norm-preserving maps between the corresponding positive definite cones. Indeed, one can follow the proof Theorem 25 given in [35] and use the following characterization of order in an arbitrary C^*-algebra $\mathcal{A}$: for any $a, b \in \mathcal{A}^{++}$ we have $a \leq b$ if and only if $\|xax\| \leq \|xbx\|$ holds for all $x \in \mathcal{A}^{++}$. To verify the nontrivial part of this assertion, i.e., the sufficiency, we choose $x = b^{-1/2}$ and see that $\|b^{-1/2}ab^{-1/2}\| \leq 1$ which gives us that $b^{-1/2}ab^{-1/2} \leq 1$ and hence that $a \leq b$.

The case of additively norm-preserving maps is definitely more difficult and we have a result only in the setting of von Neumann algebras and concerning maps between their positive semidefinite cones, see Theorem 2.7 in [35]. It reads as follows.

THEOREM 26. *Let $\mathcal{A}, \mathcal{B}$ be von Neumann algebras. The surjective function $\phi: \mathcal{A}^+ \to \mathcal{B}^+$ satisfies*

$$\|\phi(a) + \phi(b)\| = \|a + b\|, \quad a, b \in \mathcal{A}^+$$

if and only if there is a Jordan *-isomorphism $J: \mathcal{A} \to \mathcal{B}$ which extends ϕ.

The proof rests on the next lemma which provides us with a certain additional characterization of the order.

LEMMA 27. *Let $\mathcal{A}$ be a von Neumann algebra and $a, b \in \mathcal{A}^+$. The following assertions are equivalent:*

(i) $a \leq b$;
(ii) $\|pap\| \leq \|pbp\|$ *holds for all projections $p \in \mathcal{A}$;*
(iii) $\|a + \lambda p\| \leq \|b + \lambda p\|$ *holds for all projections $p \in \mathcal{A}$ and all positive real numbers λ.*

The key steps of the proof of Theorem 26 given in [35] have been to verify first that an additively norm-preserving bijection between the positive semidefinite cones of von Neumann algebras is an order isomorphism (there we used Lemma 27). After that we have shown that such a map necessarily preserves zero products, then that it preserves projections and finally that it is positive homogeneous. After that we applied Corollary 9 to finish the necessity part of the proof.

We believe that the conclusion in Theorem 26 is valid also for C^*-algebras and their positive definite cones. Obviously, the approach what we have followed in [35] cannot be used (we only refer to the existence of nontrivial projections), some significantly different argument should be invented to solve the following problem.

PROBLEM 5. Is Theorem 26 valid for C^*-algebras? Does it hold also on positive definite cones?[2]

We mention that in [10] Gaál studied a similar problem and obtained positive result for the p-norms (corresponding to faithful traces) on the positive definite cones in C^*-algebras in the case where $1 < p < \infty$. His main tool was the differentiability property of those norms which is certainly not valid for the original C^*-norm what we are interested in here. Nevertheless, the positive result in [10] gives further motivation to study the problem formulated right above.

---

[2]The problem has recently been solved and will be published in a forthcoming paper. The answers to both questions are affirmative.

4. Additional non-linear characterizations on the self-adjoint and positive definite parts of the full operator algebra $B(H)$

In the previous section we have applied the structural result Corollary 9 on positive homogeneous order isomorphisms between positive definite cones in C^*-algebras for various preserver problems that have provided characterizations of Jordan *-isomorphisms between the underlying algebras.

In what follows H stands for a complex Hilbert space and we denote by $B(H)$ the C^*-algebra of all bounded linear operators on H. In the paper [26] we obtained that in the particular case of the C^*-algebra $B(H)$, we even do not need to assume the homogeneity of order automorphisms in order to obtain a structural result like in Corollary 9. Among others we proved the following result (also see Section 2.5 in [29]).

THEOREM 28. *Suppose that* $\dim H \geq 2$. *Let* $\phi : B(H)_s \to B(H)_s$ *be a surjective map with the property that*

$$a \leq b \iff \phi(a) \leq \phi(b)$$

holds whenever $a, b \in B(H)_s$. *Then there exists an invertible operator* $t \in B(H)$ *and an element* $x \in B(H)_s$ *such that* ϕ *is of one of the following two forms:*

$$\phi(a) = tat^* + x, \quad a \in B(H)_s, \quad \text{or} \quad \phi(a) = ta^T t^* + x, \quad a \in B(H)_s.$$

Here ^{T} denotes transpose with respect to a given orthonormal basis in H.

We proved analogous statements for the order automorphisms of the positive definite cone $B(H)^{++}$ of $B(H)$ and also for the order automorphisms of the positive semidefinite cone $B(H)^+$, see [26] and [31]. We mention that recently P. Šemrl has made a fundamental contribution to those results of ours by completing them with the precise descriptions of all order isomorphisms between any operator intervals in $B(H)$. For his nice deep results, see [41] and [42].

After this, the natural question arises if those results could be extended for more general operator algebras. In particular, we pose the following problem to start investigations in that direction.

PROBLEM 6. *Let* $\mathcal{A}, \mathcal{B}$ *be nontrivial (not one-dimensional) von Neumann factors. Is it true that any order isomorphism* $\phi : \mathcal{A}^{++} \to \mathcal{B}^{++}$ *is of the form* $\phi(a) = cJ(a)c$ *with some* $c \in \mathcal{B}^{++}$ *and Jordan *-isomorphism* $J : \mathcal{A} \to \mathcal{B}$?[3]

Let us recall an interesting application of Theorem 28 for the characterization of the so-called logarithmic product

$$a \odot b = e^{\log a + \log b}, \quad a, b \in B(H)^{++}.$$

This concept originally emerged from computational geometry but soon after serious applications were found in the differential geometry of spaces of positive definite operators which is a large and active area of research in present days. In particular, important applications concern medical imaging with DT-MRI.

---

[3]The problem has recently been solved, the answer to the question is affirmative. In fact, even more general results (extensions of P. Šemrl's results mentioned above for the setting of von Neumann algebras), have been presented in the paper M. Mori, Order isomorphisms of operator intervals in von Neumann algebras, Integral Equations Operator Theory **91** (2019), no. 2, Art. 11, 26 pp.

To formulate our result we need the concept of the so-called chaotic order $\preceq$ on $B(H)^{++}$ which is defined as follows: for any $a, b \in \mathcal{A}^{++}$ we write $a \preceq b$ if and only if $\log a \leq \log b$. In [31] we proved the following

THEOREM 29. *Suppose that* $\dim H \geq 2$ *and* $\circ$ *is a binary operation on* $B(H)^{++}$ *that makes it a commutative group which is ordered under the chaotic order. Then we have*
$$a \circ b = e^{\log a + \log b - \log e}, \quad a, b \in B(H)^{++}$$
where e *is the unit of the group* $(B(H)^{++}, \circ)$.

One can now obviously ask about the usual order $\leq$. In the same paper [31] we proved the following possibly somewhat surprising fact.

THEOREM 30. *If* $\dim H \geq 2$, *then there is no binary operation on* $B(H)^{++}$ *which makes it an ordered commutative group under the usual order* $\leq$.

To put this into a wider perspective, we pose the following question.

PROBLEM 7. If the positive definite cone of a C^*-algebra can be made a commutative group with an operation which is ordered under the usual order, does it imply that the algebra is necessarily commutative (with respect to the original product)?

We conjecture an affirmative answer but have no ideas concerning a possible proof.

We have already mentioned the problem of commutativity preserving mappings in the linear context. In fact, such problems played a fundamental role in the development of the theory of preservers. Apparently, if we omit linearity, the picture concerning those maps becomes much more complicated. For example, maps sending the elements in an algebra to certain polynomials of those elements (the polynomials may be varying from elements to elements) are in general highly nonlinear maps preserving commutativity. As for the space of all self-adjoint operators on a separable Hilbert space we have the following result concerning the converse. In [28] we proved the next theorem which is the last result we present in this paper. The main reason to exhibit it here is to demonstrate that we do not always have linearity (or affinity) of our preservers for free, the situation can sometimes be very much different from what we have seen in the previous part of the paper.

THEOREM 31. *Let* H *be a complex separable Hilbert space with* $\dim H \geq 3$ *and let* $\phi : B(H)_s \to B(H)_s$ *be a bijective transformation which preserves commutativity in both directions. Then there exists a unitary operator* u *on* H *and for every operator* $a \in B(H)_s$ *there is a real valued bounded Borel function* f_a *on* $\sigma(a)$ *such that*
$$\phi(a) = u f_a(a) u^*, \quad a \in B(H)_s$$
or
$$\phi(a) = u (f_a(a))^T u^*, \quad a \in B(H)_s.$$

It is again natural to consider the problem in the setting of more general operator algebras. We close the paper with following

PROBLEM 8. Let $\mathcal{A}, \mathcal{B}$ be von Neumann factors not of type I_1 or I_2 acting on separable Hilbert spaces and let $\phi : \mathcal{A}_s \to \mathcal{B}_s$ be a bijective map which preserves commutativity in both directions. Does it follow that there is a Jordan

*-isomorphism $J : \mathcal{A} \to \mathcal{B}$ with the property that for any $a \in \mathcal{A}_s$ there is a real valued bounded Borel function f_a on $\sigma(a)$ such that

$$\phi(a) = J(f_a(a)), \quad a \in \mathcal{A}_s?$$

References

[1] P. Ara and M. Mathieu, *Local multipliers of C^*-algebras*, Springer Monographs in Mathematics, Springer-Verlag London, Ltd., London, 2003. MR1940428

[2] B. Aupetit, *Spectrum-preserving linear mappings between Banach algebras or Jordan-Banach algebras*, J. London Math. Soc. (2) **62** (2000), no. 3, 917–924, DOI 10.1112/S0024610700001514. MR1794294

[3] R. Bhatia, T. Jain, and Y. Lim, *On the Bures-Wasserstein distance between positive definite matrices*, Expo. Math. **37** (2019), no. 2, 165–191, DOI 10.1016/j.exmath.2018.01.002. MR3992484

[4] A. Bourhim, J. Mashreghi, and A. Stepanyan, *Maps between Banach algebras preserving the spectrum*, Arch. Math. (Basel) **107** (2016), no. 6, 609–621, DOI 10.1007/s00013-016-0960-9. MR3571153

[5] M. Brešar and P. Šemrl, *Mappings which preserve idempotents, local automorphisms, and local derivations*, Canad. J. Math. **45** (1993), no. 3, 483–496, DOI 10.4153/CJM-1993-025-4. MR1222512

[6] M. Brešar, M. A. Chebotar, and W. S. Martindale III, *Functional identities*, Frontiers in Mathematics, Birkhäuser Verlag, Basel, 2007. MR2332350

[7] J.-T. Chan, C.-K. Li, and N.-S. Sze, *Mappings preserving spectra of products of matrices*, Proc. Amer. Math. Soc. **135** (2007), no. 4, 977–986, DOI 10.1090/S0002-9939-06-08568-6. MR2262897

[8] D. Farenick, S. Jaques, and M. Rahaman, *The fidelity of density operators in an operator-algebraic framework*, J. Math. Phys. **57** (2016), no. 10, 102202, 15, DOI 10.1063/1.4965876. MR3564319

[9] D. Farenick and M. Rahaman, *Bures contractive channels on operator algebras*, New York J. Math. **23** (2017), 1369–1393. MR3723514

[10] M. Gaál, *Norm-additive maps on the positive definite cone of a C^*-algebra*, Results Math. **73** (2018), no. 4, Art. 151, 7, DOI 10.1007/s00025-018-0916-4. MR3871574

[11] M. Gaál, *Maps between positive cones of operator algebras preserving a measure of the difference between arithmetic and geometric means*, Positivity **23** (2019), no. 2, 461–467, DOI 10.1007/s11117-018-0617-y. MR3928821

[12] M. Gaál and S. Nayak, *On a class of determinant preserving maps for finite von Neumann algebras*, J. Math. Anal. Appl. **464** (2018), no. 1, 317–327, DOI 10.1016/j.jmaa.2018.04.006. MR3794091

[13] L. A. Harris and R. V. Kadison, *Affine mappings of invertible operators*, Proc. Amer. Math. Soc. **124** (1996), no. 8, 2415–2422, DOI 10.1090/S0002-9939-96-03445-4. MR1340389

[14] O. Hatori and L. Molnár, *Isometries of the unitary groups and Thompson isometries of the spaces of invertible positive elements in C^*-algebras*, J. Math. Anal. Appl. **409** (2014), no. 1, 158–167, DOI 10.1016/j.jmaa.2013.06.065. MR3095026

[15] O. Hatori, S. Lambert, A. Luttman, T. Miura, T. Tonev, and R. Yates, *Spectral preservers in commutative Banach algebras*, Function spaces in modern analysis, Contemp. Math., vol. 547, Amer. Math. Soc., Providence, RI, 2011, pp. 103–123, DOI 10.1090/conm/547/10812. MR2856485

[16] I. N. Herstein, *Jordan homomorphisms*, Trans. Amer. Math. Soc. **81** (1956), 331–341, DOI 10.2307/1992920. MR76751

[17] J. Hou, C.-K. Li, and N.-C. Wong, *Jordan isomorphisms and maps preserving spectra of certain operator products*, Studia Math. **184** (2008), no. 1, 31–47, DOI 10.4064/sm184-1-2. MR2365474

[18] N. Jacobson and C. E. Rickart, *Jordan homomorphisms of rings*, Trans. Amer. Math. Soc. **69** (1950), 479–502, DOI 10.2307/1990495. MR38335

[19] R. V. Kadison, *Isometries of operator algebras*, Ann. Of Math. (2) **54** (1951), 325–338, DOI 10.2307/1969534. MR0043392

[20] R. V. Kadison, *A generalized Schwarz inequality and algebraic invariants for operator algebras*, Ann. of Math. (2) **56** (1952), 494–503, DOI 10.2307/1969657. MR51442
[21] I. Kaplansky, *Algebraic and analytic aspects of operator algebras*, American Mathematical Society, Providence, R.I., 1970. Conference Board of the Mathematical Sciences Regional Conference Series in Mathematics, No. 1. MR0312283
[22] E.-Y. Lee, *A matrix reverse Cauchy-Schwarz inequality*, Linear Algebra Appl. **430** (2009), no. 2-3, 805–810, DOI 10.1016/j.laa.2008.09.026. MR2473187
[23] M. Mathieu, *Towards a non-selfadjoint version of Kadison's theorem*, Ann. Math. Inform. **32** (2005), 87–94. MR2264870
[24] C. R. Miers, *Commutativity preserving maps of factors*, Canad. J. Math. **40** (1988), no. 1, 248–256, DOI 10.4153/CJM-1988-011-1. MR928222
[25] T. Miura and D. Honma, *A generalization of peripherally-multiplicative surjections between standard operator algebras*, Cent. Eur. J. Math. **7** (2009), no. 3, 479–486, DOI 10.2478/s11533-009-0033-4. MR2534467
[26] L. Molnár, *Order-automorphisms of the set of bounded observables*, J. Math. Phys. **42** (2001), no. 12, 5904–5909, DOI 10.1063/1.1413224. MR1866695
[27] L: Molnár, *Some characterizations of the automorphisms of $B(H)$ and $C(X)$*, Proc. Amer. Math. Soc. **130** (2002), no. 1, 111–120, DOI 10.1090/S0002-9939-01-06172-X. MR1855627
[28] L. Molnár and P. Šemrl, *Nonlinear commutativity preserving maps on self-adjoint operators*, Q. J. Math. **56** (2005), no. 4, 589–595, DOI 10.1093/qmath/hah058. MR2182468
[29] L. Molnár, *Selected preserver problems on algebraic structures of linear operators and on function spaces*, Lecture Notes in Mathematics, vol. 1895, Springer-Verlag, Berlin, 2007. MR2267033
[30] L. Molnár, *Thompson isometries of the space of invertible positive operators*, Proc. Amer. Math. Soc. **137** (2009), no. 11, 3849–3859, DOI 10.1090/S0002-9939-09-09963-8. MR2529894
[31] L. Molnár, *Order automorphisms on positive definite operators and a few applications*, Linear Algebra Appl. **434** (2011), no. 10, 2158–2169, DOI 10.1016/j.laa.2010.12.007. MR2781684
[32] L. Molnár, *General Mazur-Ulam type theorems and some applications*, Operator semigroups meet complex analysis, harmonic analysis and mathematical physics, Oper. Theory Adv. Appl., vol. 250, Birkhäuser/Springer, Cham, 2015, pp. 311–342, DOI 10.1007/978-3-319-18494-4_21. MR3468225
[33] L. Molnár, *Bures isometries between density spaces of C^*-algebras*, Linear Algebra Appl. **557** (2018), 22–33, DOI 10.1016/j.laa.2018.07.008. MR3848260
[34] L. Molnár, *Quantum Rényi relative entropies on density spaces of C^*-algebras: their symmetries and their essential difference*, J. Funct. Anal. **277** (2019), no. 9, 3098–3130, DOI 10.1016/j.jfa.2019.06.009. MR3997630
[35] L. Molnár, *Spectral characterization of Jordan-Segal isomorphisms of quantum observables*, J. Operator Theory, **83** (2020), no. 1, 179–195. DOI: http://dx.doi.org/10.7900/jot.2018aug31.2207. MR4043710
[36] M. Nagisa, M. Ueda, and S. Wada, *Commutativity of operators*, Nihonkai Math. J. **17** (2006), no. 1, 1–8. MR2241356
[37] E. Nelson, *Notes on non-commutative integration*, J. Funct. Anal. **15** (1974), 103–116, DOI 10.1016/0022-1236(74)90014-7. MR0355628
[38] T. W. Palmer, *Banach algebras and the general theory of $*$-algebras. Vol. I*, Encyclopedia of Mathematics and its Applications, vol. 49, Cambridge University Press, Cambridge, 1994. Algebras and Banach algebras. MR1270014
[39] B. Russo and H. A. Dye, *A note on unitary operators in C^*-algebras*, Duke Math. J. **33** (1966), 413–416. MR193530
[40] I. E. Segal, *Postulates for general quantum mechanics*, Ann. of Math. (2) **48** (1947), 930–948, DOI 10.2307/1969387. MR22652
[41] P. Šemrl, *Order isomorphisms of operator intervals*, Integral Equations Operator Theory **89** (2017), no. 1, 1–42, DOI 10.1007/s00020-017-2395-5. MR3712250
[42] P. Šemrl, *Groups of order automorphisms of operator intervals*, Acta Sci. Math. (Szeged) **84** (2018), no. 1-2, 125–136. MR3792768
[43] A. R. Sourour, *Invertibility preserving linear maps on $\mathcal{L}(X)$*, Trans. Amer. Math. Soc. **348** (1996), no. 1, 13–30, DOI 10.1090/S0002-9947-96-01428-6. MR1311919

University of Szeged, Interdisciplinary Excellence Centre, Bolyai Institute, H-6720 Szeged, Aradi vértanúk tere 1., Hungary and Institute of Mathematics, Budapest University of Technology and Economics, H-1521 Budapest, Hungary

Email address: molnarl@math.u-szeged.hu

URL: http://www.math.u-szeged.hu/~molnarl

Multiplicatively pseudo spectrum-preserving maps

Zine El Abidine Abdelali and Hamid Nkhaylia

ABSTRACT. Let $\mathcal{B}(X)$ be the algebra of all bounded linear operators on a complex Banach space X with $\dim X \geq 3$, and let $\mathcal{A}, \mathcal{B}$ be two subsets of $\mathcal{B}(X)$ containing all operators of rank at most one. For $\varepsilon > 0$, the ε-pseudo spectrum of any $A \in \mathcal{B}(X)$ is defined by

$$\sigma_\varepsilon(A) := \left\{\lambda \in \mathbb{C} : \|(\lambda I - A)^{-1}\| > \frac{1}{\varepsilon}\right\},$$

with the convention that $\|(\lambda I - A)^{-1}\| = \infty$ if $\lambda I - A$ is not invertible. We determine the structure of surjective maps $\phi, \psi : \mathcal{A} \longrightarrow \mathcal{B}$ satisfying

$$\phi(A)\psi(B) = 0 \iff AB = 0 \quad \text{for all} \quad A, B \in \mathcal{A}.$$

Moreover, in case when $X = \langle$ is a complex Hilbert space, we characterize surjective maps $\phi, \psi : \mathcal{A} \longrightarrow \mathcal{B}$ satisfying

$$\sigma_\varepsilon(\phi(A)\psi(B)) = \sigma_\varepsilon(AB) \quad \text{for all} \quad A, B \in \mathcal{A}.$$

Furthermore, some known results are obtained as an immediate consequences of our main results.

1. Introduction

Nonlinear preserver problems demand the characterization of maps on subsets of algebras that preserve various spectral quantities or subsets or relations but without assuming any algebraic condition like linearity or multiplicativity. Over the last few decades, these problems have been studied by numerous authors and the first nonlinear preserver problem was considered by Kowalski and Słodkowski who proved in [37] that a complex-valued function f on a Banach algebra $\mathcal{A}$ is linear and multiplicative provided that $f(0) = 0$ and $f(x) - f(y)$ lies in the spectrum of $x - y$ for all x and y in $\mathcal{A}$. This generalizes the well-known theorem of Gleason–Kahane–Żelazko in the theory of Banach algebra [28, 35]. Since then, a number of techniques have been developed to treat nonlinear preserver problems and many results have been obtained mainly in matrix theory and in operator theory; see for instance [7, 8, 11, 15–21, 25, 27, 29, 30, 32, 33, 39, 41, 42, 44]. In [8], Bhatia, Šemrl and Sourour described the form of all surjective maps on the algebra $\mathcal{M}_n(\mathbb{C})$ of all complex $n \times n$-matrices preserving the spectral radius of the difference of matrices, and thus, in particular, they provided an extension of Marcus and Moyls' result [40] in the absence of the linearity. In [42], Molnár studied maps preserving the spectrum of operator or matrix products and showed, in particular, that a

---

2010 *Mathematics Subject Classification*. Primary 47B49; Secondary 47A10, 47A11.

Key words and phrases. pseudo spectrum, pseudo spectral radius, nonlinear preserver.

surjective map φ on $\mathcal{B}(\mathcal{H})$, the algebra of all bounded linear operators on an infinite-dimensional complex Hilbert space $\mathcal{H}$, preserves the spectrum of operator products if and only if φ is an automorphism or an automorphism multiplied by -1. His results have been extended in several directions, and, in particular, there has been interest in studying maps ϕ on matrices or operators satisfying $F(\phi(A) \bullet \phi(B)) = F(A \bullet B)$. Here, $F(\cdot)$ is a spectral function or a spectral set such as the spectrum, the numerical range, the ε-pseudo spectral radius, the ε-pseudo spectrum and the local spectrum and $A \bullet B$ stands for different kinds of products such as the usual product AB, the triple product ABA, the Jordan product $AB+BA$, the skew product A^*B, the skew triple product AB^*A, and the skew-Jordan product $AB^* + B^*A$; see for instance [1-6, 9, 10, 12-14, 22-24, 31, 38] and the references therein.

Let X be a complex Banach space with $\dim X \geq 3$, and $\mathcal{B}(X)$ be the Banach algebra of all bounded linear operators on X. Let $\mathcal{A}$ and $\mathcal{B}$ be two subsets of $\mathcal{B}(X)$ containing all operators in $\mathcal{B}(X)$ of rank at most one. In this paper, we first characterize all surjective maps $\phi, \psi : \mathcal{A} \longrightarrow \mathcal{B}$ satisfying

$$\phi(A)\psi(B) = 0 \iff AB = 0 \text{ for all } A, B \in \mathcal{A}.$$

If $\varepsilon > 0$ and X is a Hilbert space, we then we use this characterization to describe all surjective maps $\phi, \psi : \mathcal{A} \longrightarrow \mathcal{B}$ such that the ε-pseudo spectrum of $\phi(A)\psi(B)$ coincides with that of AB for all $A, B \in \mathcal{A}$.

2. Main results

Throughout this paper, let $\mathcal{B}(X)$ be the Banach algebra of all bounded linear operators on a complex Banach space X with $\dim X \geq 3$, and denote its identity operator by I. Recall that the natural pairing between X and its dual space X^* is defined by $\langle \, , \, \rangle : X \times X^* \to \mathbb{C}, (x, f) \mapsto f(x)$, and that the adjoint of any operator $A \in \mathcal{B}(X)$ is defined by $A^* : X^* \to X^*$; $f \mapsto f \circ A$ so that $\langle Ax, f \rangle = \langle x, A^*f \rangle$ for all $(x, f) \in X \times X^*$. Note that if $A : X \to X$ is a bounded conjugate linear operator on X (i.e., A is additive and $A(\lambda x) = \overline{\lambda} A(x)$ for all $\lambda \in \mathbb{C}$ and $x \in X$), then its adjoint A^* is the bounded conjugate linear operator on X^* defined by $A^*(f) = \overline{f} \circ A$ for all $f \in X^*$. Thus $\langle Ax, f \rangle = \overline{\langle x, A^*f \rangle}$ for all $(x, f) \in X \times X^*$. For any $x \in X$ and $f \in X^*$, let $x \otimes f$ be the operator in $\mathcal{B}(X)$ of rank at most one defined by $(x \otimes f)(y) := \langle y, f \rangle x = f(y)x$ for all $y \in X$ and note that every operator in $\mathcal{B}(X)$ of rank at most one has this form. Let $\mathcal{F}_1(X)$ be the set of all operators of rank at most one in $\mathcal{B}(X)$. This means that $\mathcal{F}_1(X) := \{x \otimes f \; : \; x \in X \text{ and } f \in X^*\}$.

Let $\mathcal{A}$ and $\mathcal{B}$ be two subsets of $\mathcal{B}(X)$ containing all operators in $\mathcal{B}(X)$ of rank at most one. Our first result characterizes all surjective maps $\phi, \psi : \mathcal{A} \longrightarrow \mathcal{B}$ satisfying

(2.1) $\qquad \phi(A)\psi(B) = 0 \iff AB = 0 \quad (\forall A, B \in \mathcal{A}).$

This result was motivated by a theorem concerning the characterization of bijectives transformations on $\{P \in \mathcal{F}_1(X) \; : \; P^2 = P\}$ preserving zero product in both directions due to Molnár [43]. Note that the result of Molnár was given for infinite dimensional case. Here we will cover also the finite dimensional case and we are influenced by the approach, due to Šemrl [46], using the fundamental theorem of projective geometry (see [46, Theorems 1.1 and 1.2]) we will also use some ideas of [21, Lemma 2.2]. Finally, note that if $\dim(X) = n$, then X and X^* will be identified with $\mathbb{C}^n$ and $\mathcal{B}(X)$ with the algebra $\mathcal{M}_n(\mathbb{C})$ of all $n \times n$ complex matrices. Accordingly, for every $x = (x_1, ..., x_n)$ and $y = (y_1, ..., y_n)$ in $\mathbb{C}^n$, we have $\langle x, y \rangle = x_1 y_1 + \cdots x_n y_n$ and $x \otimes y = x^t y$, where x^t stands for the transpose of x.

Finally, we denote $x_\tau = (\tau(x_1), \tau(x_2), ..., \tau(x_n))$ for all ring automorphisms τ of $\mathbb{C}$ and all vectors $x = (x_1, ..., x_n) \in \mathbb{C}^n$.

THEOREM 2.1. *Suppose that $\phi, \psi : \mathcal{A} \longrightarrow \mathcal{B}$ are two surjective maps. Then ϕ and ψ satisfy (2.1) if and only if the following statements hold.*

(1) *If $\dim(X) = \infty$, then there exist two maps $k : X \times X^* \longrightarrow X$, $(x, f) \mapsto k_f(x)$ and $h : X \times X^* \longrightarrow X^*, (x, f) \mapsto h_x(f)$ and a bounded invertible linear or conjugate linear operator U on X such that*

$$\phi(x \otimes f) = k_f(x) \otimes (U^{-1})^* f \text{ and } \psi(x \otimes f) = Ux \otimes h_x(f)$$

for all $(x, f) \in X \times X^*$.

(2) *If $X = \mathbb{C}^n$ is a finite-dimensional space with $n \geq 3$, then there exist two maps $k, h : \mathbb{C}^n \times \mathbb{C}^n \longrightarrow \mathbb{C}^n$, $(x, y) \mapsto k_y(x), (x, y) \mapsto h_x(y)$, a nonsingular matrix $U \in \mathcal{M}_n(\mathbb{C})$, and a ring automorphism τ of $\mathbb{C}$ such that*

$$\phi(x \otimes y) = k_y(x) \otimes y_\tau U^{-1} \text{ and } \psi(x \otimes y) = Ux_\tau \otimes h_x(y)$$

for all $x, y \in \mathbb{C}^n$.

Our next main theorem is given in Hilbert space context. Thus, by $(\mathcal{H}, \langle\ ,\ \rangle)$ we denote a complex Hilbert space with $\dim(\mathcal{H}) \geq 3$, and by $\mathcal{A}$ and $\mathcal{B}$ two subsets of $\mathcal{B}(\mathcal{H})$ containing all operators in $\mathcal{H}$ of rank at most one. For every $x, y \in \mathcal{H}$, we denote by $x \otimes y$ the operator on $\mathcal{H}$ of rank at most one defined on $\mathcal{H}$ by $x \otimes y(z) = \langle z, y \rangle x$ for all $z \in \mathcal{H}$. For every bounded linear (resp. conjugate linear) operator A on $\mathcal{H}$, the adjoint A^* of A is the bounded linear (resp. conjugate linear) operator on $\mathcal{H}$ defined by $\langle x, A^* y \rangle = \langle Ax, y \rangle$ (resp. $\langle x, A^* y \rangle = \overline{\langle Ax, y \rangle}$) for all $(x, y) \in \mathcal{H}^2$. Moreover, if $\dim(\mathcal{H}) = \mathbb{C}^n$ for some integer $n \geq 3$, then $\mathcal{H}$ will be identified with the Hilbert space $\mathbb{C}^n$ and $\mathcal{B}(\mathcal{H})$ with the algebra $\mathcal{M}_n(\mathbb{C})$. Note that, Unlike the above identification, for every $x = (x_1, ..., x_n)$ and $y = (y_1, ..., y_n)$ in $\mathbb{C}^n$, we have $\langle x, y \rangle = x_1 \overline{y_1} + \cdots x_n \overline{y_n}$ and $x \otimes y = x^t \overline{y}$. Accordingly, we provide, without proof, the following restatement of Theorem 2.1 in the Hilbert space context. This version will be needed for the proof of Theorem 2.3.

THEOREM 2.2. *Suppose that $\phi, \psi : \mathcal{A} \longrightarrow \mathcal{B}$ are two surjective maps. Then ϕ and ψ satisfying (2.1) if and only if the following statements hold.*

(1) *If $\dim(\mathcal{H}) = \infty$, then there exist two maps*

$$h, k : \mathcal{H} \times \mathcal{H} \to \mathcal{H},\ (x, y) \mapsto k_y(x), (x, y) \mapsto h_x(y)$$

and a bounded invertible linear or conjugate linear operator U on $\mathcal{H}$ such that

(2.2) $$\phi(x \otimes y) = k_y(x) \otimes (U^{-1})^* y \text{ and } \psi(x \otimes y) = Ux \otimes h_x(y)$$

for all $(x, y) \in \mathcal{H} \times \mathcal{H}$.

(2) *If $\mathcal{H} = \mathbb{C}^n$ is a finite-dimensional Hilbert space with $n \geq 3$, then there exist two maps $k, h : \mathbb{C}^n \times \mathbb{C}^n \longrightarrow \mathbb{C}^n$, $(x, y) \mapsto k_y(x), (x, y) \mapsto h_x(y)$, a nonsingular matrix $U \in \mathcal{M}_n(\mathbb{C})$, and two ring automorphisms τ_1 and τ_2 of $\mathbb{C}$ satisfying $\tau_2(\mu) = \overline{\tau_1(\overline{\mu})}$ for all $\mu \in \mathbb{C}$, such that*

(2.3) $$\phi(x \otimes y) = k_y(x) \otimes y_{\tau_2} U^{-1} \text{ and } \psi(x \otimes y) = Ux_{\tau_1} \otimes h_x(y)$$

for all $x, y \in \mathbb{C}^n$.

Now, let ε be a fixed positive real number. The ε-pseudo spectrum of an operator $A \in \mathcal{B}(\mathcal{H})$ is defined by

$$\sigma_\varepsilon(A) := \left\{ \lambda \in \mathbb{C} : \|(\lambda I - A)^{-1}\| > \frac{1}{\varepsilon} \right\},$$

with the convention that $\|(\lambda I - A)^{-1}\| = \infty$ if $\lambda I - A$ is not invertible. It is a subset of $\mathbb{C}$ and contains $\sigma(A)$, the spectrum of A. The ε-pseudo spectral radius of A, is defined by

$$r_\varepsilon(A) := \sup \{|z| : z \in \sigma_\varepsilon(A)\}.$$

For more information about these notions, one may consult [51]. We describe the form of all surjective maps $\phi, \psi : \mathcal{A} \longrightarrow \mathcal{B}$ satisfying

(2.4) $$\sigma_\varepsilon(\phi(A)\psi(B)) = \sigma_\varepsilon(AB)$$

for all $A, B \in \mathcal{A}$. Mainly, we obtain the following theorem.

THEOREM 2.3. *Two surjective maps $\phi, \psi : \mathcal{A} \longrightarrow \mathcal{B}$ satisfy (2.4) if and only if there exist a bounded invertible linear operator U on $\mathcal{H}$ and a unitary operator V on $\mathcal{H}$ such that $\phi(A) = \mu V A U^{-1}$ and $\psi(A) = \nu U A V^*$ for all $A \in \mathcal{A}$, where $\mu, \nu \in \mathbb{C}$ such that $\mu\nu = 1$.*

The rest of this paper is organized as follows. In Section 3, we present some useful results on the ε-pseudo spectrum and the ε-pseudo spectral radius. These results are needed in Sections 4, 5 and 6 which will be devoted to the proof of the above theorems.

3. Preliminaries

First, let us fix some notations which will be needed in the sequel. For a linear subspace M of X (resp, X^*), we denote $M^\perp = \{f \in X^* : \langle x, f \rangle = 0, x \in M\}$ (resp, $M^\perp = \{x \in X : \langle x, f \rangle = 0, f \in M\}$). For any $A \in \mathcal{B}(X)$, we denote by $\mathrm{ran}(A)$ and $\mathrm{rank}(A)$ respectively the range of A and the dimension of $\mathrm{ran}(A)$. For a subset $\mathcal{A}$ of $\mathcal{B}(X)$, and $A \in \mathcal{A}$ we denote

$$\{A\}^R := \{B \in \mathcal{A} \setminus \{0\} : BA = 0\},$$

and

$$\{A\}^L := \{B \in \mathcal{A} \setminus \{0\} : AB = 0\}.$$

The set $\{A\}^R$(resp, $\{A\}^L$) is said to be maximal, if for any operator $N \in \mathcal{A}$, $\{A\}^R \subseteq \{N\}^R \Rightarrow \{A\}^R = \{N\}^R$ (resp, $\{A\}^L \subseteq \{N\}^L \Rightarrow \{A\}^L = \{N\}^L$). For any $x \in X$ and $f \in X^*$, let

$$(x \otimes f)(y) := \langle y, f \rangle x, \ (y \in X)$$

be the operator of rank at most one. For all $x \in X$ and $f \in X^*$, we denote

$$R_f := \{y \otimes f : y \in X\}$$

and

$$L_x := \{x \otimes g : g \in X^*\}.$$

For any $a \in \mathbb{C}$ and $r > 0$, let $D(a, r)$ be the open disc of $\mathbb{C}$ centered at a and of radius r.

Now we collect some preliminary results which are needed in the proof of our main theorems. The first one collects some known properties of the ε-pseudo spectrum.

PROPOSITION 3.1 ([**23, 51**]). *Let $\mathcal{H}$ be a complex Hilbert space. Then, for an operator $A \in \mathcal{B}(\mathcal{H})$ the following statements hold.*

(1) $\sigma(A) + D(0, \varepsilon) \subseteq \sigma_\varepsilon(A)$.
(2) *If A is normal, then* $\sigma_\varepsilon(A) = \sigma(A) + D(0, \varepsilon)$.
(3) *For any $c \in \mathbb{C}$, we have* $\sigma_\varepsilon(A + cI) = c + \sigma_\varepsilon(A)$.
(4) *For any nonzero $c \in \mathbb{C}$, we have* $\sigma_\varepsilon(cA) = c\sigma_{\frac{\varepsilon}{|c|}}(A)$.
(5) *For every unitary operator $U \in \mathcal{B}(\mathcal{H})$, we have* $\sigma_\varepsilon(UAU^*) = \sigma_\varepsilon(A)$.
(6) *For every conjugate unitary operator U, we have* $\sigma_\varepsilon(UAU^*) = \sigma_\varepsilon(A^*) = \overline{\sigma_\varepsilon(A)}$.
(7) *For any $a \in \mathbb{C}$, we have* $\sigma_\varepsilon(A) = D(a, \varepsilon)$ *if and only if* $A = aI$.
(8) *If $a \in \mathbb{C}$ is a nonzero scalar, then* $\sigma_\varepsilon(A) = D(0, \varepsilon) \cup D(a, \varepsilon)$ *if and only if there exists a nontrivial orthogonal projection $P \in \mathcal{B}(\mathcal{H})$ such that $A = aP$.*

The second result provides an explicit formula of the ε-pseudo spectral radius of any operator of rank at most one.

PROPOSITION 3.2 ([**24**]). *Let $\mathcal{H}$ be a complex Hilbert space. If x and y are two vectors in $\mathcal{H}$, then*

$$r_\varepsilon(x \otimes y) = \frac{1}{2}\left(\sqrt{|\langle x, y \rangle|^2 + 4\varepsilon^2 + 4\varepsilon\|x\|\|y\|} + |\langle x, y \rangle|\right).$$

The following result has been proved in [**24**, Proposition 2.1].

PROPOSITION 3.3. *Let $\mathcal{H}$ be a complex Hilbert space. If x and y are two vectors in $\mathcal{H}$, then*

$$\sigma_\varepsilon(x \otimes y) = \left\{z \in \mathbb{C} : \begin{array}{l}\sqrt{(|z| + |z - \langle x, y \rangle|)^2 + \|x\|^2\|y\|^2 - |\langle x, y \rangle|^2} \\ -\sqrt{(|z| - |z - \langle x, y \rangle|)^2 + \|x\|^2\|y\|^2 - |\langle x, y \rangle|^2}\end{array} < 2\varepsilon\right\}.$$

The following lemma is a new and natural observation which is consists of characterizing of nilpotent rank one operators, see Corollary 3.5.

LEMMA 3.4. *Let $\mathcal{H}$ be a complex Hilbert space, and let x and y be two vectors in $\mathcal{H}$. If $\omega \in \mathbb{C}$, then*

$$\omega + \frac{\langle x, y \rangle}{2} \in \sigma_\varepsilon(x \otimes y) \iff -\omega + \frac{\langle x, y \rangle}{2} \in \sigma_\varepsilon(x \otimes y).$$

PROOF. Let x and y be two vectors in $\mathcal{H}$, and consider the function defined by

$$h_{x,y}(z) := \sqrt{(|z| + |z - \langle x, y \rangle|)^2 + \|x\|^2\|y\|^2 - |\langle x, y \rangle|^2} \\ -\sqrt{(|z| - |z - \langle x, y \rangle|)^2 + \|x\|^2\|y\|^2 - |\langle x, y \rangle|^2}$$

for all $z \in \mathbb{C}$. Observe that for every complex number ω, we have

$$h_{x,y}\left(\omega + \frac{\langle x, y \rangle}{2}\right) = h_{x,y}\left(-\omega + \frac{\langle x, y \rangle}{2}\right)$$

and by the characterization of the ε-pseudo spectrum given in Proposition 3.3, we conclude the lemma. □

COROLLARY 3.5. *Let $\mathcal{H}$ be a complex Hilbert space, and let x and y be two vectors in $\mathcal{H}$. Then the following statements are equivalent.*

(1) $x \otimes y$ *is a nilpotent operator.*

(2) $\sigma_\varepsilon(x \otimes y)$ is an open disc of $\mathbb{C}$ centered at zero.

(3)
$$\sigma_\varepsilon(x \otimes y) = D\left(0, \sqrt{\varepsilon^2 + \varepsilon\|x\|\|y\|}\right).$$

PROOF. Since (3) $\Rightarrow$ (2) is obvious, we only have to prove that (2) $\Rightarrow$ (1) $\Rightarrow$ (3).

First, for (1) $\Rightarrow$ (3), let $R := x \otimes y \in \mathcal{F}_1(X)$, and note that Proposition 3.3 entails that
$$\langle x, y \rangle = 0 \Longrightarrow \sigma_\varepsilon(x \otimes y) = D\left(0, \sqrt{\varepsilon^2 + \varepsilon\|x\|\|y\|}\right).$$

Now, for (2) $\Rightarrow$ (1), let x and y be two vectors in $\mathcal{H}$ such that $\langle x, y \rangle \neq 0$. By Lemma 3.4, we note that $\dfrac{\langle x, y \rangle}{2}$ is a symmetry point of $\sigma_\varepsilon(x \otimes y)$, since $\sigma_\varepsilon(x \otimes y)$ is an open disc of $\mathbb{C}$ centered at zero. Hence, $\langle x, y \rangle = 0$ and this contradiction shows that (2) $\Rightarrow$ (1). $\square$

For the proof of Theorem 2.1, we need the following lemma (see [36, Lemma 2, Corollary, page 2] and [45, Lemma B]) that tells us that a bijective semi-linear transformation S between infinite-dimensional complex normed linear spaces X and Y is cccS is automatically bicontinuous and linear or conjugate linear provided that S and S^{-1} carry closed hyperplanes to closed hyperplanes. Recall that a semi-linear map $A: X \to Y$ is an additive map for which there is τ is a fixed ring automorphism of $\mathbb{C}$ such that

$$A(\lambda x) = \tau(\lambda)A(x)$$

for all $\lambda \in \mathbb{C}$ and $x \in X$.

LEMMA 3.6. *If $S: X \longrightarrow Y$ is a bijective semi-linear transformation between infinite-dimensional complex normed linear spaces X and Y such that S and S^{-1} carry closed hyperplanes to closed hyperplanes, then S is bicontinuous and linear or conjugate linear.*

4. Proof of Theorem 2.1

The proof breaks down into four assertions.

Assertion 1. Let $\mathcal{A}$ be a subset of $\mathcal{B}(X)$ containing $\mathcal{F}_1(X)$, and let $A \in \mathcal{A}$ be a nonzero operator. Then A is an operator of rank one if and only if the set $\{A\}^R$ is maximal.

PROOF. Let $A \in \mathcal{A} \setminus \{0\}$ such that the set $\{A\}^R$ is maximal. Suppose on the contrary that rank$(A) \geq 2$, then there are two vectors x_1 and x_2 in X such that Ax_1 and Ax_2 are linearly independent. We choose $f_1, f_2 \in X^*$ such that $\langle Ax_i, f_j \rangle = \delta_{ij}$ (Kroneckor symbol), $i, j = 1, 2$. Let $P = Ax_1 \otimes f_1 \in \mathcal{A}$, then $\{A\}^R \subseteq \{P\}^R$. Since the set $\{A\}^R$ is maximal, we have $\{P\}^R = \{A\}^R$. Now, let $B = Ax_2 \otimes f_2 \in \mathcal{A}$, we have
$$BA = Ax_2 \otimes f_2 A = Ax_2 \otimes A^* f_2 \neq 0,$$
and
$$BP = (Ax_2 \otimes f_2)(Ax_1 \otimes f_1) = \langle Ax_1, f_2 \rangle Ax_2 \otimes f_1 = 0.$$
Therefore, $B \in \{P\}^R$ and $B \notin \{A\}^R$, and thus $\{P\}^R \neq \{A\}^R$, which is a contradiction. Hence, rank$(A) = 1$.

Conversely, assume that rank$(A) = 1$, and set $A = x \otimes f$ for some $x \in X$ and $f \in X^*$. If $N \in \mathcal{A}$ satisfies $\{A\}^R \subseteq \{N\}^R$, then for any operator of rank one $u \otimes h$ with $\langle x, h \rangle = 0$, we have
$$u \otimes h A = (u \otimes h)(x \otimes f) = \langle x, h \rangle u \otimes f = 0.$$
Therefore, $u \otimes h \in \{A\}^R$, and thus $u \otimes h \in \{N\}^R$. Hence $\ker(N^*) \supseteq [h]$ whenever $\langle x, h \rangle = 0$, which implies that rank$(N^*) \leq 1$. Now, we show that rank$(N) \leq 1$. Suppose on the contrary that rank$(N) \geq 2$, then there are two vectors x_1 and x_2 in X such that Nx_1 and Nx_2 are linearly independent. We choose $f_1, f_2 \in X^*$ satisfying $\langle Nx_i, f_j \rangle = \delta_{ij}$, $i, j = 1, 2$. If $g_i = f_i \circ N$, then $\langle x_i, g_j \rangle = \delta_{ij}$, $i, j = 1, 2$. Hence rank$(N^*) \geq 2$, which is a contradiction. Thus, rank$(N) \leq 1$. So, let $N = y \otimes g$ for some $(y, g) \in X \times X^*$. We show that x and y are linearly dependent. Suppose on the contrary that x and y are linearly independent and choose $l \in X^*$ such that $\langle x, l \rangle = 0$ and $\langle y, l \rangle = 1$. Then $x \otimes l \in \{A\}^R$ and $x \otimes l \notin \{N\}^R$. This contradicts the fact $\{A\}^R \subseteq \{N\}^R$ and shows that x and y are linearly dependent. Therefore, $\{N\}^R = \{A\}^R$, and the set $\{A\}^R$ is maximal. $\square$

Assertion 2. Let $\mathcal{A}$ be a subset of $\mathcal{B}(X)$ containing $\mathcal{F}_1(X)$, and let $A \in \mathcal{A}$ be a nonzero operator. Then A is an operator of rank one if and only if the set $\{A\}^L$ is maximal.

PROOF. Let $A \in \mathcal{A} \setminus \{0\}$ such that the set $\{A\}^L$ is maximal. Suppose on the contrary that rank$(A) \geq 2$, then there are two vectors x_1 and x_2 in X such that Ax_1 and Ax_2 are linearly independent. We choose $f \in X^*$ satisfying $\langle Ax_2, f \rangle = 0$ and $\langle Ax_1, f \rangle = 1$. Let $P := x_1 \otimes A^*f \in \mathcal{A}$, and note that $\{A\}^L \subseteq \{P\}^L$. Since the set $\{A\}^L$ is maximal, then $\{P\}^L = \{A\}^L$. Now, if $B = x_2 \otimes A^*f \in \mathcal{A}$, then
$$AB = Ax_2 \otimes A^*f \neq 0,$$
and
$$PB = (x_1 \otimes A^*f)(x_2 \otimes A^*f) = \langle Ax_2, f \rangle x_1 \otimes A^*f = 0.$$
Therefore, $B \in \{P\}^L$ and $B \notin \{A\}^L$. Thus $\{P\}^L \neq \{A\}^L$, which is a contradiction. Hence, rank$(A) = 1$.

Conversely, assume that rank$(A) = 1$, and let $A = x \otimes f$. If $N \in \mathcal{A}$ satisfies $\{A\}^L \subseteq \{N\}^L$, then for any operator of rank one $u \otimes h$ with $\langle u, f \rangle = 0$, we have
$$Au \otimes h = (x \otimes f)(u \otimes h) = \langle u, f \rangle x \otimes h = 0.$$
Therefore, $u \otimes h \in \{A\}^L$, thus $u \otimes h \in \{N\}^L$, and $\ker(N) \supseteq [u]$ whenever $\langle u, f \rangle = 0$. This implies that rank$(N) \leq 1$, and $N = y \otimes g$ for some $(y, g) \in X \times X^*$. We show that f and g are linearly dependent. Suppose on the contrary that f and g are linearly independent. Choose $z \in X$ such that $\langle z, f \rangle = 0$ and $\langle z, g \rangle = 1$. Then $z \otimes f \in \{A\}^L$ and $z \otimes f \notin \{N\}^L$, which is a contradiction. Thus f and g are linearly dependent and $\{N\}^L = \{A\}^L$, so that the set $\{A\}^L$ is maximal. $\square$

Assertion 3. Let $\phi, \psi : \mathcal{A} \longrightarrow \mathcal{B}$ be two surjective maps satisfying (2.1). Then, ϕ and ψ preserves the operators of rank at most one in both directions.

PROOF. By the surjectivity of ϕ, ψ and (2.1), we have $\phi(0) = 0$ and $\psi(0) = 0$. Now, it suffices to prove that ϕ and ψ preserve operators of rank one in both directions. Let $A \in \mathcal{A} \setminus \{0\}$, and let us first show that rank$(A) = 1$ if and only if

$\operatorname{rank}(\psi(A)) = 1$. For any $N \in \mathcal{A} \setminus \{0\}$, we note that (2.1) and the surjectivity of ϕ imply that

$$\{A\}^R \subseteq \{N\}^R \iff \{\psi(A)\}^R \subseteq \{\psi(N)\}^R,$$

and

$$\{A\}^R = \{N\}^R \iff \{\psi(A)\}^R = \{\psi(N)\}^R.$$

This together with Assertion 1 and the surjectivity of ϕ entail that

$\operatorname{rank}(A) = 1 \iff$ the set $\{A\}^R$ is maximal
$\iff$ for any $N \in \mathcal{A} \setminus \{0\}$, we have $\{A\}^R \subseteq \{N\}^R \Rightarrow \{A\}^R = \{N\}^R$
$\iff$ for any $N \in \mathcal{B} \setminus \{0\}$, we have $\{\psi(A)\}^R \subseteq \{N\}^R \Rightarrow \{\psi(A)\}^R = \{N\}^R$
$\iff$ the set $\{\psi(A)\}^R$ is maximal
$\iff \operatorname{rank}(\psi(A)) = 1$.

Similarly, we show that $\operatorname{rank}(A) = 1$ if and only if $\operatorname{rank}(\phi(A)) = 1$. By (2.1), Assertion 2, and the surjectivity of ψ, we have

$\operatorname{rank}(A) = 1 \iff$ the set $\{A\}^L$ is maximal
$\iff$ for any $N \in \mathcal{A} \setminus \{0\}$, we have $\{A\}^L \subseteq \{N\}^L \Rightarrow \{A\}^L = \{N\}^L$
$\iff$ for any $N \in \mathcal{B} \setminus \{0\}$, we have $\{\phi(A)\}^L \subseteq \{N\}^L \Rightarrow \{\phi(A)\}^L = \{N\}^L$
$\iff$ the set $\{\phi(A)\}^L$ is maximal
$\iff \operatorname{rank}(\phi(A)) = 1$.

The proof of Assertion 3 is now complete. $\square$

In the following assertion we will use essentially some notions and results given in [44, §2.2]. For every nonzero vector $x \in X$, the ray generated by x will be denoted by $\underline{x}$, this is the set $\{\lambda x : \lambda \in \mathbb{C} \setminus \{0\}\}$. The set of all rays in X will be denoted by $\underline{X}$, this means that $\underline{X} := \{\underline{x} : x \in X \setminus \{0\}\}$. We say that the rays $\underline{x}, \underline{y} \in \underline{X}$ are orthogonal to each other, in notation $\langle \underline{x}, \underline{y} \rangle = 0$, if we have $\langle x, y \rangle = 0$. Observe that $\langle x, y \rangle = 0$ if and only if $\langle u, v \rangle = 0$ for all $u \in \underline{x}$ and $v \in \underline{y}$. We well also need the notation $\underline{x} + \underline{y} := \{\alpha x + \beta y : \alpha, \beta \in \mathbb{C}\} \setminus \{0\}$ for all nonzero vectors $x, y \in X$.

Assertion 4. Let $\phi, \psi : \mathcal{A} \longrightarrow \mathcal{B}$ be two surjective maps satisfying (2.1). Then there exist a bijective semilinear map $U : X \longrightarrow X$, a bijective semilinear map $V : X^* \longrightarrow X^*$ and two maps $k : X \times X^* \longrightarrow X$, $(x, f) \mapsto k_f(x)$ and $h : X \times X^* \longrightarrow X^*$, $(x, f) \mapsto h_x(f)$ such that

(4.1) $\qquad \phi(x \otimes f) = k_f(x) \otimes (Vf)$ and $\psi(x \otimes f) = (Ux) \otimes h_x(f)$

for all $(x, f) \in X \times X^*$. Furthermore, we have

(4.2) $\qquad \langle Ux, Vf \rangle = 0 \iff \langle x, f \rangle = 0$

for all $(x, f) \in X \times X^*$.

PROOF. Keep in mind that ϕ and ψ are surjective maps preserving operators of rank at most one in both directions, and let $R = x \otimes f$ and $Q = y \otimes g$ be two operators of rank one. Let $\psi(R) = x' \otimes f'$ and $\psi(Q) = y' \otimes g'$, and note that (2.1) entails that

$$
\begin{aligned}
x = \lambda y \text{ for some } \lambda &\iff \text{for any } P \in \mathcal{F}_1(X), \text{ we have } PR = 0 \Leftrightarrow PQ = 0 \\
&\iff \text{for any } P \in \mathcal{F}_1(X), \text{ we have } \phi(P)\psi(R) = 0 \Leftrightarrow \phi(P)\psi(Q) = 0 \\
&\iff \text{for any } P \in \mathcal{F}_1(X), \text{ we have } \phi(P)(x' \otimes f') = 0 \Leftrightarrow \phi(P)(y' \otimes g') = 0 \\
&\iff \text{for any } z \otimes h \in \mathcal{F}_1(X), \text{ we have } (z \otimes h)(x' \otimes f') = 0 \Leftrightarrow (z \otimes h)(y' \otimes g') = 0 \\
&\iff \text{for any } h \in X^*, \text{ we have } \langle x', h \rangle = 0 \Leftrightarrow \langle y', h \rangle = 0 \\
&\iff x' = \lambda' y' \text{ for some } \lambda'.
\end{aligned}
$$

This implies that for every nonzero $x \in X$ there exists a nonzero $z \in X$ such that $\psi(L_x) = L_z$.

Now, we define a map $\varphi : \underline{X} \longrightarrow \underline{X}$ by

$$\varphi(\underline{x}) = \underline{y} \text{ if and only if } \psi(L_x) = L_y.$$

Sinse ϕ and ψ preserve operators of rank at most one in both directions, φ is a bijective map. We show that for every nonzero vectors $x, u, v \in X$, we have

$$\underline{x} \subseteq \underline{u} + \underline{v} \iff \varphi(\underline{x}) \subseteq \varphi(\underline{u}) + \varphi(\underline{v}).$$

Let $(x, u, v) \in X^3$ and let $\varphi(\underline{x}) = \underline{x}'$, $\varphi(\underline{v}) = \underline{v}'$, $\varphi(\underline{u}) = \underline{u}'$. We have

$$
\begin{aligned}
\underline{x} \subseteq \underline{u} + \underline{v} &\iff \text{for any } P \in \mathcal{F}_1(X), \text{ we have } (PL_u = \{0\} \text{ and } PL_v = \{0\}) \Rightarrow PL_x = \{0\} \\
&\iff \text{for any } P \in \mathcal{F}_1(X), \text{ we have } (\phi(P)\psi(L_u) = \{0\} \text{ and } \phi(P)\psi(L_v) = \{0\}) \\
&\qquad \Rightarrow \phi(P)\psi(L_x) = \{0\} \\
&\iff \text{for any } Q \in \mathcal{F}_1(X), \text{ we have } (QL'_u = \{0\} \text{ and } QL'_v = \{0\}) \Rightarrow QL'_x = \{0\} \\
&\iff \varphi(\underline{x}) \subseteq \varphi(\underline{u}) + \varphi(\underline{v}).
\end{aligned}
$$

Thus, using the fundamental theorem of projective geometry as in [**46**, Proof of Theorems 1.1 and 1.2], we conclude that the map φ is induced by a semilinear bijective map $U : X \longrightarrow X$, this means that $\varphi(\underline{x}) = \underline{Ux}$ for all $x \in X$. Therefore, for every operator of rank at most one $x \otimes f \in \mathcal{F}_1(X)$ there exists a map $h : X \times X^* \longrightarrow X^*$, $(x, f) \mapsto h_x(f)$ such that

$$\psi(x \otimes f) = Ux \otimes h_x(f)$$

for all $(x, f) \in X \times X^*$.

Applying a similar process to $\Psi : \underline{X}^* \longrightarrow \underline{X}^*$ defined by

$$\Psi(\underline{f}) = \underline{g} \text{ if and only if } \phi(R_f) = R_g,$$

we prove that there is a semilinear bijective map $V : X^* \longrightarrow X^*$ such that $\Psi(\underline{f}) = \underline{Vg}$. Hence, there exists a map $k : X \times X^* \longrightarrow X$, $(x, f) \mapsto k_f(x)$ such that

$$\phi(x \otimes f) = k_f(x) \otimes Vf$$

for all $(x, f) \in X \times X^*$. To complete the proof of the assertion, observe that by (2.1) we have

$$
\begin{aligned}
\langle Uy, Vf \rangle = 0 &\iff (k_f(x) \otimes Vf)(Uy \otimes h_y(f)) = 0 \\
&\iff \phi(x \otimes f)\psi(y \otimes f) = 0 \\
&\iff (x \otimes f)(y \otimes f) = 0 \\
&\iff \langle y, f \rangle = 0
\end{aligned}
$$

for all $y \in X$ and $f \in X^*$. This finish the proof of assertion 4. □

Assertion 5. Let $\phi, \psi : \mathcal{A} \longrightarrow \mathcal{B}$ be two surjective maps satisfying (2.1) and suppose that $\dim(X) = \infty$, then there exist two maps

$$k : X \times X^* \longrightarrow X; (x, f) \mapsto k_f(x)$$

and

$$h : X \times X^* \longrightarrow X^*; (x, f) \mapsto h_x(f)$$

and a bounded invertible linear or conjugate linear operator U on X such that

$$\phi(x \otimes f) = k_f(x) \otimes (U^{-1})^* f \text{ and } \psi(x \otimes f) = Ux \otimes h_x(f)$$

for all $(x, f) \in X \times X^*$.

PROOF. Keep in mind that ϕ and ψ satisfying the conditions (4.1) of Assertion 4, and let us show that U is continuous and linear or conjugate linear. We will prove that U and U^{-1} carry every closed hyperplane of X to a closed hyperplane. Pick a nonzero $x \in X$, by (4.2) we know that

$$\langle Uy, Vf \rangle = 0 \iff \langle y, f \rangle = 0$$

for all nonzero $y \in X$ and nonzero $f \in X^*$. Since U and V are bijective we deduce that

(4.3) $$\langle U^{-1}z, V^{-1}f \rangle = 0 \iff \langle z, f \rangle = 0$$

for all nonzero $z \in X$ and nonzero $f \in X^*$. Now, let $H \subseteq X$ be a closed hyperplane of X. Then, there is a nonzero $f_0 \in X^*$ such that $H = \{x \in X : \langle x, f_0 \rangle = 0\}$. By (4.2) and (4.3), we have

$$U(H) = \{x \in X : \langle x, Vf_0 \rangle = 0\},$$

and

$$U^{-1}(H) = \{x \in X : \langle x, V^{-1}f_0 \rangle = 0\}.$$

From Lemma 3.6, we conclude that U is continuous and linear or conjugate linear. Therefore, from the definition of the adjoint U^* and (4.2) we have

(4.4) $$\langle y, U^*Vf \rangle = 0 \iff \langle Uy, Vf \rangle = 0 \iff \langle y, f \rangle = 0$$

for all $y \in X$ and $f \in X^*$. Let us show that $(U^{-1})^*$ and V are locally linearly dependent. Assume by way of contradiction that they are locally linearly independent, and let f be a vector in X^* such that U^*Vf and f are linearly independent. Then there exists a nonzero vector $z \in X$ such that $\langle z, U^*Vf \rangle = 0$ and $\langle z, f \rangle = 1$, this contradicts (4.4). Thus U^*Vf and f are linearly dependent for all $f \in X^*$. Thus, there exists a functional $\mu : X^* \mapsto \mathbb{C} \setminus \{0\}, f \mapsto \mu_f$ such that $Vf = \mu_f(U^{-1})^*f$ for all $f \in X^*$. Hence $\phi(x \otimes f) = \mu_f k_f(x) \otimes (U^{-1})^*f$ and $\psi(x \otimes f) = Ux \otimes h_x(f)$ for all $(x, f) \in X \times X^*$. Thus, Assertion 5 is proved. $\square$

Assertion 6. Let $\phi, \psi : \mathcal{A} \longrightarrow \mathcal{B}$ be two surjective maps satisfying (2.1) and suppose that $X = \mathbb{C}^n$ is a finite-dimensional space with $n \geq 3$, then there exist two maps $k, h : \mathbb{C}^n \times \mathbb{C}^n \longrightarrow \mathbb{C}^n$, $(x, y) \mapsto k_y(x), (x, y) \mapsto h_x(y)$, a nonsingular matrix $U \in \mathcal{M}_n(\mathbb{C})$, and a ring automorphism τ of $\mathbb{C}$ such that

$$\phi(x \otimes y) = k_y(x) \otimes y_\tau U^{-1} \text{ and } \psi(x \otimes y) = Ux_\tau \otimes h_x(y)$$

for all $x, y \in \mathbb{C}^n$.

PROOF. Note that Assertion 4, tells us that
$$\phi(x \otimes y) = k_y(x) \otimes Vy \text{ and } \psi(x \otimes y) = Ux \otimes h_x(y),$$
where U and V are two semilinear mapping on $\mathbb{C}$. Thus, there exist two ring automorphisms τ_1 and τ_2 of $\mathbb{C}$ and two matrices B and C such that
$$Ux = Cx_{\tau_1} \text{ and } Vx = Bx_{\tau_2}$$
for all $x \in \mathbb{C}$. First, we show that $\tau_1 = \tau_2$. Let $\mu \in \mathbb{C}$ and $x, y \in X$ be two unit vectors such that $\langle x, y \rangle = 0$. Clearly, $\langle \mu x - y, x + \mu y \rangle = 0$ and by (4.2), we have
$$\begin{aligned} 0 &= \langle U(\mu x - y), V(x + \mu y) \rangle \\ &= \tau_1(\mu)\langle Ux, Vx \rangle - \tau_2(\mu)\langle Uy, Vy \rangle + \tau_1(\mu)\tau_2(\mu)\langle Ux, Vy \rangle - \langle Uy, Vx \rangle \\ &= \tau_1(\mu)\langle Ux, Vx \rangle - \tau_2(\mu)\langle Uy, Vy \rangle. \end{aligned}$$
Therefore,
$$(4.5) \qquad \tau_1(\mu)\langle Cx_{\tau_1}, Bx_{\tau_2} \rangle - \tau_2(\mu)\langle Cy_{\tau_1}, By_{\tau_2} \rangle = 0.$$
for all $\mu \in \mathbb{C}$. Choosing $\mu = 1$ in the above equality, we obtain that
$$(4.6) \qquad \langle Cx_{\tau_1}, Bx_{\tau_2} \rangle = \langle Cy_{\tau_1}, By_{\tau_2} \rangle.$$
From (4.5) and (4.6), we conclude that
$$(\tau_1(\mu) - \tau_2(\mu))\langle Cx_{\tau_1}, Bx_{\tau_2} \rangle = 0,$$
for all $\mu \in \mathbb{C}$ and all unit vectors $x \in X$. Again by (4.2) and the fact that $\langle x, x \rangle \neq 0$, we have $\langle Cx_{\tau_1}, Bx_{\tau_2} \rangle = \langle Ux, Vx \rangle \neq 0$. Then
$$\tau_1(\mu) = \tau_2(\mu) \text{ for all } \mu \in \mathbb{C}.$$
In the rest of the proof of Assertion 6, we set $\tau := \tau_1 = \tau_2$. Now, let us show that B^t and C^{-1} are linearly dependent. By (2.1), we have
$$\begin{aligned} \langle y_\tau, x_\tau \rangle = 0 &\iff \langle y, x \rangle = 0 \\ &\iff \langle Uy, Vx \rangle = 0 \\ &\iff \langle Cy_{\tau_1}, Bx_{\tau_2} \rangle = 0 \\ &\iff \langle B^* Cy_\tau, x_\tau \rangle = 0 \\ &\iff \langle (B^t C)y_\tau, x_\tau \rangle = 0 \end{aligned}$$
for all $x, y \in \mathbb{C}^n$. Hence
$$\langle y, x \rangle = 0 \iff \langle (B^t C)y, x \rangle = 0 \text{ for all } x, y \in \mathbb{C}^n.$$
Then $(B^t C)y$ and y are linearly independent for all $y \in \mathbb{C}^n$, thus $B^t = \lambda C^{-1}$ for some nonzero scalar $\lambda \in \mathbb{C}$. Therefore,
$$\phi(x \otimes y) = k_y(x) \otimes Vy = k_y(x) \otimes By_\tau = k_y(x) \otimes y_\tau B^t = (\lambda k_y(x)) \otimes y_\tau C^{-1}$$
for all $x, y \in \mathbb{C}^n$, and
$$\psi(x \otimes y) = Ux \otimes h_x(y) = Cx_\tau \otimes h_x(y)$$
for all $x, y \in \mathbb{C}^n$. By writing U instead of C and $k_y(x)$ instead of $\lambda k_y(x)$ we obtain the desired forms of ϕ and ψ. □

5. Proof of Theorem 2.3

We only need to prove the "*only if*" part. So, assume that $\phi, \psi : \mathcal{A} \longrightarrow \mathcal{B}$ are surjective maps such that
$$\sigma_\varepsilon(\phi(A)\psi(B)) = \sigma_\varepsilon(AB)$$
for all $A, B \in \mathcal{A}$. This and the statement (7) of Proposition 3.1 imply that
$$\phi(A)\psi(B) = 0 \iff AB = 0$$
for all $A, B \in \mathcal{A}$. Thus we can assume that ϕ and ψ have the forms given in Theorem 2.2. We will proceed by checking five assertions.

Assertion 1: If x is a unit vector in $\mathcal{H}$, we have
$$(5.1) \qquad \langle u, v \rangle = 0 \iff \langle k_x(u), h_x(v) \rangle = 0$$
for all $u, v \in \mathcal{H}$.

PROOF. Let x, u and v be three nonzero vectors in $\mathcal{H}$ such that $\|x\| = 1$. By (2.4), (2.2) and (2.3), we get
$$
\begin{aligned}
\sigma_\varepsilon(u \otimes v) &= \sigma_\varepsilon((u \otimes x)(x \otimes v)) \\
&= \sigma_\varepsilon(\phi(u \otimes x)\psi(x \otimes v)) \\
&= \sigma_\varepsilon(k_x(u) \otimes h_x(v)).
\end{aligned}
\qquad (5.2)
$$
Assume that $\langle u, v \rangle = 0$, and note that Corollary 3.5 entails that $\sigma_\varepsilon(u \otimes v) = D\left(0, \sqrt{\varepsilon^2 + \varepsilon\|u\|\|v\|}\right)$. This together with equation (5.2) imply that
$$\sigma_\varepsilon(k_x(u) \otimes h_x(v)) = D\left(0, \sqrt{\varepsilon^2 + \varepsilon\|u\|\|v\|}\right).$$
Thus, by Corollary 3.5, we have $\langle k_x(u), h_x(v) \rangle = 0$. Hence,
$$\langle u, v \rangle = 0 \implies \langle k_x(u), h_x(v) \rangle = 0.$$
By a similar way as in the above discussion, we get
$$\langle u, v \rangle = 0 \impliedby \langle k_x(u), h_x(v) \rangle = 0,$$
and hence
$$\langle u, v \rangle = 0 \iff \langle k_x(u), h_x(v) \rangle = 0.$$
Thus, the proof of Assertion 1 is complete. $\square$

In the following assertion we discuss some properties of the maps
$$h_x(\cdot) : \begin{array}{c} \mathcal{H} \\ y \end{array} \begin{array}{c} \to \\ \mapsto \end{array} \begin{array}{c} \mathcal{H} \\ h_x(y) \end{array} \quad \text{and} \quad k_x(\cdot) : \begin{array}{c} \mathcal{H} \\ y \end{array} \begin{array}{c} \to \\ \mapsto \end{array} \begin{array}{c} \mathcal{H} \\ k_x(y) \end{array}$$
for each $x \in \mathcal{H} \setminus \{0\}$.

Assertion 2: For each $x \in \mathcal{H} \setminus \{0\}$, we have that $h_x(\cdot)$ and $k_x(\cdot)$ are a bijective maps satisfying
$$(5.3) \qquad \langle u, v \rangle = 0 \iff \langle h_x(u), k_x(v) \rangle = 0$$
for all $u, v \in \mathcal{H}$.

PROOF. For a nonzero vector x in $\mathcal{H}$, we show that k_x is surjective. It suffices to show that the following equality

(5.4) $$\psi(\{x \otimes y : \ y \in \mathcal{H}\}) = \{Ux_\tau \otimes y : \ y \in \mathcal{H}\}$$

holds, where
$$x_\tau = \begin{cases} x_{\tau_1} & \text{if} \quad \mathcal{H} = \mathbb{C}^n \\ x & \text{if} \quad \dim(\mathcal{H}) = \infty. \end{cases}$$

Since the inclusion $\psi(\{x \otimes y : \ y \in \mathcal{H}\}) \subseteq \{Ux_\tau \otimes y : \ y \in \mathcal{H}\}$ is obvious, we only have to prove the reverse inclusion. For $z \in \mathcal{H}$, since ψ preserves operators of rank at most one in both directions, then there exists $u \otimes v \in \mathcal{F}_1(\mathcal{H})$ such that

$$Ux_\tau \otimes z = \phi(u \otimes v) = Uu_\tau \otimes k_u(v).$$

Then Ux_τ and Uu_τ are linearly dependent, and so $u = \alpha x$ for some nonzero scalar α. Thus, $\phi(x \otimes \overline{\alpha}v) = Ux_\tau \otimes z$, and the equality (5.4) is proved. The surjectivity of k_x immediately follows. By a similar argument as above, one can conclude that $h_x(\cdot)$ is surjective as well.

Now, we show that $\underline{h_x}(\cdot)$ and $\underline{k_x}(\cdot)$ are injective and well defined. Let $\underline{y}, \underline{y'} \in \underline{\mathcal{H}}$, by Assertion 1 and the surjectivity of $k_x(\cdot)$, we have

$$\begin{aligned}
\underline{y} = \underline{y'} &\iff \langle v, y \rangle = 0 \quad \text{if and only if} \quad \langle v, y' \rangle = 0 \quad \text{for all } v \in \mathcal{H} \\
&\iff \langle k_x(v), h_x(y) \rangle = 0 \quad \text{if and only if} \quad \langle k_x(v), h_x(y') \rangle = 0 \quad \text{for all } v \in \mathcal{H} \\
&\iff \langle w, h_x(y) \rangle = 0 \quad \text{if and only if} \quad \langle w, h_x(y')) \rangle = 0 \quad \text{for all } w \in \mathcal{H} \\
&\iff \underline{h_x(y)} = \underline{h_x(y')} \\
&\iff \underline{h_x}\left(\underline{y}\right) = \underline{h_x}\left(\underline{y'}\right).
\end{aligned}$$

Consequently, $\underline{h_x}(\cdot)$ is injective and well defined. By a similar argument, one shows that $\underline{k_x}(\cdot)$ is injective and well defined. Furthermore, since $k_x(\cdot)$ is surjective, it follows that $\underline{k_x}(\cdot)$ must be surjective too. Therefore, $\underline{k_x}(\cdot)$ is bijective. Now, from Assertion 1, we have

$$\begin{aligned}
\langle \underline{u}, \underline{v} \rangle = 0 &\iff \langle u, v \rangle = 0 \\
&\iff \langle h_x(u), k_x(v) \rangle = 0 \\
&\iff \langle \underline{h_x(u)}, \underline{k_x(v)} \rangle = 0 \\
&\iff \langle \underline{h_x}\left(\underline{u}\right), \underline{k_x}\left(\underline{v}\right) \rangle = 0
\end{aligned}$$

for all $u, v \in \mathcal{H}$. □

The following assertion is a consequence of [44, Corollary 2.2.2].

Assertion 3. There exist a unitary or conjugate unitary operator W and a functional $d : \mathcal{H} \setminus \{0\} \times \mathcal{H} \setminus \{0\} \to \mathbb{C} \setminus \{0\}$, $(x, y) \mapsto d_x(y)$ such that

$$h_x(y) = k_x(y) = d_x(y)Wy \quad \text{for all} \quad (x, y) \in \mathcal{H} \setminus \{0\} \times \mathcal{H}.$$

PROOF. We first show that $h_u(x)$ and $k_u(x)$ are linearly dependent for all $x, y, u \in \mathcal{H}$ such that $\|u\| = 1$. Let x, y and u be three nonzero vectors in $\mathcal{H}$ such that $\|u\| = 1$. By equations (2.4), (2.2) and (2.3), we get

$$\begin{aligned}
\sigma_\varepsilon(x \otimes x) &= \sigma_\varepsilon((x \otimes u)(u \otimes v)) \\
&= \sigma_\varepsilon(\phi(x \otimes u)\psi(u \otimes x)) \\
&= \sigma_\varepsilon(k_u(x) \otimes h_u(x)).
\end{aligned}$$

By Proposition 3.1, we conclude that $k_u(x) \otimes h_u(x)$ is an orthogonal projection. Hence there exists a functional $\lambda_u : \mathcal{H} \longrightarrow \mathbb{R} \setminus \{0\}$ such that $k_u(x) = \lambda_u(x) h_u(x)$.

Now, let x be a nonzero vector in $\mathcal{H}$. From Assertion 2, we know that $\underline{h_x}$ and $\underline{k_x}$ are two bijective transformations on $\underline{\mathcal{H}}$ satisfying (5.3). Then in light of [**44**, Corollary 2.2.2], we have the following two possibilities:

(1) $\dim(\mathcal{H}) = \infty$, and there exists an invertible bounded linear or conjugate-linear operator V_x on $\mathcal{H}$ such that

$$\underline{k_x}(\underline{u}) = \underline{h_x}(\underline{u}) = \underline{V_x(u)} \text{ for all } u \in \mathcal{H}; \tag{5.5}$$

(2) $\mathcal{H} = \mathbb{C}^n$, for a given integer $n \geq 3$, and there exist a nonsingular matrix $V_x \in \mathcal{M}_n(\mathbb{C})$ and a ring automorphism τ_x of $\mathbb{C}$ such that

$$\underline{k_x}(\underline{u}) = \underline{h_x}(\underline{u}) = \underline{V_x(u_{\tau_x})} \text{ for all } u \in \mathcal{H}. \tag{5.6}$$

Thus, in the case $\mathcal{H} = \mathbb{C}^n$, by (5.5) and (5.6), there exist two maps $d'_x(\cdot), l'_x(\cdot) : \mathcal{H} \longrightarrow \mathbb{C} \setminus \{0\}$, a nonsingular matrix $V_x \in \mathcal{M}_n(\mathbb{C})$ and a ring automorphism τ_x of $\mathbb{C}$ such that

$$h_x(u) = d'_x(u) V_x(u_{\tau_x}) \text{ and } k_x(u) = l'_x(u) V_x(u_{\tau_x}) \tag{5.7}$$

for all $u \in \mathcal{H}$. Now, let us show that τ_x is either the identity or the complex conjugation. Observe that by (5.1) and (5.6) we get

$$\langle u, v \rangle = 0 \iff \langle u_{\tau_x}, v_{\tau_x} \rangle = 0 \tag{5.8}$$

for all nonzero vectors $u, v \in \mathcal{H}$. Choose $u = (1, \frac{1}{\mu}, 0, ..., 0)$ and $v = (1, -\overline{\mu}, 0, ..., 0)$ with $\mu \in \mathbb{C} \setminus \{0\}$. Therefore, by equation (5.8), we get that $\tau_x(\mu) = \overline{\tau_x(\overline{\mu})}$, $\mu \in \mathbb{C} \setminus \{0\}$. Thus, τ_x maps the real numbers into real numbers, and consequently, the restriction of τ_x to $\mathbb{R}$ is the identity map. Clearly, $\tau_x(i) = \pm i$, and thus τ_x is either the identity or the complex conjugation.

Therefore, in the both cases $\dim(\mathcal{H}) = \infty$ and $\dim(\mathcal{H}) < \infty$, there exist two maps $d'_x(\cdot), l'_x(\cdot) : \mathcal{H} \longrightarrow \mathbb{C} \setminus \{0\}$, such that

$$h_x(u) = d'_x(u) W'_x(u) \text{ and } k_x(u) = l'_x(u) W'_x(u) \text{ for all } u \in \mathcal{H}, \tag{5.9}$$

where W'_x is an invertible bounded linear or conjugate-linear operator on $\mathcal{H}$.

Now, we will prove that there exist a nonzero scalar β_x depending only on x, and a unitary or conjugate unitary operator W' independent of x such that $W'_x = \beta_x W$. Indeed, assume by the way of contradiction that $(W'_x)^* W'_y$ is not a scalar operator for some non orthogonal vectors $x, y \in \mathcal{H}$. Then there exists a nonzero vector $v \in \mathcal{H}$ such that v and $(W'_x)^* W'_y v$ are linearly independent. In this case, one can find another nonzero vector $u \in \mathcal{H}$ such that $\langle u, (W'_x)^* W'_y v \rangle \neq 0$ and $\langle u, v \rangle = 0$. Therefore, by Assertion 1, we have

$$\begin{aligned} \langle u, (W'_x)^* W'_y v \rangle &= \langle d'_x(u) W'_x u, l'_y(v) W'_y v \rangle \\ &= \langle h_x(u), k_y(v) \rangle \\ &= 0 \end{aligned}$$

which is a contradiction. Then $(W'_x)^* W'_y$ is a scalar operator and hence W'_x and W'_y are linearly dependent for all $x, y \in \mathcal{H}$ with $\langle x, y \rangle \neq 0$. In the other case, if $x, y \in \mathcal{H} \setminus \{0\}$ such that $\langle x, y \rangle = 0$, then there exists $z \in \mathcal{H}$ such that $\langle x, z \rangle \neq 0$ and $\langle y, z \rangle \neq 0$. Then W'_x and W'_z are linearly dependent and W'_y and W'_z are also linearly dependent. Therefore, W'_x and W'_y are linearly dependent for all $x, y \in \mathcal{H} \setminus \{0\}$.

Then, we conclude that there exist a unitary or conjugate unitary operator W on $\mathcal{H}$ and a functional $\beta : \mathcal{H} \to \mathbb{C} \setminus \{0\}; x \mapsto \beta_x$ such that

$$W'_x = \beta_x W \text{ for all } x \in \mathcal{H}.$$

Denote $d_x(u) = \beta_x d'_x(u)$ and $l_x(u) = \beta_x l'_x(u)$ for all $x, u \in \mathcal{H}$, we conclude that

$$h_x(u) = d_x(u)W(u) \text{ and } k_x(u) = l'_x(u)W(u)$$

for all $x, u \in \mathcal{H}$. This completes the proof of Assertion 3. $\square$

In the rest of this section U denotes the invertible linear or conjugate linear operator on $\mathcal{H}$ given in Theorem 2.2, and W is the unitary or conjugate unitary operator on $\mathcal{H}$ given in the previous assertion.

Assertion 4. There are two nonzero scalars $r, s \in \mathbb{C}$ with $rs = 1$ such that

$$\phi(A) = \mu W A U^{-1} \text{ and } \psi(A) = \nu U A W^*$$

for all operators of rank one $A \in \mathcal{F}_1(\mathcal{H})$.

PROOF. Note that by Theorem 2.2 and Assertion 3, we know that there exist two functionals $h_1, h_2 : \mathcal{F}_1(\mathcal{H}) \longrightarrow \mathbb{C}$ such that

(5.10) $\quad \phi(x \otimes u) = h_1(x \otimes u) W x \otimes (U^{-1})^* u_{\tau_3}$ and $\psi(x \otimes u) = h_2(x \otimes u) U x_\tau \otimes u W^*$

for all $x \otimes u \in \mathcal{F}_1(\mathcal{H})$, where

$$x_{\tau_3} = \begin{cases} x_{\tau_2} & \text{if } \mathcal{H} = \mathbb{C}^n \\ x & \text{if } \dim(\mathcal{H}) = \infty. \end{cases}$$

for all $x \in \mathcal{H}$. We also denote

(5.11) $\quad \eta(z) := \begin{cases} z & \text{if } \dim(\mathcal{H}) = \infty \text{ and } U \text{ is linear} \\ \overline{z} & \text{if } \dim(\mathcal{H}) = \infty \text{ and } U \text{ is conjugate linear} \\ \tau_1(z) & \text{if } \dim(\mathcal{H}) < \infty. \end{cases}$

for all $z \in \mathbb{C}$. To complete the proof of the assertion, we will proceed in two steps.

Step 1. We have to show that $h_1, h_2 : \mathcal{F}_1(\mathcal{H}) \setminus \{0\} \longrightarrow \mathbb{C} \setminus \{0\}$ are in fact constant. Let $x \otimes u, y \otimes v \in \mathcal{F}_1(\mathcal{H}) \setminus \{0\}$ with $\langle y, u \rangle \neq 0$. By equations (2.4) and (5.10), we have

$$\begin{aligned} r_\varepsilon(h_1(x \otimes u) h_2(y \otimes v) \eta(\langle y, u \rangle) x \otimes v) &= r_\varepsilon(\phi(x \otimes u) \psi(y \otimes v)) \\ &= r_\varepsilon((x \otimes u)(y \otimes v)) \\ &= r_\varepsilon(\langle y, u \rangle x \otimes v), \end{aligned}$$

Thus, by Proposition 3.2, we have

$$r_\varepsilon(\langle y, u\rangle x \otimes v)$$
$$= \frac{1}{2}\left(\sqrt{|\langle y,u\rangle|^2|\langle x,v\rangle|^2 + 4\varepsilon^2 + 4\varepsilon|\langle y,u\rangle|\|x\|\|v\|} + |\langle y,u\rangle||\langle x,v\rangle|\right)$$
$$r_\varepsilon(h_1(x \otimes u)h_2(y \otimes v)\eta(\langle y,u\rangle)x \otimes v)$$
$$= \frac{1}{2}\sqrt{|\lambda_{x,y,u,v}(h_1,h_2)|^2|\langle x,v\rangle|^2 + 4\varepsilon^2 + 4\varepsilon|\lambda_{x,y,u,v}(h_1,h_2)|\|x\|\|v\|}$$
$$+ \frac{1}{2}|\lambda_{x,y,u,v}(h_1,h_2)||\langle x,v\rangle|,$$

where
$$\lambda_{x,y,u,v}(h_1, h_2) = h_1(x \otimes u)h_2(y \otimes v)\eta(\langle y, u\rangle).$$

Thus, we conclude that

$$\left(\sqrt{|\langle y,u\rangle|^2|\langle x,v\rangle|^2 + 4\varepsilon^2 + 4\varepsilon|\langle y,u\rangle|\|x\|\|v\|} + |\langle y,u\rangle||\langle x,v\rangle|\right)$$
$$=$$
$$\sqrt{|\lambda_{x,y,u,v}(h_1,h_2)|^2|\langle x,v\rangle|^2 + 4\varepsilon^2 + 4\varepsilon|\lambda_{x,y,u,v}(h_1,h_2)|\|x\|\|v\|} + |\lambda_{x,y,u,v}(h_1,h_2)||\langle x,v\rangle|.$$

Therefore

$$|\langle y,u\rangle|^2|\langle x,v\rangle|^2 + 4\varepsilon^2 + 4\varepsilon|\langle y,u\rangle|\|x\|\|v\|$$
$$= (\gamma_{x,y,u,v}(h_1,h_2) + |\lambda_{x,y,u,v}(h_1,h_2)||\langle x,v\rangle| - |\langle y,u\rangle||\langle x,v\rangle|)^2$$
$$= \gamma_{x,y,u,v}^2(h_1,h_2) + |\lambda_{x,y,u,v}(h_1,h_2)|^2\langle x,v\rangle|^2 + |\langle y,u\rangle|^2|\langle x,v\rangle|^2$$
$$- 2\gamma_{x,y,u,v}(h_1,h_2)|\langle y,u\rangle||\langle x,v\rangle|$$
$$+ 2\gamma_{x,y,u,v}(h_1,h_2)|\lambda_{x,y,u,v}(h_1,h_2)\langle x,v\rangle|$$
$$- 2|\langle y,u\rangle||\lambda_{x,y,u,v}(h_1,h_2)\langle x,v\rangle|^2,$$

where
$$\gamma_{x,y,u,v}(h_1,h_2) = \sqrt{|\lambda_{x,y,u,v}(h_1,h_2)|^2|\langle x,v\rangle|^2 + 4\varepsilon^2 + 4\varepsilon|\lambda_{x,y,u,v}(h_1,h_2)|\|x\|\|v\|}.$$

Hence

$$4\varepsilon\|x\|\|v\|\left(|\langle y,u\rangle| - |\lambda_{x,y,u,v}(h_1,h_2)|\right) = 2|\lambda_{x,y,u,v}(h_1,h_2)|^2|\langle x,v\rangle|^2$$
$$- 2\gamma_{x,y,u,v}(h_1,h_2)|\langle y,u\rangle||\langle x,v\rangle|$$
$$+ 2\gamma_{x,y,u,v}(h_1,h_2)|\lambda_{x,y,u,v}(h_1,h_2)\langle x,v\rangle|$$
$$- 2|\langle y,u\rangle||\lambda_{x,y,u,v}(h_1,h_2)\langle x,v\rangle|^2.$$

A straightforward computation shows that

$$|\langle x,v\rangle|\left(\lambda_{x,y,u,v}(h_1,h_2)|\langle x,v\rangle| + \gamma_{x,y,u,v}(h_1,h_2)\right)\left(|\lambda_{x,y,u,v}(h_1,h_2)| - |\langle y,u\rangle|\right)$$
$$=$$
$$2\varepsilon\|x\|\|v\|\left(|\langle y,u\rangle| - |\lambda_{x,y,u,v}(h_1,h_2)|\right).$$

Hence,
$$\left(\frac{|\langle x,v\rangle|}{2\varepsilon\|x\|\|v\|}\left(\lambda_{x,y,u,v}(h_1,h_2)|\langle x,v\rangle| + \gamma_{x,y,u,v}(h_1,h_2)\right) + 1\right)\left(|\lambda_{x,y,u,v}(h_1,h_2)| - |\langle y,u\rangle|\right) = 0.$$

Since
$$\frac{|\langle x,v \rangle|}{2\varepsilon \|x\|\|v\|} \left(\lambda_{x,y,u,v}(h_1,h_2)|\langle x,v \rangle| + \gamma_{x,y,u,v}(h_1,h_2) \right) + 1 > 0,$$
we obtain
$$|\lambda_{x,y,u,v}(h_1,h_2)| = |\langle y,u \rangle|.$$
Thus $|h_1(x \otimes u)h_2(y \otimes v)| = \frac{|\langle y,u \rangle|}{|\eta(\langle y,u \rangle)|} = 1$ if $\dim(\mathcal{H}) = \infty$. Now, assume that $\mathcal{H} = \mathbb{C}^n$ for some integer $n \geq 3$, clearly we have

(5.12) $$|h_1(x \otimes u)h_2(y \otimes v)| = \frac{|\langle y,u \rangle|}{|\eta(\langle y,u \rangle)|} = 1$$

for all $x,y,u,v \in \mathbb{C}^n \setminus \{0\}$ with $|\langle y,u \rangle|$ is a positive integer. Next, assume that $x,y,u,v \in \mathbb{C}^n \setminus \{0\}$ and $|\langle y,u \rangle| \neq 1$, we have two cases to discuss:

Case 1. If y and u are linearly independent, then
$$\{z \in \mathcal{H} : \langle z,u \rangle = 1\} \text{ and } \{z \in \mathcal{H} : \langle z,y \rangle = 1\}$$
are two non-parallel affine hyperplanes of $\mathbb{C}^n$. So
$$\{z \in \mathcal{H} : \langle z,u \rangle = 1\} \cap \{z \in \mathcal{H} : \langle z,y \rangle = 1\}$$
is a nonempty affine hyperplane, hence there exists $z \in \mathcal{H}$ such that $\langle z,z \rangle$ is a positive integer and
$$\langle z,u \rangle = 1 \text{ and } \langle z,y \rangle = 1.$$
Therefore, we have
$$|h_1(x \otimes u)h_2(z \otimes v)| = 1 \text{ and } |h_1(x \otimes z)h_2(y \otimes v)| = 1.$$
Then
$$|h_1(x \otimes u)h_2(z \otimes v)||h_1(x \otimes z)h_2(y \otimes v)| = 1 \text{ and } |h_1(x \otimes z)h_2(z \otimes v)| = 1,$$
hence $|h_1(x \otimes u)h_2(y \otimes v)| = 1$.

Case 2. If y and u are linearly dependent. Choose $z \in \mathcal{H}$ such $\langle z,z \rangle$ is a positive integer, $\langle z,y \rangle \langle z,u \rangle \neq 0$ and u,y,z are linearly independent. By the previous case we conclude that
$$|h_1(x \otimes u)h_2(z \otimes v)||h_1(x \otimes z)h_2(y \otimes v)| = 1 \text{ and } |h_1(x \otimes z)h_2(z \otimes v)| = 1,$$
hence $|h_1(x \otimes u)h_2(y \otimes v)| = 1$.

We conclude that if $\mathcal{H}$ is a finite or infinite dimensional Hilbert space, we have
$$|h_1(x \otimes u)h_2(y \otimes v)| = 1$$
for all $x,y,u,v \in \mathcal{H} \setminus \{0\}$ with $\langle y,u \rangle \neq 0$. Finally, if $x,y,u,v \in \mathcal{H} \setminus \{0\}$ with $\langle y,u \rangle = 0$, one can choose $z \in \mathbb{C}^n$ such that $\langle z,u \rangle \langle z,y \rangle \neq 0$, then
$$|h_1(x \otimes u)h_2(z \otimes v)| = 1 \,, \, |h_1(x \otimes z)h_2(y \otimes v)| = 1 \text{ and } |h_1(x \otimes z)h_2(z \otimes v)| = 1.$$
This implies that
$$|h_1(x \otimes u)h_2(y \otimes v)| = 1$$
for all $x,y,u,v \in \mathcal{H} \setminus \{0\}$. Consequently, there exist two nonzero scalars μ and ν such that $|\mu\nu| = 1$ and

(5.13) $$h_1(x \otimes u) = \mu \text{ and } h_2(x \otimes u) = \nu$$

for all nonzero $x \otimes u \in \mathcal{F}_1(\mathcal{H})$.

Step 2. The operators U and W are linear and $\mu\nu = 1$. Moreover, if $\dim(\mathcal{H}) < \infty$ then τ_1 is the identity of $\mathbb{C}$.

The equalities (5.13) together with (2.4) and (5.10) tells us that

(5.14) $$\sigma_\varepsilon(\mu\nu\eta(\langle y, u \rangle)x \otimes v) = \sigma_\varepsilon(\langle y, u \rangle x \otimes v)$$

for all nonzero $x, v \in \mathcal{H}$. Therefore, if $x \in \mathcal{H}$ is a unit vector, then by Proposition 3.1, we have

$$\begin{aligned}
D(0,\varepsilon) \cup D(\alpha,\varepsilon) &= \sigma_\varepsilon(\alpha x \otimes x) \\
&= \sigma_\varepsilon((\alpha x \otimes x)(x \otimes x)) \\
&= \sigma_\varepsilon(\phi(\alpha x \otimes x)\psi(x \otimes x)) \\
&= \sigma_\varepsilon\left((\mu W \alpha x \otimes (U^{-1})^* x_{\tau_3})(\nu U x_\tau \otimes x W^*)\right) \\
&= \sigma_\varepsilon\left((W \mu^W \alpha x \otimes (U^{-1})^* x_{\tau_3})(\nu U x_\tau \otimes x W^*)\right) \\
&= \sigma_\varepsilon\left(\mu^W \nu \eta(\langle x, x \rangle)\alpha x \otimes x\right)^W \\
&= \sigma_\varepsilon\left(\mu^W \nu \alpha x \otimes x\right)^W \\
&= \left(D(0,\varepsilon) \cup D(\mu^W \nu \alpha, \varepsilon)\right)^W \\
&= D(0,\varepsilon) \cup D(\mu(\nu\alpha)^W, \varepsilon)
\end{aligned}$$

for all $\alpha \in \mathbb{C} \setminus \{0\}$, where $\lambda^W = \lambda$ (resp. $\overline{\lambda}$) if W is unitary (resp. conjugate unitary) for all scalars λ. This implies that $\mu(\nu\alpha)^W = \alpha$ for all $\alpha \in \mathbb{C} \setminus \{0\}$. For $\alpha = 1$ we obtain $\mu\nu^W = 1$, hence $\alpha^W = \alpha$ for all $\alpha \in \mathbb{C} \setminus \{0\}$. Thus W is an unitary operator and $\mu\nu = 1$.

Next, if $x \in \mathcal{H}$ is a unit vector, then by Proposition 3.1, we have

$$\begin{aligned}
D(0,\varepsilon) \cup D(\alpha,\varepsilon) &= \sigma_\varepsilon(\alpha x \otimes x) \\
&= \sigma_\varepsilon((x \otimes \overline{\alpha}x)(x \otimes x)) \\
&= \sigma_\varepsilon(\phi(x \otimes \overline{\alpha}x)\psi(x \otimes x)) \\
&= \sigma_\varepsilon\left((\mu W x \otimes (U^{-1})^*(\overline{\alpha}x)_{\tau_3})(\nu U x_\tau \otimes x W^*)\right) \\
&= \sigma_\varepsilon(\eta(\alpha) x \otimes x) \\
&= D(0,\varepsilon) \cup D(\eta(\alpha), \varepsilon)
\end{aligned}$$

for all $\alpha \in \mathbb{C} \setminus \{0\}$. Hence $\eta(\alpha) = \alpha$ for all $\alpha \in \mathbb{C}$ and so U is a bounded linear operator. Thus, we obtain the desired forms of ϕ and ψ and the proof of Assertion 4 is complete. $\square$

Keeping in mind the notations of the previous assertion, the following assertion shows that ϕ and ψ have the desired forms on $\mathcal{A}$.

Assertion 5. We have $\phi(A) = \mu W A U^{-1}$ and $\psi(A) = \nu U A W^*$ for all $A \in \mathcal{A}$.

PROOF. Let A be an arbitrary nonzero operator in $\mathcal{A}$. For any unit vector $x \in \mathcal{H}$, we have

$$\begin{aligned}
\sigma_\varepsilon(\nu\phi(A) U x \otimes W A x) &= \sigma_\varepsilon(\phi(A)\psi(x \otimes Ax)) \\
&= \sigma_\varepsilon(A(x \otimes Ax)) \\
&= D(0,\varepsilon) \cup D(\|Ax\|^2, \varepsilon).
\end{aligned}$$

If $Ax \neq 0$, then from Proposition 3.1 we obtain that $\nu\phi(A)Ux \otimes WAx = \|Ax\|^2 w \otimes w$ for a unit vector $w \in \mathcal{H}$. Therefore, $w = \frac{1}{\|WAx\|} WAx$ and we have

$$\begin{aligned}
\nu\phi(A)Ux \otimes WAx &= \|Ax\|^2 \left(\frac{WAx}{\|WAx\|}\right) \otimes \left(\frac{WAx}{\|WAx\|}\right) \\
&= \|WAx\|^2 \left(\frac{WAx}{\|WAx\|}\right) \otimes \left(\frac{WAx}{\|WAx\|}\right) \\
&= WAx \otimes WAx.
\end{aligned}$$

Hence
$$\nu\phi(A)Ux = WAx$$
for all unit vectors $x \in \mathcal{H}$ with $Ax \neq 0$. Thus

(5.15) $$\phi(A)y = \frac{1}{\nu} WAU^{-1}y$$

for all vectors $y \in \mathcal{H}$ such that $AU^{-1}y \neq 0$. Now if $AU^{-1}y = 0$ for some y one can choose any vector y_0 such that $AU^{-1}y_0 \neq 0$, then (5.15) holds for y_0 and $y - y_0$. Consequently,

$$\phi(A) = \frac{1}{\nu} WAU^{-1} = \mu WAU^{-1}$$

for all $A \in \mathcal{A}$.

Similarly, we have

$$\begin{aligned}
\sigma_\varepsilon\left(\mu WA^*x \otimes xU^{-1}\psi(A)\right) &= \sigma_\varepsilon\left(\phi(A^*x \otimes x)\psi(A)\right) \\
&= \sigma_\varepsilon\left(A^*x \otimes A^*x\right) \\
&= D(0, \varepsilon) \cup D(\|A^*x\|^2, \varepsilon)
\end{aligned}$$

for all unit vectors $x \in \mathcal{H}$ and $A \in \mathcal{A}$. Thus,

$$\mu WA^*x \otimes xU^{-1}\psi(A) = \|WAx\|^2 \left(\frac{WA^*x}{\|WA^*x\|}\right) \otimes \left(\frac{WA^*x}{\|WA^*x\|}\right)$$

Hence
$$\overline{\mu}\psi(A)^*(U^*)^{-1}x = WA^*x$$
for all unit vectors $x \in \mathcal{H}$ and $A \in \mathcal{A}$. Thus
$$\overline{\mu}\psi(A)^*(U^*)^{-1} = WA^*$$
for all $A \in \mathcal{A}$. Therefore
$$\psi(A) = \frac{1}{\mu} UAW^* = \nu UAW^*$$
for all $A \in \mathcal{A}$.

This completes the proof of Theorem 2.3. $\square$

6. Concluding remarks

In this section, we provide some applications of our main results. We start with the following theorem which is a direct consequence of Theorem 2.1.

THEOREM 6.1. *Suppose that $\phi, \psi : \mathcal{A} \longrightarrow \mathcal{B}$ are two surjective maps satisfying*

(6.1) $$AB = 0 \iff \phi(A)\psi(B) = 0$$

for all $A, B \in \mathcal{A}$, and

(6.2) $$AB = 0 \iff \psi(A)\phi(B) = 0$$

for all $A, B \in \mathcal{A}$. Then the following statements hold.
(1) If $\dim(X) = \infty$, then there exist two maps $\alpha, \beta : X \times X^* \longrightarrow X$, $(x, f) \mapsto \alpha_{x,f}$, $(x, f) \mapsto \beta_{x,f}$ and two bounded invertible linear or conjugate linear operators U and V on X such that
$$\phi(x \otimes f) = \alpha_{x,f} Vx \otimes (U^{-1})^* f \text{ and } \psi(x \otimes f) = \beta_{x,f} Ux \otimes (V^{-1})^* f$$
for all $(x, f) \in X \times X^*$.
(2) If $X = \mathbb{C}^n$ is a finite-dimensional space with $n \geq 3$, then there exist two maps $k, h : \mathbb{C}^n \times \mathbb{C}^n \longrightarrow \mathbb{C}^n$, $(x, y) \mapsto k_y(x), (x, y) \mapsto h_x(y)$, two nonsingular matrices $U, V \in \mathcal{M}_n(\mathbb{C})$, and two ring automorphisms τ_1 and τ_2 of $\mathbb{C}$ such that
$$\phi(x \otimes y) = \alpha_{x,y} V x_{\tau_1} \otimes y_{\tau_2} U^{-1} \text{ and } \psi(x \otimes y) = \beta_{x,y} U x_{\tau_2} \otimes y_{\tau_1} V^{-1}$$
for all $x, y \in \mathbb{C}^n$.

PROOF. Assume first that $\dim(X) = \infty$. Applying Theorem 2.1, (6.2) and (6.1) we conclude that there exist four X-valued maps k, k', h and h' on $X \times X^*$ and two bounded invertible linear or conjugate linear operators U and V on X such that

(6.3) $\quad \begin{aligned} \phi(x \otimes f) &= k_f(x) \otimes (U^{-1})^* f &= Vx \otimes h'_x(f) \\ \psi(x \otimes f) &= Ux \otimes h_x(f) &= k'_f(x) \otimes (V^{-1})^* f \end{aligned}$

for all $(x, f) \in X \times X^*$. Therefore, there exists two maps α and β from $X \times X^*$ to $\mathbb{C} \setminus \{0\}$ such that
$$k'_f(x) = \nu(x, f) Ux \text{ and } h'_f(x) = \nu(x, f)(V^{-1})^* fx$$
for all $(x, f) \in X \times X^*$. This allows us to conclude that there exist two maps α and β from $X \times X^*$ to $\mathbb{C} \setminus \{0\}$ and two bounded invertible linear or conjugate linear operators U and V on X such that
$$\phi(x \otimes f) = \alpha_{x,f} Vx \otimes (U^{-1})^* f \text{ and } \psi(x \otimes f) = \beta_{x,f} Ux \otimes (V^{-1})^* f$$
for all $(x, f) \in X \times X^*$.

Now, if $X = \mathbb{C}^n$, for a given integer $n \geq 3$, then by Theorem 2.1 there exist four $\mathbb{C}^n$-valued maps $k, , k', h, h'$ on $\mathbb{C}^n \times \mathbb{C}^n$, two nonsingular matrices $U, V \in \mathcal{M}_n(\mathbb{C})$, and two ring automorphisms τ_1 and τ_2 of $\mathbb{C}$ such that

(6.4) $\quad \begin{aligned} \phi(x \otimes y) &= k_y(x) \otimes y_{\tau_1} U^{-1} &= V x_{\tau_2} \otimes h'_x(y) \\ \psi(x \otimes y) &= U x_{\tau_1} \otimes h_x(y) &= k'_y(x) \otimes y_{\tau_2} V^{-1} \end{aligned}$

for all $x, y \in \mathbb{C}^n$. By a similar argument as above, there exist two functionals α and β on $\mathbb{C}^n \times \mathbb{C}^n$, two nonsingular matrices $U, V \in \mathcal{M}_n(\mathbb{C})$, and two ring automorphisms τ_1 and τ_2 of $\mathbb{C}$ such that
$$\phi(x \otimes y) = \alpha_{x,y} V x_{\tau_1} \otimes y_{\tau_2} U^{-1} \text{ and } \psi(x \otimes y) = \beta_{x,y} U x_{\tau_2} \otimes y_{\tau_1} V^{-1}$$
for all $x, y \in \mathbb{C}^n$. This completes the proof of Theorem 6.1. $\square$

Now, we will apply Theorem 6.1 to characterize maps preserving zero products and zero skew products. Let $\mathcal{A}$ and $\mathcal{B}$ are two subsets of $\mathcal{B}(X)$ which contain all operators of rank one, and note that surjective maps $\phi : \mathcal{A} \longrightarrow \mathcal{B}$ satisfying

(6.5) $\quad \phi(A)\phi(B) = 0 \iff AB = 0 \quad (A, B \in \mathcal{A}).$

were described in [**21**, Lemma 2.2]. Such a characterization is now a consequence of Theorem 6.1.

COROLLARY 6.2. *Let X be a complex Banach space with $\dim X \geq 3$, and let $\mathcal{A}, \mathcal{B}$ be two subsets of $\mathcal{B}(X)$ containing all operators in $\mathcal{B}(X)$ of rank at most one. Suppose that $\phi : \mathcal{A} \longrightarrow \mathcal{B}$ is a surjective map satisfying (6.5), then there exists a functional $l : X \otimes X^* \longrightarrow \mathbb{C} \setminus \{0\}$ and*

(1) *if $\dim(X) = \infty$, there exists a bounded invertible linear or conjugate linear operator U on X such that*

$$\phi(x \otimes f) = l(x, f)U(x \otimes f)U^{-1}$$

for all rank one operator $(x, f) \in X \times X^*$;

(2) *if $X = \mathbb{C}^n$, there exist a nonsingular matrix $U \in \mathcal{M}_n(\mathbb{C})$ and a ring automorphism τ of $\mathbb{C}$ such that*

$$\phi(x \otimes y) = l(x, y)U x_\tau \otimes y_\tau U^{-1}$$

for all rank one matrix $x \otimes y = x^t y \in \mathbb{C}^n$.

PROOF. Applying Theorem 6.1 in the case when $\psi = \phi$, we conclude that:

If $\dim(X) = \infty$, then there exist two maps $\alpha, \beta : X \times X^* \longrightarrow X$, $(x,f) \mapsto \alpha_{x,f}$, $(x,f) \mapsto \beta_{x,f}$ and two bounded invertible linear or conjugate linear operators U and V on X such that

$$\alpha_{x,f} V x \otimes (U^{-1})^* f = \phi(x \otimes f) = \psi(x \otimes f) = \beta_{x,f} U x \otimes (V^{-1})^* f$$

for all $(x, f) \in X \times X^*$. Therefore, $\beta_{x,y} U x$ and $\alpha_{x,y} V x$ are linearly dependent for all $x \in X$. Hence, U and V are linearly dependent, and thus, there exist a functional $l : X \times X^* \longrightarrow \mathbb{C} \setminus \{0\}$ and a bounded invertible linear or conjugate linear operator U on X such that

$$\phi(x \otimes f) = l(x, f)U x \otimes (U^{-1})^* f = l(x, f)U(x \otimes f)U^{-1}$$

for every rank one operator $x \otimes f \in \mathcal{A}$.

If $X = \mathbb{C}^n$ is a finite-dimensional space with $n \geq 3$, then there exist two maps $k, h : \mathbb{C}^n \times \mathbb{C}^n \longrightarrow \mathbb{C}^n$, $(x,y) \mapsto k_y(x), (x,y) \mapsto h_x(y)$, two nonsingular matrices $U, V \in \mathcal{M}_n(\mathbb{C})$, and two ring automorphisms τ_1 and τ_2 of $\mathbb{C}$ such that

$$\alpha_{x,y} V x_{\tau_1} \otimes y_{\tau_2} U^{-1} = \phi(x \otimes y) = \psi(x \otimes y) = \beta_{x,y} U x_{\tau_2} \otimes y_{\tau_1} V^{-1}$$

for all $x, y \in \mathbb{C}^n$. Therefore, $\beta_{x,y} U x_{\tau_2}$ and $\alpha_{x,y} V x_{\tau_1}$ are linearly dependent for all $x \in \mathbb{C}^n$. Hence, Ux and Vx are linearly dependent for all $x = (x_1, ..., x_n) \in \{0,1\}^n$, thus $U = \lambda V$ for some scalar λ and $\tau_1 = \tau_2$. Therefore, there exists a functional $l : \mathcal{A} \longrightarrow \mathbb{C} \setminus \{0\}$ and there exist a nonsingular matrix $U \in \mathcal{M}_n(\mathbb{C})$ and a ring automorphism τ of $\mathbb{C}$ such that $\phi(A) = l(A)U(x_\tau \otimes y_\tau)U^{-1}$ for every all rank one matrices $x \otimes y$. □

Next, let $\mathcal{H}$ be a complex Hilbert space with $\dim(\mathcal{H}) \geq 3$, and $\mathcal{A}$ and $\mathcal{B}$ be two subsets of $\mathcal{B}(\mathcal{H})$ containing all operators in $\mathcal{B}(\mathcal{H})$ of rank at most one. Let $J \in \mathcal{B}(\mathcal{H})$ be an invertible self-adjoint operator and denote $A^\dagger = J^{-1} A^* J$ for all $A \in \mathcal{B}(\mathcal{H})$. Surjective maps $\phi : \mathcal{A} \longrightarrow \mathcal{B}$ satisfying

(6.6) $$\phi(A)\phi(B)^\dagger = 0 \Longleftrightarrow AB^\dagger = 0 \quad (A, B \in \mathcal{A}),$$

and

(6.7) $$\phi(A)^\dagger \phi(B) = 0 \Longleftrightarrow A^\dagger B = 0 \quad (A, B \in \mathcal{A}),$$

were described in [**31**, Theorem 2.1]. Such a characterization is now a consequence of Theorem 6.1. Here, we state this result, in a slightly different way, since the original version in [**31**, Theorem 2.1] requires a minor correction; see the discussion after the proof of the following result.

COROLLARY 6.3. *Let $\mathcal{H}$ be a complex Hilbert space with $\dim(\mathcal{H}) \geq 3$. Let $\mathcal{A}$, $\mathcal{B}$ be two subsets of $\mathcal{B}(\mathcal{H})$ containing all operators in $\mathcal{B}(\mathcal{H})$ of rank at most one. Suppose that $\phi : \mathcal{A} \longrightarrow \mathcal{B}$ is a surjective map satisfying (6.6) and (6.7), then there exist a functional $h : \mathcal{H} \times \mathcal{H} \longrightarrow \mathbb{C} \setminus \{0\}$, two nonzero real scalars c and d, and linear or conjugate linear bounded invertible operators U and V on $\mathcal{H}$ such that $U^*JU = cJ$, $V^*JV = dJ$ and*

$$\phi(x \otimes y) = h(x,y) Ux \otimes V^*y \quad (x, y \in \mathcal{H}).$$

PROOF. Set
$$\psi(A) := J^{-1} \left(\phi(J^{-1}A^*J) \right)^* J$$
for all $A \in \mathcal{A}$, and note that ϕ and ψ satisfy (6.1) and (6.2). Therefore, Theorem 6.1 applies, and thus two cases will be discussed.

If $\dim(\mathcal{H}) = \infty$, then there exist two maps α and β from $\mathcal{H}^2$ to $\mathbb{C} \setminus \{0\}$, and two bounded invertible linear or conjugate linear operators U and V on $\mathcal{H}$ such that
$$\phi(x \otimes y) = \alpha_{x,y} Vx \otimes (U^{-1})^*y \text{ and } \psi(x \otimes y) = \beta_{x,y} Ux \otimes (V^{-1})^*y$$
for all $(x, y) \in \mathcal{H}^2$. Then, for every $x \in \mathcal{H}$, we have
$$\begin{aligned}
\beta_{x,x} Ux \otimes (V^{-1})^*x &= \psi(x \otimes x) \\
&= J^{-1} \left(\phi(J^{-1}x \otimes Jx) \right)^* J \\
&= J^{-1} \left(\alpha_{x,x} VJ^{-1}x \otimes (U^{-1})^*Jx \right)^* J \\
&= \overline{\alpha_{x,x}} J^{-1}(U^{-1})^* Jx \otimes JVJ^{-1}x.
\end{aligned}$$
Thus, there exists two functionals α' and β' on $\mathcal{H}$ such that
$$J^{-1}(U^{-1})^* Jx = \beta'_x Ux$$
and
$$JVJ^{-1}x = (\alpha'_x)(V^{-1})^*x.$$
Therefore, there exist a nonzero scalars c and d such that $U^*JU = cJ$ and $V^*JV = dJ$. Thus c and d are in fact reals. Hence, there exists a functional $h : \mathcal{H} \times \mathcal{H} \longrightarrow \mathbb{C} \setminus \{0\}$ and two bounded invertible linear or conjugate linear operators U and V on $\mathcal{H}$ such that $\phi(x \otimes y) = h(x,y) Vx \otimes U^*y$, for all rank one operators $x \otimes y \in \mathcal{B}(\mathcal{H})$.

If $\mathcal{H} = \mathbb{C}^n$ is a finite-dimensional Hilbert space with $n \geq 3$, then there exist two functionals α and β on $\mathbb{C}^n \times \mathbb{C}^n$, two nonsingular matrices $U, V \in \mathcal{M}_n(\mathbb{C})$, and two ring automorphisms τ_1 and τ_2 of $\mathbb{C}$ such that
$$\phi(x \otimes y) = \alpha_{x,y} Vx_{\tau_1} \otimes y_{\tau_2^*} U^{-1} \text{ and } \psi(x \otimes y) = \beta_{x,y} Ux_{\tau_2} \otimes y_{\tau_1^*} V^{-1}$$
for all $x, y \in \mathbb{C}^n$, where $\tau_i^*(\mu) = \overline{\tau_i(\overline{\mu})}$. Thus, there exists two maps $\alpha', \beta' : \mathcal{H} \longrightarrow \mathcal{H}$, $x \mapsto \alpha'_x$, $x \mapsto \beta'_x$ such that
$$J^{-1}(U^{-1})^* J^{-1} x_{\tau_2^*} = \beta'_x Ux_{\tau_2}$$
and
$$JVJx_{\tau_1} = \alpha'_x (V^{-1})^* x_{\tau_1^*}.$$

Thus $\tau_i^* = \tau_i$ and hence τ_i is either the identity or the complex conjugation map on $\mathbb{C}$, $i = 1, 2$. By similar arguments as above, we obtain the desired form of ϕ. □

Note that in the statement of the previous result, U and V need not be simultaneously linear or simultaneously conjugate linear. For example, let $\mathcal{H} = \mathbb{C}^n$ and $J = I$. If
$$\phi(x \otimes y) = x \otimes \overline{y}$$
for all $x, y \in \mathbb{C}^n$, then ϕ satisfies (6.6) and (6.7). While there is no simultaneously linear or simultaneously conjugate linear operators U and V in $\mathcal{B}(\mathcal{H})$ such that ϕ has the form $x \otimes y \mapsto U(x \otimes y)V$. In fact the equality
$$h_1(x \otimes f)Ux \otimes V^*f = h_1(x \otimes f)U(x \otimes f)V$$
claimed in [**31**, page 2245, line 3, in the proof of Theorem 2.1] is not correct when exactly one of operators U and V is conjugate linear. For this reason, we also give the following restatement of [**31**, Corollary 2.3].

COROLLARY 6.4. *Let $\mathcal{H}$ be a complex Hilbert space with* $\dim(\mathcal{H}) \geq 3$, *and let $\mathcal{A}$ and $\mathcal{B}$ be two subsets of $\mathcal{B}(\mathcal{H})$ containing all operators of rank at most one. Suppose that $\phi : \mathcal{A} \longrightarrow \mathcal{B}$ is a surjective map satisfying*

(6.8) $$\phi(A)\phi(B)^* = 0 \iff AB^* = 0$$

for all $A, B \in \mathcal{A}$, and

(6.9) $$\phi(A)^*\phi(B) = 0 \iff A^*B = 0$$

for all $A, B \in \mathcal{A}$. Then there exist unitary or conjugate unitary operators U and V and a functional $h : \mathcal{H}^2 \longrightarrow \mathbb{C} \setminus \{0\}$ such that
$$\phi(x \otimes y) = h(x, y)Ux \otimes V^*y$$
for all $x, y \in \mathcal{H}$.

In the following two corollaries, we will apply Theorem 2.3 to characterize maps preserving ε-pseudo spectrum of the product AB, and maps preserving the ε-pseudo spectrum of the skew product AB^*. Surjective maps ϕ from $\mathcal{B}(\mathcal{H})$ into itself satisfying

(6.10) $$\sigma_\varepsilon(\phi(A)\phi(B)) = \sigma_\varepsilon(AB), \quad (A, B \in \mathcal{A})$$

were described in [**23**, Theorem 4.1] and [**24**, Theorem 3.3]. Such characterizations are now a consequence of Theorem 2.3.

COROLLARY 6.5. *Let $\mathcal{H}$ be a complex Hilbert space with $\dim(\mathcal{H}) \geq 3$, and let $\mathcal{A}$ and $\mathcal{B}$ be two subsets of $\mathcal{B}(\mathcal{H})$ containing all operators in $\mathcal{B}(\mathcal{H})$ of rank at most one. A surjective map $\phi : \mathcal{A} \longrightarrow \mathcal{B}$ satisfies (6.10) if and only if there exist a nonzero scalar λ with $\lambda^2 = 1$ and a unitary operator V on $\mathcal{H}$ such that $\phi(A) = \lambda V A V^*$, for all $A \in \mathcal{A}$.*

PROOF. Applying Theorem 2.3 for ϕ and $\psi = \phi$, we conclude that there exist a bounded invertible linear operator U on $\mathcal{H}$ and a unitary operator V on $\mathcal{H}$ such that
$$\mu V A U^{-1} = \phi(A) = \psi(A) = \nu U A V^*$$
for all $A \in \mathcal{A}$, where $\mu, \nu \in \mathbb{C}$ such that $\mu\nu = 1$. Therefore, for every $x \otimes u \in \mathcal{F}_1(\mathcal{H})$, we have $\mu V x \otimes (U^{-1})^* u = \nu U x \otimes V u$, then there exists $\beta \in \mathbb{C}$ such that $U x = \beta V x$

for all $x \in \mathcal{H}$. Thus, $U = \beta V$, so, $\phi(A) = \lambda VAV^*$, with $\lambda = \nu\beta$. Now, we show that $\lambda^2 = 1$. By equation (6.10) and Proposition 3.1, for a unit vector $x \in \mathcal{H}$, we have

$$\begin{aligned} D(0,\varepsilon) \cup D(\lambda^2,\varepsilon) &= \sigma_\varepsilon(\lambda^2 x \otimes x) \\ &= \sigma_\varepsilon(\lambda^2(x \otimes x)(x \otimes x)) \\ &= \sigma_\varepsilon(\lambda Vx \otimes xV^* \lambda Vx \otimes xV^*) \\ &= \sigma_\varepsilon(\phi(x \otimes x)\phi(x \otimes x)) \\ &= \sigma_\varepsilon((x \otimes x)(x \otimes x)) \\ &= \sigma_\varepsilon(x \otimes x) \\ &= D(0,\varepsilon) \cup D(1,\varepsilon), \end{aligned}$$

and then $\lambda^2 = 1$. □

Surjective maps ϕ from $\mathcal{B}(\mathcal{H})$ into itself satisfying

(6.11) $\qquad \sigma_\varepsilon(\phi(A)^*\phi(B)) = \sigma_\varepsilon(A^*B) \quad (\forall A, B \in \mathcal{B}(\mathcal{H}))$,

were described in [22, Corollary 5.4]. Such a characterizations is now a consequence of Theorem 2.3.

COROLLARY 6.6. *Let $\mathcal{H}$ be a complex Hilbert space with $\dim(\mathcal{H}) \geq 3$, and let $\mathcal{A}$ and $\mathcal{B}$ be two subsets of $\mathcal{B}(\mathcal{H})$ containing all operators in $\mathcal{B}(\mathcal{H})$ of rank at most one. A surjective map $\phi : \mathcal{A} \longrightarrow \mathcal{B}$ satisfies*

(6.12) $\qquad \sigma_\varepsilon(\phi(A)^*\phi(B)) = \sigma_\varepsilon(A^*B), \quad (A, B \in \mathcal{A})$

if and only if there exist a complex unit scalar λ and unitary operators $U, V \in \mathcal{A}$ such that $\phi(A) = \lambda VAU$ for all $A \in \mathcal{A}$.

PROOF. Applying Theorem 2.3 for $\psi(A) := (\phi(A^*))^*$ for all $A \in \mathcal{A}$, clearly, ϕ and ψ satisfy (2.4). Then there exist a bounded invertible linear operator U on $\mathcal{H}$ and a unitary operator V on $\mathcal{H}$ such that $\phi(A) = \mu VAU^{-1}$ and $\psi(A) = \nu UAV^*$ for all $A \in \mathcal{A}$, where $\mu, \nu \in \mathbb{C}$ such that $\mu\nu = 1$. Thus

(6.13) $\qquad \overline{\mu}(U^{-1})^*AV^* = (\phi(A^*))^* = \psi(A) = \nu UAV^*$

for every $A \in \mathcal{A}$. For $x \in \mathcal{H}$, we choose $A = x \otimes xV$, then, equation (6.13) implies that $\overline{\mu}(U^{-1})^*x \otimes x = \nu Ux \otimes x$, hence $\left(\overline{\mu}(U^{-1})^* - \nu U\right)x = 0$, for all $x \in \mathcal{H}$. Thus, U is a λ multiple of a unitary operator W, where λ is a complex unit scalar. Therefore, $\phi(A) = \lambda WAV^*$ for all $A \in \mathcal{A}$. □

References

[1] Z. E. A. Abdelali and A. Bourhim, *Maps preserving the local spectrum of quadratic products of matrices*, Acta Sci. Math. (Szeged) **84** (2018), no. 1-2, 49–64. MR3792765

[2] Z. Abdelali, A. Bourhim, and M. Mabrouk, *Lie product and local spectrum preservers*, Linear Algebra Appl. **553** (2018), 328–361, DOI 10.1016/j.laa.2018.05.013. MR3809383

[3] Z. E. A. Abdelali and H. Nkhaylia, *Maps preserving the pseudo spectrum of skew triple product of operators*, Linear Multilinear Algebra **67** (2019), no. 11, 2297–2306, DOI 10.1080/03081087.2018.1490690. MR4002350

[4] Z. Abdelali, A. Achchi, and R. Marzouki, *Maps preserving the local spectrum of skew-product of operators*, Linear Algebra Appl. **485** (2015), 58–71, DOI 10.1016/j.laa.2015.07.019. MR3394138

[5] Z. E. A. Abdelali, A. Achchi, and R. Marzouki, *Maps preserving the local spectrum of some matrix products*, Oper. Matrices **12** (2018), no. 2, 549–562, DOI 10.7153/oam-2018-12-34. MR3812190

[6] A. Achchi, M. Mabrouk, and R. Marzouki, *Maps preserving the local spectrum of the skew Jordan product of operators*, Oper. Matrices **11** (2017), no. 1, 133–146, DOI 10.7153/oam-11-10. MR3602635
[7] L. Baribeau and T. Ransford, *Non-linear spectrum-preserving maps*, Bull. London Math. Soc. **32** (2000), no. 1, 8–14, DOI 10.1112/S0024609399006426. MR1718765
[8] R. Bhatia, P. Šemrl, and A. R. Sourour, *Maps on matrices that preserve the spectral radius distance*, Studia Math. **134** (1999), no. 2, 99–110. MR1688218
[9] A. Bourhim and J. Mashreghi, *Maps preserving the local spectrum of triple product of operators*, Linear Multilinear Algebra **63** (2015), no. 4, 765–773, DOI 10.1080/03081087.2014.898299. MR3291562
[10] A. Bourhim and J. Mashreghi, *Maps preserving the local spectrum of product of operators*, Glasg. Math. J. **57** (2015), no. 3, 709–718, DOI 10.1017/S0017089514000585. MR3395343
[11] A. Bourhim, J. Mashreghi, and A. Stepanyan, *Nonlinear maps preserving the minimum and surjectivity moduli*, Linear Algebra Appl. **463** (2014), 171–189, DOI 10.1016/j.laa.2014.09.002. MR3262395
[12] A. Bourhim and M. Mabrouk, *Maps preserving the local spectrum of Jordan product of matrices*, Linear Algebra Appl. **484** (2015), 379–395, DOI 10.1016/j.laa.2015.06.034. MR3385068
[13] A. Bourhim and J. Mashreghi, *A survey on preservers of spectra and local spectra*, Invariant subspaces of the shift operator, Contemp. Math., vol. 638, Amer. Math. Soc., Providence, RI, 2015, pp. 45–98, DOI 10.1090/conm/638/12810. MR3309349
[14] M. Bendaoud, A. Benyouness, and M. Sarih, *Condition spectra of special operators and condition spectra preservers*, J. Math. Anal. Appl. **449** (2017), no. 1, 514–527, DOI 10.1016/j.jmaa.2016.12.022. MR3595215
[15] J.-T. Chan, C.-K. Li, and N.-S. Sze, *Mappings preserving spectra of products of matrices*, Proc. Amer. Math. Soc. **135** (2007), no. 4, 977–986, DOI 10.1090/S0002-9939-06-08568-6. MR2262897
[16] M. A. Chebotar, Y. Fong, and P.-H. Lee, *On maps preserving zeros of the polynomial $xy - yx$*, Linear Algebra Appl. **408** (2005), 230–243, DOI 10.1016/j.laa.2005.06.015. MR2166866
[17] C. Costara, *On nonlinear maps preserving the reduced minimum modulus on differences of matrices*, Linear Algebra Appl. **507** (2016), 288–299, DOI 10.1016/j.laa.2016.06.016. MR3536959
[18] C. Costara, *Maps on matrices that preserve the spectrum*, Linear Algebra Appl. **435** (2011), no. 11, 2674–2680, DOI 10.1016/j.laa.2011.04.026. MR2825274
[19] C. Costara and D. Repovš, *Nonlinear mappings preserving at least one eigenvalue*, Studia Math. **200** (2010), no. 1, 79–89, DOI 10.4064/sm200-1-5. MR2720208
[20] C. Costara and T. Ransford, *On local irreducibility of the spectrum*, Proc. Amer. Math. Soc. **135** (2007), no. 9, 2779–2784, DOI 10.1090/S0002-9939-07-08779-5. MR2317952
[21] J. Cui and J. Hou, *Maps leaving functional values of operator products invariant*, Linear Algebra Appl. **428** (2008), no. 7, 1649–1663, DOI 10.1016/j.laa.2007.10.010. MR2388647
[22] J. Cui, C.-K. Li, and N.-S. Sze, *Unitary similarity invariant function preservers of skew products of operators*, J. Math. Anal. Appl. **454** (2017), no. 2, 716–729, DOI 10.1016/j.jmaa.2017.04.072. MR3658795
[23] J. Cui, C.-K. Li, and Y.-T. Poon, *Pseudospectra of special operators and pseudospectrum preservers*, J. Math. Anal. Appl. **419** (2014), no. 2, 1261–1273, DOI 10.1016/j.jmaa.2014.05.041. MR3225433
[24] J. Cui, V. Forstall, C.-K. Li, and V. Yannello, *Properties and preservers of the pseudospectrum*, Linear Algebra Appl. **436** (2012), no. 2, 316–325, DOI 10.1016/j.laa.2011.03.044. MR2854873
[25] G. Dolinar and B. Kuzma, *General preservers of quasi-commutativity*, Canad. J. Math. **62** (2010), no. 4, 758–786, DOI 10.4153/CJM-2010-041-x. MR2674700
[26] P. A. Fillmore and W. E. Longstaff, *On isomorphisms of lattices of closed subspaces*, Canad. J. Math. **36** (1984), no. 5, 820–829, DOI 10.4153/CJM-1984-048-x. MR762744
[27] A. Fošner and B. Kuzma, *Preserving zeros of Lie product on alternate matrices*, Spec. Matrices **4** (2016), 80–100, DOI 10.1515/spma-2016-0009. MR3451272
[28] A. M. Gleason, *A characterization of maximal ideals*, J. Analyse Math. **19** (1967), 171–172, DOI 10.1007/BF02788714. MR213878
[29] A. E. Guterman and B. Kuzma, *Preserving zeros of a polynomial*, Comm. Algebra **37** (2009), no. 11, 4038–4064, DOI 10.1080/00927870802545687. MR2573234

[30] J. Hou and Q. Di, *Maps preserving numerical ranges of operator products*, Proc. Amer. Math. Soc. **134** (2006), no. 5, 1435–1446, DOI 10.1090/S0002-9939-05-08101-3. MR2199190

[31] J. Hou, K. He, and X. Zhang, *Nonlinear maps preserving numerical radius of indefinite skew products of operators*, Linear Algebra Appl. **430** (2009), no. 8-9, 2240–2253, DOI 10.1016/j.laa.2008.12.002. MR2503969

[32] J. Hou, C.-K. Li, and N.-C. Wong, *Jordan isomorphisms and maps preserving spectra of certain operator products*, Studia Math. **184** (2008), no. 1, 31–47, DOI 10.4064/sm184-1-2. MR2365474

[33] J. Hou, C.-K. Li, and N.-C. Wong, *Maps preserving the spectrum of generalized Jordan product of operators*, Linear Algebra Appl. **432** (2010), no. 4, 1049–1069, DOI 10.1016/j.laa.2009.10.018. MR2577648

[34] T. Jari, *Nonlinear maps preserving the inner local spectral radius*, Rend. Circ. Mat. Palermo (2) **64** (2015), no. 1, 67–76, DOI 10.1007/s12215-014-0181-7. MR3324374

[35] J.-P. Kahane and W. Żelazko, *A characterization of maximal ideals in commutative Banach algebras*, Studia Math. **29** (1968), 339–343, DOI 10.4064/sm-29-3-339-343. MR226408

[36] S. Kakutani and G. W. Mackey, *Ring and lattice characterization of complex Hilbert space*, Bull. Amer. Math. Soc. **52** (1946), 727–733, DOI 10.1090/S0002-9904-1946-08644-9. MR16534

[37] S. Kowalski and Z. Słodkowski, *A characterization of multiplicative linear functionals in Banach algebras*, Studia Math. **67** (1980), no. 3, 215–223, DOI 10.4064/sm-67-3-215-223. MR592387

[38] J. Hou, C.-K. Li, and N.-C. Wong, *Jordan isomorphisms and maps preserving spectra of certain operator products*, Studia Math. **184** (2008), no. 1, 31–47, DOI 10.4064/sm184-1-2. MR2365474

[39] C.-K. Li, P. Šemrl, and N.-S. Sze, *Maps preserving the nilpotency of products of operators*, Linear Algebra Appl. **424** (2007), no. 1, 222–239, DOI 10.1016/j.laa.2006.11.013. MR2324385

[40] M. Marcus and B. N. Moyls, *Linear transformations on algebras of matrices*, Canadian J. Math. **11** (1959), 61–66, DOI 10.4153/CJM-1959-008-0. MR99996

[41] J. Mashreghi and A. Stepanyan, *Nonlinear maps preserving the reduced minimum modulus of operators*, Linear Algebra Appl. **493** (2016), 426–432, DOI 10.1016/j.laa.2015.12.010. MR3452747

[42] L. Molnár, *Some characterizations of the automorphisms of $B(H)$ and $C(X)$*, Proc. Amer. Math. Soc. **130** (2002), no. 1, 111–120, DOI 10.1090/S0002-9939-01-06172-X. MR1855627

[43] L. Molnár, *Orthogonality preserving transformations on indefinite inner product spaces: generalization of Uhlhorn's version of Wigner's theorem*, J. Funct. Anal. **194** (2002), no. 2, 248–262, DOI 10.1006/jfan.2002.3970. MR1934603

[44] L. Molnár, *Selected preserver problems on algebraic structures of linear operators and on function spaces*, Lecture Notes in Mathematics, vol. 1895, Springer-Verlag, Berlin, 2007. MR2267033

[45] G. W. Mackey, *Isomorphisms of normed linear spaces*, Ann. of Math. (2) **43** (1942), 244–260, DOI 10.2307/1968868. MR6604

[46] P. Šemrl, *Applying projective geometry to transformations on rank one idempotents*, J. Funct. Anal. **210** (2004), no. 1, 248–257, DOI 10.1016/j.jfa.2003.07.009. MR2052121

[47] P. Šemrl, *Order and spectrum preserving maps on positive operators*, Canad. J. Math. **69** (2017), no. 6, 1422–1435, DOI 10.4153/CJM-2016-039-0. MR3715017

[48] P. Šemrl, *Maps on matrix and operator algebras*, Jahresber. Deutsch. Math.-Verein. **108** (2006), no. 2, 91–103. MR2248755

[49] P. Šemrl, *Maps on matrix spaces*, Linear Algebra Appl. **413** (2006), no. 2-3, 364–393, DOI 10.1016/j.laa.2005.03.011. MR2198941

[50] P. Šemrl, *Jordan ∗-derivations of standard operator algebras*, Proc. Amer. Math. Soc. **120** (1994), no. 2, 515–518, DOI 10.2307/2159889. MR1186136

[51] L. N. Trefethen and M. Embree, *Spectra and pseudospectra: The behavior of nonnormal matrices and operators*, Princeton University Press, Princeton, NJ, 2005. MR2155029

Department of Mathematics, Mathematical Research Center of Rabat, Laboratory of Mathematics, Statistics and Applications, Faculty of Sciences, Mohammed-V University in Rabat, Rabat, Morocco.

Email address: zineelabidineabdelali@gmail.com

Department of Mathematics, Mathematical Research Center of Rabat, Laboratory of Mathematics, Statistics and Applications, Faculty of Sciences, Mohammed-V University in Rabat, Rabat, Morocco.

Email address: nkhaylia.hamid@gmail.com

On algebraic characterizations of advertibly complete algebras

Martin Weigt and Ioannis Zarakas

ABSTRACT. In this paper, we deal with the question as to whether advertible completeness for general topological algebras can be characterized via the characters of the algebra, in a manner which resembles the known case of a unital commutative locally m-convex algebra. We provide certain results and stimulus for further research around the previous question, in the setting of Gelfand-Mazur GB*-algebras.

1. Introduction

A topological algebra $A[\tau]$ with identity element 1 is said to be *advertibly complete* if every Cauchy net (x_α) of A, having the property that there exists $x \in A$ such that $xx_\alpha \to 1$ and $x_\alpha x \to 1$, converges in A. It is well known that a unital commutative locally m-convex algebra $A[\tau]$ is advertibly complete if and only if $G_A = \{x \in A : \phi(x) \neq 0 \text{ for all } \phi \in \text{hom }(A)\}$ [**12**]. Here, G_A denotes the group of invertible elements of A and hom(A) denotes the set of all nonzero continuous characters of A. Advertible completeness, in the case of unital commutative locally m-convex algebras, is therefore characterized by fairly algebraic means (although not entirely through algebraic means, as the characters are assumed to be continuous and advertible completeness is a topological property). We recall that every unital topological algebra, for which G_A is an open set (called a Q-algebra), is advertibly complete [**12**, Theorem 6.5]. In this regard, we mention that one can characterize Q-algebras amongst the class of locally convex algebras in terms of concepts which are of a fairly algebraic nature [**19**]. All of this provides some motivation to find a characterization of advertibly complete locally convex algebras $A[\tau]$ in terms of constructs which are algebraic to a fair extent. Here, it must be emphasized that we are not looking for any such characterization involving an enlargement of the algebra A, such as its completion (should it exist as an algebra).

A Gelfand-Mazur algebra is a topological algebra which has a lot in common with a commutative locally m-convex algebra. More precisely, a *Gelfand-Mazur algebra* (over $\mathbb{C}$) is a topological algebra $A[\tau]$ such that $A/M \cong \mathbb{C}$ up to topological isomorphism, for all $M \in \mathcal{M}(A)$, where $\mathcal{M}(A)$ denotes the set of all closed regular two-sided ideals of A which are maximal as a left or right ideal of A [**2**]. If

---

2010 *Mathematics Subject Classification.* Primary 46H05, 46J05, 46K05.
Key words and phrases. Advertibly complete algebra, Gelfand-Mazur algebra, GB*-algebra.

$\mathcal{M}(A) = \emptyset$, then $A[\tau]$ is trivially a Gelfand-Mazur algebra. If $A[\tau]$ is a Gelfand-Mazur algebra with $\mathcal{M}(A) \neq \emptyset$, then $M \in \mathcal{M}(A)$ if and only if it is the kernel of some nonzero continuous character of A [2]. One is therefore led to the problem of characterizing advertible completeness amongst the class of Gelfand-Mazur algebras using fairly algebraic means (although not entirely algebraic). We specifically look at partial results whereby the property of advertible completeness amongst the class of Gelfand-Mazur algebras is characterized in terms of characters in a similar manner to that of commutative locally m-convex algebras mentioned above.

In this paper, we provide in Section 3 an overview of existing results in this direction for general topological algebras, as can, for example, be found in [5]. In Sections 3 and 4, we also give some new results within the framework of Gelfand-Mazur algebras.

This work was inspired by a recent problem we were working on, concerning the domains of closed unbounded $*$-derivations of GB*-algebras, which are $*$-algebras of unbounded linear operators on a Hilbert space which generalize C*-algebras [20]. More specifically, if $A[\tau]$ is a complete GB*-algebra, and $\delta : D(\delta) \to A$ is a closed $*$-derivation of A, then partial results are given in [20], whereby one has that $\lambda 1 - x$ is invertible in A if and only if $\lambda 1 - x$ is invertible in $D(\delta)$ for all $x \in D(\delta)$, where $\lambda \in \mathbb{C}$. The question, whether this is true in general, is still open, and the latter property is equivalent to $D(\delta)$ being advertibly complete. This provides the main source of motivation for considering the problem mentioned above.

In Section 2, we give the necessary background, which is needed for Sections 3 and 4. We include in this section also the background material on GB*-algebras, which is relevant to us.

2. Preliminaries

Throughout the paper, we consider all vector spaces to be over the field $\mathbb{C}$ of complex numbers and all topological spaces are assumed to be Hausdorff.

The term *topological algebra* refers to an algebra, which is also a topological vector space such that the multiplication is separately continuous. If the underlying topological vector space of a topological algebra is metrizable and complete, then the algebra is called a *Fréchet topological* algebra. A topological algebra, endowed with a continuous involution $*$, is called a topological $*$-algebra. A topological $*$-algebra, which is also a locally convex space, is called a *locally convex $*$-algebra*. In particular, a topological algebra whose topology is defined by a family $\Gamma = \{p\}$ of m-seminorms (i.e. $p(xy) \leq p(x)p(y)$, for all $x, y \in A, p \in \Gamma$) is called a locally m-convex algebra.

With the symbol $A[\tau]$, we denote a topological ($*$-)algebra A endowed with a given topology τ. For an algebra A with identity 1, the symbol G_A denotes the group of invertible elements in A. Furthermore, for $x \in A$, the *spectrum* of x, denoted by Sp (x), is defined to be the set

$$\text{Sp}\,(x) = \{\lambda \in \mathbb{C} : x - \lambda 1 \notin G_A\}.$$

DEFINITION 2.1. ([6]) Let $A[\tau]$ be a unital topological $*$-algebra and let $\mathcal{B}_A^*$ denote a collection of subsets B of A with the following properties:

(i) B is absolutely convex, closed and bounded;
(ii) $1 \in B$, $B^2 \subset B$ and $B^* = B$.

For every $B \in \mathcal{B}^*$, denote by $A[B]$ the linear span of B, which is a normed algebra under the gauge function $\|\cdot\|_B$ of B. If $A[B]$ is complete for every $B \in \mathcal{B}^*$, then $A[\tau]$ is called *pseudo-complete*.

An element $x \in A$ is called *bounded* if, for some nonzero complex number λ, the set $\{(\lambda x)^n : n = 1, 2, 3, ...\}$ is bounded in A. We denote by A_0 the set of all bounded elements in A.

A unital topological $*$-algebra $A[\tau]$ is called *symmetric* if, for every $x \in A$, the element $(1 + x^*x)^{-1}$ exists and belongs to A_0.

DEFINITION 2.2. [6] A unital symmetric pseudo-complete locally convex $*$-algebra $A[\tau]$, such that the collection $\mathcal{B}_A^*$ has a greatest member, denoted by B_0, is called a *GB^*-algebra* over B_0.

If A is commutative, then $A_0 = A[B_0]$ [6, p. 94]. In general, A_0 is not a $*$-subalgebra of A, and $A[B_0]$ contains all normal elements of A_0, i.e., all $x \in A$ such that $xx^* = x^*x$ [6, p. 94].

PROPOSITION 2.3. [6, Theorem 2.6] *If $A[\tau]$ is a GB^*-algebra, then the Banach $*$-algebra $A[B_0]$ is a C^*-algebra, which is sequentially dense in A. Moreover, $(1 + x^*x)^{-1} \in A[B_0]$ for every $x \in A$ and B_0 is the unit ball of $A[B_0]$.*

THEOREM 2.4. [6, Theorem 3.9] *Any commutative GB^*-algebra A is algebraically $*$-isomorphic to an algebra $N(X)$ of continuous $\mathbb{C}^*$-valued functions on a compact Hausdorff space X, which are allowed to take the value infinity on at most a nowhere dense subset of X (by $\mathbb{C}^*$, we mean the one-point compactification of $\mathbb{C}$). This isomorphism extends the Gelfand isomorphism of the C^*-algebra $A[B_0]$ onto the corresponding C^*-algebra $C(X)$, the $*$-algebra of all complex valued continuous functions on X.*

For a topological algebra A with unit 1, an element $x \in A$ is called *topologically-invertible* if there exists a net (x_α) of elements of A such that $x_\alpha x \to 1$ and $xx_\alpha \to 1$. The set of all topologically-invertible elements in A is denoted by Tinv$A$.

DEFINITION 2.5. [3] Let $A[\tau]$ be a topological algebra with identity 1. The algebra A is called an *invertive topological algebra* if Tinv$A = G_A$. The algebra A is called an *advertibly complete algebra* if every Cauchy net (x_α) in A, with the property that there exists $x \in A$ such that $x_\alpha x \to 1$ and $xx_\alpha \to 1$, converges in A.

Let $A[\tau]$ be a topological algebra, which does not necessarily have an identity element. Let $x \circ y = x + y - xy$ for all $x, y \in A$. We say that $A[\tau]$ is advertibly complete if every Cauchy net (x_α) in A, with the property that there exists $x \in A$ such that $x_\alpha \circ x \to 0$ and $x \circ x_\alpha \to 0$, converges in A [3, p. 15]. If $A[\tau]$ has an identity element, then it is easily verified that this notion of advertible completeness is equivalent to that in Definition 2.5. Note that, in [3], an advertibly complete topological algebra with identity element (as defined in Definition 2.5) is called *invertibly complete*.

We recall that, for a general algebra A, a two-sided ideal I of A is called *regular* if there exists a $y \in A$ such that $x - xy \in I$ and $x - yx \in I$ for every $x \in A$. The concepts of left or right regular ideals are similarly defined. In the case of a unital algebra, all ideals are regular.

DEFINITION 2.6. [3] A topological algebra $A[\tau]$ is called *simplicial* if every closed regular left (right or two-sided) ideal of A is contained in some closed maximal regular left (resp. right, two-sided) ideal of A.

3. Characterizations of advertible completeness for general topological algebras

We recall that an *unbounded derivation* of a topological algebra $A[\tau]$ is a linear map $\delta : D(\delta) \to A$ such that $\delta(xy) = \delta(x)y + x\delta(y)$ for all $x,y \in D(\delta)$, where $D(\delta)$ is the domain of the derivation δ, which is taken to be a dense subalgebra of A. In case A is a topological $*$-algebra, a derivation δ in A is said to be a *$*$-derivation* if $a \in D(\delta)$ implies that $a^* \in D(\delta)$ and $\delta(a^*) = \delta(a)^*$. A derivation δ in A is said to be *closed* if, for any net (x_α) in $D(\delta)$ such that $x_\alpha \to x$ and $\delta(x_\alpha) \to y$, we have that $x \in D(\delta)$ and $y = \delta(x)$.

Proposition 3.2 (similar to [20, Proposition 3.11]) generalizes a result in [17] that the domain of every closed unbounded derivation of a Banach algebra is a Q-algebra, and hence, advertibly complete. The proof of (iii) $\Rightarrow$ (i) of Proposition 3.2 relies on [12, Proposition 6.2, p. 70]. For convenience of the reader, the latter result is recorded, for the case of a unital algebra, in the following Proposition 3.1.

PROPOSITION 3.1. [12, Proposition 6.2] *Let $A[\tau]$ be a unital advertibly complete topological algebra, whose completion $\tilde{A}[\tilde{\tau}]$ is also supposed to be a topological algebra. Then, for an element $x \in A$, $x \in G_A$ if and only if $x \in G_{\tilde{A}}$.*

PROPOSITION 3.2. *Let $\delta : D(\delta) \to A$ be a closed unbounded derivation of a unital complete locally convex algebra $A[\tau]$, with $1 \in D(\delta)$. The following statements are equivalent.*

   (i) $G_{D(\delta)} = G_A \cap D(\delta)$, *where $G_{D(\delta)}$ and G_A denote the groups of invertible elements in $D(\delta)$ and A respectively.*
   (ii) $\mathrm{Sp}_{D(\delta)}(x) = \mathrm{Sp}_A(x)$ *for all $x \in A$.*
   (iii) $D(\delta)$ *is advertibly complete with respect to the relative topology inherited from τ.*

PROOF. (i) $\Rightarrow$ (iii): Let (x_α) be a Cauchy net in $D(\delta)$, for which there exists $x \in D(\delta)$ such that $xx_\alpha \to 1$ and $x_\alpha x \to 1$. Since (x_α) is a Cauchy net in $D(\delta)$ and A is the completion of $D(\delta)$, we have that $x_\alpha \to y$ for some $y \in A$. Therefore, $xx_\alpha \to xy$ and $x_\alpha x \to yx$. Hence, $x \in G_A \cap D(\delta) = G_{D(\delta)}$ (by (i)) and $y = x^{-1}$. Therefore, $y = x^{-1} \in D(\delta)$.

(iii) $\Rightarrow$ (i): This is a straight consequence of Proposition 3.1. Clearly, (i) and (ii) are equivalent statements. □

The above result also holds *for any dense subalgebra of A* and gives a purely algebraic characterization of advertible completeness. The only problem is that it is in terms of an enlargement of the algebra, namely, its completion. Can we characterize advertible completeness in an algebra A fairly algebraically (if not purely algebraically) so that it does not rely on some enlargement of A?

We attempt to solve here the problem of finding a fairly algebraic characterization of those locally convex algebras which are advertibly complete.

The following results appear to be well known and trivial, although no references could be found in the literature. Therefore, we give the proofs for sake of completeness.

PROPOSITION 3.3. *Let $A[\tau]$ be a unital topological algebra such that x is invertible in A if and only if $x \notin M$ for all closed maximal one-sided ideals M of A. Suppose also that the multiplication on A is jointly continuous. Then A is advertibly complete.*

PROOF. Let (x_α) be a Cauchy net in A for which there exists $z \in A$ such that $zx_\alpha \to 1$ and $x_\alpha z \to 1$. We show that (x_α) converges to some element in A.

Since A has jointly continuous multiplication, we know that the completion $\widetilde{A}$ of A is a topological algebra (into which A is embedded as a dense subalgebra). Since (x_α) is a Cauchy net in $A \subseteq \widetilde{A}$, we know that $x_\alpha \to y$ for some $y \in \widetilde{A}$. By continuity of multiplication, $zx_\alpha \to zy$ and $x_\alpha z \to yz$. By uniqueness of limits, $zy = yz = 1$. Therefore, $z^{-1} = y$.

If $z \in M$ for some closed maximal left ideal M of A, we get that $x_\alpha z \in M$ for all α. Therefore, $1 = yz \in M$, which is a contradiction. Thus, $z \notin M$ for all closed maximal left ideals M of A. Similarly, $z \notin M$ for all closed maximal right ideals M of A. Therefore, by hypothesis, z is an invertible element of A. This means that z^{-1} is not only in $\widetilde{A}$, but also in A. Therefore, $y = z^{-1} \in A$. Hence A is advertibly complete. □

To obtain our characterization, it would be nice if we can prove the converse of this proposition. This motivates the next two trivial observations, which are noteworthy to recall and not lose sight of.

PROPOSITION 3.4. *Let A be a unital algebra. Then x is invertible in A if and only if x is in no maximal one-sided ideal of A.*

PROOF. If x is invertible, then, clearly, x is in no maximal one-sided ideal of A. Now suppose that x is in no maximal one-sided ideal of A. Let $I = Ax$. Then I is a left ideal of A. If I is a *proper* left ideal of A, then I is contained in a maximal left ideal M of A. Therefore, $x \in M$. This is a contradiction, and hence, $I = A$, i.e., $Ax = A$. Hence there exists $z \in A$ such that $zx = 1$. Similarly, by considering $J = xA$, we get that there exists $w \in A$ such that $1 = xw$. Therefore, x is invertible in A. □

COROLLARY 3.5. *If $A[\tau]$ is a topological algebra, with identity, having jointly continuous multiplication, and has all its maximal one-sided ideals closed, then $A[\tau]$ is advertibly complete.*

Proving the converse of Proposition 3.3 seems to be harder. In fact, if A is a locally m-convex algebra which is advertibly complete and simplicial, we have the converse [**16**, Proposition 2.1] (we note here that the property (P) in [**16**] is equivalent to advertible completeness [**12**, Proposition 6.10 (4)]). In this regard, we also have the following Proposition 3.6, which follows directly from [**16**, Remark 2.1 and Corollary, p. 190]. With respect to the following proposition, we recall that an algebra A is *two-sided* if for $x, y \in A$, there exist $u, v \in A$ such that $xy = ux = yv$.

PROPOSITION 3.6. *The following statements are equivalent for a unital two-sided locally m-convex algebra $A[\tau]$.*

(i) *$A[\tau]$ is advertibly complete and simplicial.*
(ii) *$x \in A$ is invertible if and only if x is in no closed maximal one-sided ideal of A.*

The proof of Proposition 3.6 relies on the following result: If $A[\tau]$ is an advertibly complete algebra, which is also simplicial, then, if $x \in A$ is not quasi-invertible, then x is a unit element of A modulo some regular, maximal and closed one-sided ideal of A [**16**, Proposition 2.1]. If there is a unital two-sided locally m-convex algebra, which is advertibly complete but not simplicial, then Proposition 3.6 informs us that the converse of Proposition 3.3 is false.

If A is a unital commutative algebra, we let $M(A)$ (resp. $m(A)$) denote the set of all (resp. closed) maximal ideals of A

PROPOSITION 3.7. *Let $A[\tau]$ be a unital commutative simplicial algebra. The following statements are equivalent.*

(i) *$A[\tau]$ is an invertive algebra.*
(ii) $\bigcup_{M \in M(A)} M = \bigcup_{M \in m(A)} M.$
(iii) *$x \in G_A$ if and only if x is in no closed maximal ideal of A (i.e., $x \notin M$ for all $M \in m(A)$).*

PROOF. (i)$\Leftrightarrow$(ii): This is [**3**, Proposition 4].

(ii) $\Rightarrow$ (iii): If $x \in G_A$, then x is not in M for all $M \in m(A)$. Conversely, if $x \notin M$ for all $M \in m(A)$, then $x \notin M$ for all $M \in M(A)$, by (ii). Therefore, $x \in G_A$, by Proposition 3.4.

(iii) $\Rightarrow$ (ii): It is clear that $\bigcup_{M \in m(A)} M \subseteq \bigcup_{M \in M(A)} M$. Now, let $x \notin \bigcup_{M \in m(A)} M$. By (iii), we have that $x \in G_A$. Therefore, by Proposition 3.4, $x \notin M$ for every $M \in M(A)$, i.e., $x \notin \bigcup_{M \in M(A)} M$. Therefore, $\bigcup_{M \in M(A)} M \subseteq \bigcup_{M \in m(A)} M$. This completes the proof. □

By [**3**, Corollary 1], every invertive algebra is an advertibly complete algebra. The converse of this statement is not true in general. Therefore, in general, one cannot characterize advertible completeness fairly algebraically via statement (iii) in Proposition 3.7.

In what follows, we recall that $\hom(A)$ stands for the set of all nonzero continuous characters of A and $\ker(\phi)$ denotes the kernel of a character ϕ of A. The proof of Proposition 3.8 can be deduced from [**3**, Lemma 2, Proposition 5, Proposition 6].

PROPOSITION 3.8. *Let $A[\tau]$ be a unital commutative topological algebra with $\hom(A) \neq \emptyset$. Then the following statements are equivalent.*

(i) $\bigcup_{M \in M(A)} M = \bigcup_{\phi \in \hom(A)} \ker(\phi).$
(ii) $\mathrm{Sp}_A(x) = \{\phi(x) : \phi \in \hom(A)\}.$
(iii) $G_A = \{x \in A : \phi(x) \neq 0 \text{ for all } \phi \in \hom(A)\}.$

If, in addition, $A[\tau]$ is a simplicial Gelfand-Mazur algebra, then the above equivalent conditions are equivalent to

(iv) *$A[\tau]$ is an invertive algebra.*

Note that a commutative unital locally m-convex algebra is advertibly complete if and only if it is an invertive algebra [**3**, Proposition 11]. Any commutative unital locally m-convex algebra is simplicial [**8**]. Using these facts and Proposition 3.8, we recover Proposition 6.10(6) in [**12**], which states that a commutative unital locally m-convex algebra A is advertibly complete if and only if the following equivalence is true: An element $x \in A$ belongs to G_A if and only if $\phi(x) \neq 0$ for all $\phi \in \hom(A)$.

We recall that for a unital topological algebra, $\mathcal{M}(A)$ denotes the set of all closed two-sided ideals of A, which are maximal as a left or right ideal of A.

PROPOSITION 3.9. *Let $A[\tau]$ be a unital Gelfand-Mazur algebra with $\mathcal{M}(A) \neq \emptyset$. Let $\mathcal{M}_{os}(A)$ denote the set of all maximal one-sided ideals of A. Then*
$$G_A = \{x \in A : \phi(x) \neq 0 \text{ for all } \phi \in \hom(A)\}$$
if and only if $\bigcup_{M \in \mathcal{M}_{os}(A)} M = \bigcup_{\phi \in \hom(A)} \ker(\phi).$

PROOF. Assume that $\bigcup_{M \in \mathcal{M}_{os}(A)} M = \bigcup_{\phi \in \hom(A)} \ker(\phi)$. Clearly,
$$G_A \subseteq \{x \in A : \phi(x) \neq 0 \text{ for all } \phi \in \hom(A)\}.$$
So, let $x \in A$ be such that $\phi(x) \neq 0$ for all $\phi \in \hom(A)$. Assume that $x \notin G_A$. By Proposition 3.4, there is a maximal one-sided ideal M of A such that $x \in M$. By hypothesis, $x \in \ker(\phi)$ for some $\phi \in \hom(A)$, i.e., $\phi(x) = 0$. This is a contradiction, and therefore, $x \in G_A$.

Now, assume that $x \in G_A$ if and only if $\phi(x) \neq 0$ for all $\phi \in \hom(A)$. Let $x \in \bigcup_{M \in \mathcal{M}_{os}(A)} M$. Then $x \notin G_A$. Hence, $\phi(x) = 0$ for some $\phi \in \hom(A)$, i.e., $x \in \ker(\phi)$. The reverse inclusion is obvious. □

The proof of of the following result is a direct consequence of [3, Corollary 1 and Proposition 6].

PROPOSITION 3.10. *Let $A[\tau]$ be a unital topological algebra with $\hom(A) \neq \emptyset$. Assume that*
$$G_A = \{x \in A : \phi(x) \neq 0 \text{ for all } \phi \in \hom(A)\}.$$
Then $A[\tau]$ is advertibly complete.

A big question is now when advertible completeness implies that
$$G_A = \{x \in A : \phi(x) \neq 0 \text{ for all } \phi \in \hom(A)\}.$$

This we answer in the remainder of this section. For this quest, it would be helpful to know what is the structure of characters on a general topological algebra. In this regard, see also Remark 4.8 below. The next proposition follows immediately from [3, Corollary 7].

PROPOSITION 3.11. *Let A be a dense subalgebra of a complete topological algebra $\widetilde{A}$, with $\mathcal{M}(\widetilde{A}) \neq \emptyset$, such that*
$$G_{\widetilde{A}} = \{x \in \widetilde{A} : \phi(x) \neq 0 \text{ for all } \phi \in \hom(\widetilde{A})\}.$$
Then A is advertibly complete if and only if
$$G_A = \{x \in A : \phi(x) \neq 0 \text{ for all } \phi \in \hom(A)\}.$$

In the previous proposition, note that $\mathcal{M}(\widetilde{A}) \neq \emptyset$ if and only if $\mathcal{M}(A) \neq \emptyset$. The above motivates the question about finding the conditions under which $\mathcal{M}(A) \neq \emptyset$ for a Gelfand-Mazur algebra A. For this, we let $\text{comm}(A)$ denote the closure of the two-sided ideal of A, generated by all commutators in A.

THEOREM 3.12. [4, Proposition 1]
(i) *Let $A[\tau]$ be a Gelfand-Mazur algebra. If $\mathcal{M}(A) \neq \emptyset$, then $\text{comm}(A) \neq A$.*
(ii) *Let $A[\tau]$ be a simplicial Gelfand-Mazur algebra with identity. If $\text{comm}(A) \neq A$, then $\mathcal{M}(A) \neq \emptyset$.*

The same argument as that of Proposition 3.3 yields the following result.

PROPOSITION 3.13. *Let $A[\tau]$ be a unital topological algebra such that*
$$G_A = \{x \in A : x \notin I \text{ for all proper closed two-sided ideals } I \text{ of } A\}.$$
If $A[\tau]$ also has jointly continuous multiplication, then $A[\tau]$ is advertibly complete.

REMARK 3.14. We remark that the converse of Proposition 3.13 is not true. For instance, the algebra $M_n(\mathbb{C})$ of all square $n \times n$ complex matrices, which is identified with $B(\mathbb{C}^n)$ (all bounded linear maps on $\mathbb{C}^n$), endowed with the usual norm operator topology, is a unital Banach algebra. Therefore, it is advertibly complete since it is complete. Moreover, since $M_n(\mathbb{C})$ is simple, we have that $\{x \in M_n(\mathbb{C}) : x \notin I \text{ for all proper closed two-sided ideals } I \text{ of } M_n(\mathbb{C})\}$ coincides with the set $\{x \in M_n(\mathbb{C}) : x \neq 0\}$. Hence, in this case, we clearly have that
$$G_{M_n(\mathbb{C})} \neq \{x \in M_n(\mathbb{C}) : x \notin I \text{ for all proper closed ideals } I \text{ of } M_n(\mathbb{C})\}.$$
Note that $M_n(\mathbb{C})$ has no characters. So it would make sense to ask if the converse of Proposition 3.13 is true for Gelfand-Mazur algebras A with $\mathcal{M}(A) \neq \emptyset$.

A type of converse implication to that of Proposition 3.13, in the setting of an invertive unital commutative algebra, is given by the following.

PROPOSITION 3.15. *Let $A[\tau]$ be a unital commutative invertive algebra. Then $G_A = \{x \in A : x \notin I \text{ for all proper closed ideals } I \text{ of } A\}$.*

PROOF. In general, $G_A \subseteq \{x \in A : x \notin I \text{ for all proper closed ideals } I \text{ of } A\}$. For the reverse implication, let $x \in A$ be such that $x \notin I$ for all proper closed ideals I of A and suppose that $x \notin G_A$. Since A is invertive, we have that $x \notin \text{Tinv}A$. Therefore, by [3, Lemma 1 (b)], $x \in I$ for some proper closed ideal I of A, a contradiction. $\square$

For the following result, we recall that a *representation* of an algebra A on a vector space X is a homomorphism $\pi : A \to L(X)$, where $L(X)$ stands for all linear operators on X. A representation π of A on X is *irreducible* if it is non-trivial, i.e., $\pi(A)X \neq 0$, and the only subspaces of X, which are invariant under π, are $\{0\}$ and X.

Let $\pi : A \to L(X)$ be an irreducible representation of A on X, and let
$$D = \{a \in L(X) : \pi(y)a(\xi) = a(\pi(y)\xi), \text{ for all } y \in A \text{ and } \xi \in X\}.$$

PROPOSITION 3.16. [9, Proposition 24.6] *Let A be an algebra and let X be a vector space. If $\pi : A \to L(X)$ is an irreducible representation of A on X, then the set D, above, contains the identity operator on X and is a division subalgebra of $L(X)$.*

PROPOSITION 3.17. *Let $A[\tau]$ be a unital commutative topological algebra such that $y \mapsto \pi(y)\xi$ is continuous for every irreducible representation π of A on X and for some non-zero ξ in X, where X is a topological vector space. Then, for $x \in A$, if x is in no closed maximal ideal of A, then x is an invertible element of A.*

PROOF. Let x be an element in A such that x is in no closed maximal ideal of A. Assume that $x \notin G_A$. Following the arguments of the proof of [7, Theorem 4.2.1(iii)], we have that $\text{Sp}_A(x) = \bigcup_\pi \text{Sp}_{L(X)}(\pi(x))$, where π runs through all irreducible representations of A on some topological vector space X. Since we assumed that $x \notin G_A$, we have that $0 \in \text{Sp}_A(x)$ and thus, $0 \in \text{Sp}_{L(X)}(\pi(x))$, for some irreducible representation π of A on X.

By continuity of the map $y \mapsto \pi(y)\xi$, we have that
$$L_\xi := \{x \in A : \pi(x)\xi = 0\}$$
is closed. Moreover, by [7, Theorem 4.2.1 (i)], L_ξ is a maximal ideal of A. Hence, we get that $x \notin L_\xi$, i.e., $\pi(x)\xi \neq 0$ and thus, $\pi(x) \neq 0$.

It follows that $\pi(x) \in D$: Since A is commutative, we get that
$$\pi(y)\pi(x)\xi = \pi(x)\big(\pi(y)\xi\big)$$
for all $y \in A$ and $\xi \in X$. Therefore $\pi(x) \in D$.

It follows from Proposition 3.16 that $\pi(x)$ is an invertible element of $L(X)$, which is a contradiction. Therefore, we conclude that $x \in G_A$. $\square$

4. Characterizations of advertible completeness for GB*-algebras

Motivated by the second last paragraph of the introduction, one of the main question is to find the conditions under which $D(\delta)$ will be a Gelfand-Mazur algebra with $\mathcal{M}(D(\delta)) \neq \emptyset$. This we answer in the following proposition. Here, $A[\tau]$ is a GB*-algebra and $\delta : D(\delta) \to A$ is a closed *-derivation of A. For the proof of the following Proposition 4.1, we recall from [4] that a topological algebra A is an *exponentially galbed algebra* if, for any neighbourhood V of 0 in A, there is a neighbourhood U of 0 in A such that $\{\sum_{k=0}^n \frac{x_k}{2^k} : n \in \mathbb{N}, x_0, \cdots, x_n \in U\} \subseteq V$.

PROPOSITION 4.1. *Let $A[\tau]$ be a unital Gelfand-Mazur algebra with $\mathcal{M}(A) \neq \emptyset$. If B is a unital subalgebra of A and*
$$\mathcal{M}(B) = \{M \cap B : M \in \mathcal{M}(A)\},$$
where $\mathcal{M}(B)$ is taken to be with respect to the relative topology τ' on B induced by τ, then B is also a Gelfand-Mazur algebra.

PROOF. By [4, Theorem 1], there exists a topology τ'' on A such that $A[\tau'']$ is an exponentially galbed algebra such that every element of A is τ''-bounded and such that every $M \in \mathcal{M}(A)$ is τ''-closed. Then, for every τ''-neighbourhood V of $0 \in A$, there exists a τ''-neighbourhood U of $0 \in A$ such that
$$\left\{\sum_{k=0}^n \frac{x_k}{2^k} : n \in \mathbb{N}, x_0, \ldots, x_n \in U\right\} \subseteq V.$$

Let τ''' denote the relative topology on B, induced by τ''. Then, by hypothesis, every $M \in \mathcal{M}(B)$ is τ'''-closed. Let W be a τ'''-neighbourhood of $0 \in B$. Then $W = V \cap B$, where V is a τ''-neighbourhood of $0 \in A$. Also, $U_1 = U \cap B$ is a τ'''-neighbourhood of $0 \in B$. Then
$$\left\{\sum_{k=0}^n \frac{x_k}{2^k} : n \in \mathbb{N}, x_0, \ldots, x_n \in U_1\right\} \subseteq W,$$
i.e., $B[\tau''']$ is exponentially galbed. Also, every element of B is τ'''-bounded: This is due to the facts that every element in A is τ''-bounded and that τ''' is the relative topology on B inherited from the topology τ'' on A.

By [4, Theorem 1], $B[\tau']$ is a Gelfand-Mazur algebra with $\mathcal{M}(B) \neq \emptyset$. $\square$

Proposition 4.1 immediately implies the following result concerning unbounded derivations of GB*-algebras.

COROLLARY 4.2. *Let $A[\tau]$ be a GB^*-algebra such that $\mathcal{M}(A) \neq \emptyset$. Let $\delta : D(\delta) \to A$ be a closed unbounded $*$-derivation of A with $1 \in D(\delta)$. If $A[\tau]$ is also a Gelfand-Mazur algebra and*
$$\mathcal{M}(D(\delta)) = \{M \cap D(\delta) : M \in \mathcal{M}(A)\},$$
then so is $D(\delta)$ in the relative topology and $\mathcal{M}(D(\delta)) \neq \emptyset$.

In the setting of GB^*-algebras, we now provide partial motivation for reducing our problem of characterizing advertible completeness in the class of GB^*-algebras to the setting of commutative GB^*-algebras. We recall that the problem we are referring to is whether the fact that the algebra A is advertibly complete implies that $G_A = \{x \in A : \phi(x) \neq 0 \text{ for all } \phi \in \hom(A)\}$.

(1) Let $A[\tau]$ be a Gelfand-Mazur algebra with identity 1 such that $\mathcal{M}(A) \neq \emptyset$. Let B be a closed subalgebra of A containing 1. Let $M_B \in \mathcal{M}(B)$.

Assumption 1. $M_B = M \cap B$ for some $M \in \mathcal{M}(A)$.

(sufficient conditions for Assumption 1 to hold are given in [**1**, Theorem 3.10, p. 59] in the setting of a unital strongly topologically semisimple simplicial Hausdorff algebra over $\mathbb{C}$).

From this assumption, it follows from Proposition 4.1 that B is a Gelfand-Mazur algebra.

Now, $M_B = \ker(\phi_B)$ for some $\phi_B \in \hom(B)$ and $M = \ker(\phi_A)$ for some $\phi_A \in \hom(A)$.

Let $x \in B$. Then $\phi_B(x) = \lambda \in \mathbb{C}$ and hence $\phi_B(x - \lambda 1) = 0$. Now, $M_B \subseteq M$ and hence, $\ker(\phi_B) \subseteq \ker(\phi_A)$. So, $\phi_A(x - \lambda 1) = 0$, and hence, $\phi_A(x) = \lambda$, i.e., $\phi_A(x) = \phi_B(x)$ for all $x \in B$, i.e., $\phi_B = \phi_A|_B$.

(2) **Assumption 2.** Assume that for any commutative Gelfand-Mazur algebra C, with $\mathcal{M}(C) \neq \emptyset$, we have that C is advertibly complete if and only if
$$G_C = \{x \in C : \phi(x) \neq 0 \text{ for all } \phi \in \hom(C)\}.$$

Assume that $A[\tau]$ is an advertibly complete Gelfand-Mazur algebra with $\mathcal{M}(A) \neq \emptyset$. Then, if $x \in G_A$, then $\phi_A(x) \neq 0$ for all $\phi_A \in \hom(A)$.

Now, let $x \in A$ be such that $\phi_A(x) \neq 0$ for all $\phi_A \in \hom(A)$. Then, $x \notin \ker(\phi_A)$ for all $\phi_A \in \hom(A)$. Also, for every $\phi_A \in \hom(A)$, $\ker(\phi_A) \in \mathcal{M}(A)$.

Observe that x generates a maximal commutative subalgebra B of A. Since $\phi_A(x) \neq 0$ for all $\phi_A \in \hom(A)$, we have from (1) above that $x \notin \ker(\phi_B)$ for all $\phi_B \in \hom(B)$ (in (1) above, M_B, and hence, ϕ_A, is chosen arbitrarily). That is, $\phi_B(x) \neq 0$ for all $\phi_B \in \hom(B)$.

Since B is a maximal commutative subalgebra of A, it is closed, so B is advertibly complete, and therefore, by Assumption 2, we get that $x \in G_B$, and so, $x \in G_A$. The remainder of this article is devoted to finding a strategy for proving Assumption 2 for the case where C is, in addition, a GB^*-algebra (motivation for this is given below).

Now, let $A[\tau]$ be a complete GB^*-algebra which is a Gelfand-Mazur algebra (for example, $A[\tau]$ is such an algebra if $A[\tau]$ is a Fréchet algebra - see [**2**]) with

$\mathcal{M}(A) \neq \emptyset$. Suppose that

$$G_A = \{x \in A : \phi(x) \neq 0 \text{ for all } \phi \in \hom(A)\}.$$

It follows from Proposition 3.11 that if B is a dense subalgebra of A, then B is advertibly complete in the relative topology if and only if

$$G_B = \{x \in B : \phi(x) \neq 0 \text{ for all } \phi \in \hom(B)\}.$$

When applied to the domain $D(\delta)$ of a closed $*$-derivation $\delta : D(\delta) \to A$, one obtains necessary and sufficient conditions under which $D(\delta)$ is advertibly complete. One must therefore attempt to prove that

$$G_A = \{x \in A : \phi(x) \neq 0 \text{ for all } \phi \in \hom(A)\}$$

for any complete GB*-algebra $A[\tau]$ which is a Gelfand-Mazur algebra with $\mathcal{M}(A) \neq \emptyset$.

For this purpose, it suffices to assume that the element x in part (2) above is self-adjoint: Let $y \in A$ be such that $\phi(y) \neq 0$ for all $\phi \in \hom(A)$. Then $\phi(y^*y) = \overline{\phi(y)} \cdot \phi(y) \neq 0$ for all $\phi \in \hom(A)$. The self-adjoint element $y^*y \in A$ generates a maximal commutative $*$-subalgebra C of A. It follows that C is a complete commutative GB*-algebra containing the self-adjoint element y^*y (by [**10**, Lemma 4.10 (ii)], any maximal abelian $*$-subalgebra of a GB*-algebra is a GB*-algebra). By part (2) and Assumption 2, it follows that $y^*y \in G_A$. Similarly, $yy^* \in G_A$. Therefore, there exist $u \in A$ and $z \in A$ such that $(uy^*)y = 1$ and $y(y^*z) = 1$, implying that y has a left inverse and a right inverse in A. It follows that $y \in A$ is invertible in A.

It follows that, in Assumption 2, we may assume that C is, in addition, also a commutative GB*-algebra. Therefore, when attempting to prove that

$$G_A = \{x \in A : \phi(x) \neq 0 \text{ for all } \phi \in \hom(A)\}$$

it is worthwhile to restrict our attention to the case where A is a complete commutative GB*-algebra which is also a Gelfand-Mazur algebra with $\mathcal{M}(A) \neq \emptyset$. This is what is done in the remainder of this article.

If $A[\tau]$ is a simplicial Gelfand-Mazur algebra, then

$$G_A = \{x \in A : \phi(x) \neq 0 \text{ for all } \phi \in \hom(A)\}$$

is equivalent to A being an invertive algebra (see Proposition 3.8).

REMARK 4.3. We recall that in a commutative GB*-algebra, every $\phi \in \hom(A[B_0])$ has a unique extension to a $\mathbb{C}^*$-valued function on A, say, ϕ', where the latter map ϕ' behaves as a 'partial homomorphism' on A, and $\mathbb{C}^*$ denotes the one point compactification of $\mathbb{C}$. In particular, $\phi'(xy) = \phi'(x)\phi'(y)$, for all $x, y \in A$ such that $\phi'(x)$, $\phi'(y)$ are not $0, \infty$ in some order (for details see [**6**, Proposition 3.1]). Let $x \in A$ be such that $\phi'(x) \neq 0$ and $\phi'(x) \neq \infty$, for all $\phi \in \hom(A[B_0])$. Then, since $\phi'(x) \neq \infty$, by [**6**, Lemma 3.5 (ii)], we have that $\phi'((1+x^*x)^{-1}) \neq 0$. Moreover, $(1+x^*x)^{-1} \in A[B_0]$ (see Proposition 2.3), so, clearly,

$$\phi'((1+x^*x)^{-1}) = \phi((1+x^*x)^{-1}) \neq \infty.$$

Hence, taking into consideration that $x(1+x^*x)^{-1} \in A[B_0]$ ([**6**, Lemma 3.2]), we have that

$$\begin{aligned} \phi(x(1+x^*x)^{-1}) &= \phi'(x(1+x^*x)^{-1}) \\ &= \phi'(x)\phi'((1+x^*x)^{-1}) \\ &\neq 0 \end{aligned}$$

for every $\phi \in \hom(A[B_0])$. That is, $\widehat{x(1+x^*x)^{-1}}(\phi) \neq 0$, for every $\phi \in \hom(A[B_0])$. By the Gelfand representation of the commutative C*-algebra $A[B_0]$, we thus have that $x(1+x^*x)^{-1} \in G_{A[B_0]}$. So, we conclude that $x \in G_A$. Therefore, for a commutative GB*-algebra A, we have the following inclusions:

$$\{x \in A : \phi'(x) \neq 0 \text{ and } \phi'(x) \neq \infty \text{ for all } \phi \in \hom(A[B_0])\} \subseteq G_A$$

and

$$G_A \subseteq \{x \in A : \phi(x) \neq 0 \text{ for all } \phi \in \hom(A)\}.$$

The Stone-Weierstrass theorem is well known to hold for all locally compact (hence, compact) spaces. This theorem has a generalization to all topological spaces, and is given below. For a reference, see [**11**, Theorem XIII 3.3 and p. 283].

THEOREM 4.4 (Stone-Weierstrass). *Let X be a topological space and let $C(X)$ denote the space of all complex-valued continuous functions on X. If D is a subset of $C(X)$, which contains all nonzero constant functions and is separating, then the algebra A, generated by D, is dense in $C(X)$ with respect to the compact-open topology on $C(X)$, provided that A is also closed under complex conjugation.*

PROPOSITION 4.5. *Let $A[\tau]$ be a Gelfand-Mazur *-algebra with $\mathcal{M}(A) \neq \emptyset$ and all characters self-adjoint, having the following additional properties.*
   (i) *$\phi(x) = 0$ for all $\phi \in \hom(A)$ implies $x = 0$,*
   (ii) *the range of the Gelfand map $x \mapsto \hat{x}$ is closed in $C(\hom(A))$ with respect to the compact open topology.*

Then

$$G_A = \{x \in A : \phi(x) \neq 0 \text{ for all } \phi \in \hom A\}.$$

PROOF. The set $\hom(A)$ is a topological space with respect to the Gelfand topology. We recall that a net (ϕ_α) in $\hom(A)$ converges to $\phi \in \hom(A)$ with respect to the Gelfand topology if and only if $\phi_\alpha(x) \to \phi(x)$ for all $x \in A$. Clearly, $\hat{x}$ is a continuous complex-valued function on $\hom(A)$. The algebra

$$\hat{A} = \{\hat{x} : x \in A\}$$

is clearly closed under complex conjugation, is separating and contains all nonzero constant functions. By the Stone-Weierstrass theorem, $\hat{A}$ is dense in $C(\hom(A))$ with respect to the compact open topology. By (ii), we get that $\hat{A} = C(\hom(A))$.

If $x \in G_A$, then $\phi(x) \neq 0$ for all $\phi \in \hom(A)$ is obvious. So, suppose that $\phi(x) \neq 0$ for all $\phi \in \hom(A)$. Then $\hat{x}(\phi) = \phi(x) \neq 0$ for all $\phi \in \hom(A)$. Therefore, $\hat{x}$ is invertible in $C(\hom(A))$, and is, therefore, invertible in $\hat{A}$. By (i), the Gelfand map is injective, and thus, $x \in G_A$. □

Proposition 4.5 holds, in particular, for any complete commutative GB*-algebra $A[\tau]$, which is also a Gelfand-Mazur algebra with $\mathcal{M}(A) \neq \emptyset$, and which satisfies (i) and (ii) of Proposition 4.5. If one wishes to prove (i) in the previous result for such an algebra, then a characterization of all closed maximal ideals, and all maximal ideals, of a complete commutative GB*-algebra will be useful. In this regard, we draw the reader's attention to Remark 4.8 below.

We give an example of a commutative complete GB*-algebra $A[\tau]$, which is also a Gelfand-Mazur algebra with $\mathcal{M}(A) \neq \emptyset$, and which is not a pro-C*-algebra. We recall that a pro-C*-algebra $A[\tau]$ is a complete topological *-algebra whose topology τ is defined by a family $\Gamma = \{p\}$ of C*-seminorms, i.e., $p(x^*x) = p(x)^2$ for every $p \in \Gamma$ and $x \in A$. Moreover, recall that a commutative pro-C*-algebra is a commutative complete GB*-algebra (see [**6**, Example (3), p. 95]) which is also a Gelfand-Mazur algebra with $\mathcal{M}(A) \neq \emptyset$ (since a pro-C*-algebra is a complete locally m-convex algebra).

EXAMPLE 4.6. Let (X, Σ, μ) be a finite measure space which has an atom Y (i.e., Y is a non-empty measurable subset of X with $\mu(Y) > 0$ such that if B is a measurable subset of X with $B \subseteq Y$, then $B = \emptyset$ or $B = Y$). Assume that $\mu(Y) < \mu(X)$. Consider the Arens algebra $L^\omega(X, \Sigma, \mu)$, which consists of all equivalence classes of complex-valued measurable functions f on X such that $f \in L^p(X, \Sigma, \mu)$ for all $p = 1, 2, \ldots$, where the equivalence is given by equality almost everywhere. The topology of the algebra $L^\omega(X, \Sigma, \mu)$ is given by the family of norms $\|\cdot\|_p' = \mu(X)^{-\frac{1}{p}} \|\cdot\|_p$ for $p = 1, 2, \ldots$. Let $\phi : L^\omega(X, \Sigma, \mu) \to \mathbb{C}$ be defined by $\phi(f) = f(Y)$ for all $f \in L^\omega(X, \Sigma, \mu)$. This makes sense because measurable functions are constant on atoms. Clearly, ϕ is a (nonzero) character of $L^\omega(X, \Sigma, \mu)$, since $\phi(1) = 1$.

On a subtle note, it is essential that Y is an atom in order for ϕ to be well-defined: Recall that equality is defined almost everywhere. So, let $f, g \in L^\omega(X, \Sigma, \mu)$ be such that f and g are equal almost everywhere and let $f(Y) = \alpha_1 \in \mathbb{C}$, $g(Y) = \alpha_2 \in \mathbb{C}$ and suppose that $\alpha_1 \neq \alpha_2$. Hence, we have that
$$Y \subseteq \{x \in X : f(x) \neq g(x)\}.$$
Then $\mu(Y) \leq \mu([f \neq g]) = 0$. Thus, $\mu(Y) = 0$, a contradiction.

For the fact that ϕ is a homomorphism, we recall again that measurable functions are constant on atoms. So, $\phi(fg) = (fg)(Y)$ and, say, $(fg)(Y) = \alpha \in \mathbb{C}$. Then $(fg)(y) = \alpha$, for all $y \in Y$. If we then pick any element of Y, say y_0, we have that $f(y_0)g(y_0) = (fg)(y_0) = \alpha = (fg)(Y)$. Hence, $f(Y)g(Y) = (fg)(Y)$, since f, g are stable on Y, i.e., $f(y_0) = f(Y)$ and $g(y_0) = g(Y)$. This example does not hold for $L^\omega([0, 1])$ (where the measure is the Lebesgue measure), since $[0, 1]$ has no atoms. If, in this case, we select Y to be a one-point set (hence, not an atom), then, clearly, there are $f, g \in L^\omega([0, 1])$ such that f, g are equal everywhere, except on Y. Thus, ϕ is not well-defined.

The character ϕ above is continuous: Let (f_n) be a sequence in $L^\omega(X, \Sigma, \mu)$ with $f_n \to f \in L^\omega(X, \Sigma, \mu)$. By definition of the seminorms defining the topology on $L^\omega(X, \Sigma, \mu)$, it follows that $\|f_n - f\|_p \to 0$ as $n \to \infty$. That is, $\int |f_n - f|^p \, d\mu \to 0$ for all $p = 1, 2, \ldots$. Hence, $\int_Y |f_n - f|^p \, d\mu \to 0$ for all $p = 1, 2, \ldots$. Since measurable functions are constant on atoms, we know that $f_n(Y) = \lambda_n$ (for some $\lambda_n \in \mathbb{C}$) for all $n \in \mathbb{N}$ and $f(Y) = \lambda$ for some $\lambda \in \mathbb{C}$. Hence, $\int_Y |\lambda_n - \lambda|^p \, d\mu \to 0$ for all $p = 1, 2, \ldots$, i.e., $|\lambda_n - \lambda|^p \mu(Y) \to 0$ for all $p = 1, 2, \ldots$. Hence, $\lambda_n \to \lambda$ as $n \to \infty$,

because $\mu(Y) > 0$. Hence,
$$\phi(f_n) = f_n(Y) = \lambda_n \to \lambda = f(Y) = \phi(f)$$
as $n \to \infty$.

Now, $A = L^\omega(X, \Sigma, \mu)$ is a Fréchet GB*-algebra (see the next paragraph), and is, hence, a unital Fréchet symmetric topological *-algebra. A result of S. B. Ng and S. Warner says that any character of a unital Fréchet symmetric topological *-algebra is continuous (see [**18**, Theorem 3]). So, the continuity of ϕ also follows from the latter result.

Recall that $A = L^\omega(X, \Sigma, \mu)$ is a Fréchet (locally convex) GB*-algebra, see [**6**, p. 96], and hence, a Gelfand-Mazur algebra with $\mathcal{M}(A) \neq \emptyset$. It is metrizable, since its topology is defined by a countable number of seminorms, namely the L_p norms. The GB*-algebra $A[\tau]$ is not a pro-C*-algebra: Let $f = \chi_Y$, where χ_Y is the characteristic function of the atom Y. Since $0 < \mu(Y) < \mu(X)$, it follows easily (by direct calculation) that $\|f^*f\|'_p = \|f\|'_p = \mu(Y)^{\frac{1}{p}}\mu(X)^{-\frac{1}{p}} \neq (\|f\|'_p)^2$.

If, above, we replace $L^\omega(X, \Sigma, \mu)$ with $L^0(X, \Sigma, \mu)$, then the above construction is still valid, and one then has an example of a Dixon Fréchet GB*-algebra $A[\tau]$ (in the sense of [**10**, Definition 2.5]) which is Gelfand-Mazur (being Fréchet) with $\mathcal{M}(A) \neq \emptyset$. The topology here is the topology of convergence in measure, which is not necessarily locally convex.

REMARK 4.7. A C*-like locally convex *-algebra $A[\tau]$, with identity element 1, is a complete topological *-algebra whose topology τ is defined by a family Γ of seminorms having the following properties: For every $p \in \Gamma$, there exists $q \in \Gamma$ such that $p(xy) \leq q(x)q(y)$, $p(x^*) \leq q(x)$ and $p(x)^2 \leq q(x^*x)$ for all $x, y \in A$. Every C*-like locally convex *-algebra is a GB*-algebra (see [**15**, Theorem 2.1]).

Now, let $A[\tau]$ be a commutative C*-like locally convex *-algebra, with identity element 1, such that $x \in A$ is invertible if and only if it is in no closed maximal ideal of A.

Let X be a completely regular space. Observe that $C(X)$ is a pro-C*-algebra and hence, a C*-like locally convex *-algebra. If follows that
$$B := C(X, A) \cong C(X) \widehat{\otimes}_\epsilon A$$
is a commutative complete (and hence, advertibly complete) GB*-algebra, by [**13**, Theorem 5.2], where $C(X, A)$ denotes the *-algebra of all A-valued continuous functions on X. Let $\mathcal{M}$ be a closed maximal ideal of B. By [**5**, Theorem 2.3], there exists $x \in X$ and a maximal ideal M of A such that
$$\mathcal{M} = \{f \in B : f(x) \in M\}.$$

Conversely, if $x \in X$ and M is a closed maximal ideal of A, then it can easily be shown that $\{f \in B : f(x) \in M\}$ is a closed maximal ideal of B (see [**5**]).

Let $f \in B$ be such that f is not in any closed maximal ideal of B. From the above, it follows that, for all $x \in X$, $f(x)$ is not in any closed maximal ideal of A. Therefore, $f(x)$ is invertible in A for every $x \in X$. Does this imply that f is an invertible element in B? Since $f(x)$ is invertible in A for all $x \in X$, there exists, for all $x \in X$, an element $g(x)$ in A such that $f(x)g(x) = g(x)f(x) = 1$. The question is therefore whether g is a continuous function of X. If A has continuous inversion, then g is clearly a continuous function of X, thereby implying that f is an invertible element of B. If A does not have continuous inversion, then it would be interesting to know whether this implies that f is not necessarily an invertible element of B. If this is indeed the case, then there exists at least one element of B which is not invertible in B and which is not in any closed maximal ideal of B, thereby illustrating that the converse of Proposition 3.10 is not true in general.

The following remark is partially motivated by the first paragraph following the proof of Proposition 4.5.

REMARK 4.8. (1) The results in [14] are also true if $\mathbb{R}$ is replaced with $\mathbb{C}$, as we are working with complex algebras. In [14], $C^*(Y)$ is $C_b(Y)$, the $*$-algebra of all bounded continuous functions on Y.

Let $A[\tau]$ be a unital commutative pro-C*-algebra. Then $A \cong C(Y)$ up to $*$-algebra isomorphism, where $Y = \hom(A)$ is a completely regular Hausdorff space (see [12, Theorem 9.3 (3)]). Then $A[B_0] \cong C(X)$, where X is the maximal ideal space (i.e., the character space) of the C*-algebra $A[B_0]$. Let βY be the Stone-Čech compactification of Y.

We show that βY is homeomorphic to X. Every $f \in C_b(Y)$ has a complex valued continuous extension to βY. Therefore,
$$C(\beta Y) \cong C_b(Y) \cong A[B_0] \cong C(X).$$

Since $A[B_0]$ has an identity element, we get that X is a compact Hausdorff space. By the Banach-Stone theorem, βY is homeomorphic to X.

(2) Given a unital commutative pro-C*-algebra $A[\tau]$, where the family, say $\Gamma_\tau = \{p\}$ induces the topology τ of A, then, as in (1), we have that $A \cong C(\hom(A))$ up to $*$-algebra isomorphism. For the present case, the respective ideals of the form M_p on page 448 in [14] are the ideals
$$M_\phi = \{f \in C(\hom(A)) : f(\phi) = 0\}, \ \phi \in \hom(A).$$

One sees that there is a one to one and onto correspondence between elements of $Y = \hom(A)$ and the maximal ideals M_ϕ, $\phi \in \hom(A)$. Moreover, for every $\phi \in \hom(A)$, M_ϕ is closed with respect to the topology induced by the seminorms $\|\cdot\|_{K_p}, p \in \Gamma$, where
$$K_p = \hom(A) \cap U_p^\circ(\epsilon), \ 0 < \epsilon \leq 1,$$
and
$$\|f\|_{K_p} = \sup\{|f(\psi)| : \psi \in K_p\}.$$

Indeed, let $\phi \in \hom(A)$ and let (f_i) be a net in M_ϕ such that $f_i \to f \in C(Y)$ with respect to $\{\|\cdot\|_{K_p}\}_{p \in \Gamma}$. Then, by continuity of ϕ, there is a neighbourhood of zero, say U, in A such that $|\phi(x)| \leq 1$, for every $x \in U$. Without loss of generality, we can take U to be the set $\{x \in A : p_0(x) < \epsilon\}$, for some $p_0 \in \Gamma$ and $0 < \epsilon \leq 1$. Hence, $\phi \in K_{p_0}$. Since $f_i \overset{\|\cdot\|_{K_p}}{\to} f$, we have that $\|f_i - f\|_{K_{p_0}} \overset{i}{\to} 0$. Therefore, $\sup\{|(f_i - f)(\psi)| : \psi \in K_{p_0}\} \overset{i}{\to} 0$, which implies that $|(f_i - f)(\phi)| \overset{i}{\to} 0$. So, $f \in M_\phi$ since $f_i \in M_\phi$. By [12, Lemma 9.1 and Theorem 9.3 (4)], we have that $C(\hom(A)) \cong A[\tau]$ up to a topological $*$-isomorphism, where $C(\hom(A))$ is considered with respect to the topology induced by the seminorms $\|\cdot\|_{K_p}, p \in \Gamma$, and the isomorphism is given by the Gelfand map. Therefore, since M_ϕ is shown to be $\|\cdot\|_{K_p}$-closed, we conclude that

M_ϕ is τ closed (where actually M_ϕ is identified in its latter call with its pre-image via the Gelfand map).

Recall that Y is a dense subspace of βY, where the latter is identified with the set of all maximal ideals of $A[B_0]$. By [14, Theorem 3], the set of all τ-closed maximal ideals of A can be identified with a subset of the set of all maximal ideals of $A[B_0]$. Therefore, the set of all τ-closed maximal ideals of A can be identified with a dense subspace of βY.

(3) Let A be a commutative $*$-algebra and assume that $A[\tau_1]$ and $A[\tau_2]$ are both pro-C*-algebras. Then, since all pro-C*-algebras are GB*-algebras, we have from a result of Allan ([6, Corollary 5.6]) that τ_1 is equivalent to τ_2 (the equivalence is to be understood in the sense of Allan, i.e., if $\mathcal{B}_1^*$, $\mathcal{B}_2^*$ denote the relevant families of subsets of A, as in Definition 2.1, under the topologies τ_1 and τ_2 respectively, and if B_1 denotes the greatest member of $\mathcal{B}_1^*$, then τ_1, τ_2 are equivalent if and only if B_1 is also the greatest member of $\mathcal{B}_2^*$). Therefore, $A[B_0^1] = A[B_0^2]$ (this follows directly from the definition of equivalent GB*-topologies). Now, $A[\tau_1] \cong C(Y_1)$ and $A[\tau_2] \cong C(Y_2)$, up to algebra $*$-isomorphism, where Y_1 and Y_2 are completely regular Hausdorff spaces. So, $C(Y_1) \cong C(Y_2)$ up to algebra $*$-isomorphism, and therefore, βY_1 is homeomorphic to βY_2 (by (1), above).

(4) Now, let A be a commutative $*$-algebra and assume that $A[\tau_1]$ and $A[\tau_2]$ are both GB*-algebras (not necessarily pro-C*-algebras). By a result of Allan (as in (3), above), we know that τ_1 is equivalent to τ_2. Therefore, $A[B_0^1] = A[B_0^2]$. Hence, $A[\tau_1]$ and $A[\tau_2]$ are *algebraically* $*$-isomorphic to the *same* $*$-algebra of functions with the *same* Gelfand isomorphism $x \mapsto \hat{x}$ (see Theorem 2.4). Therefore, the presence of a GB*-topology on a commutative $*$-algebra seems only to play an *algebraic* role. Therefore, the set of all closed maximal ideals of a commutative GB*-algebra appears to be an algebraic artifact!

(5) Let A be a commutative $*$-algebra with identity. If A can be equipped with a topology τ for which $A[\tau]$ is a GB*-algebra, then it follows from Remark (4), above, that the *algebraic* structure of A is entirely determined by a compact Hausdorff space (since any unital commutative C^*-algebra is $*$-isomorphic to some $C(X)$, where X is a compact Hausdorff space).

As a final remark, we note that the article [14] reveals the full structure of all maximal ideals of a commutative pro-C*-algebra.

Acknowledgments. This work is based on part of the research for which the first named author received financial support from the National Research Foundation (NRF) of South Africa.

This work was presented by the first named author at the International Conference of Algebra and Related Topics, hosted by Mohammed V University in Rabat, Morocco, during the time period 2-5 July 2018. He thanks the organizers of this conference for their warm hospitality.

The authors would like to kindly thank the referee for his/her meticulous reading of the manuscript and his/her comments, which resulted in a significant improvement of the overall appearance of the paper. The authors also kindly thank the referee for bringing reference [1] to their attention.

References

[1] M. Abel, *Structure of Gelfand-Mazur algebras*, Dissertationes Mathematicae Universitatis Tartuensis, vol. 31, Tartu University Press, Tartu, 2003. Dissertation, University of Tartu, Tartu, 2002. MR1964191

[2] M. Abel, *Gel'fand-Mazur algebras*, Topological vector spaces, algebras and related areas (Hamilton, ON, 1994), Pitman Res. Notes Math. Ser., vol. 316, Longman Sci. Tech., Harlow, 1994, pp. 116–129. MR1319378

[3] M. Abel, *Advertive topological algebras*, General topological algebras (Tartu, 1999), Math. Stud. (Tartu), vol. 1, Est. Math. Soc., Tartu, 2001, pp. 14–24. MR1853508

[4] M. Abel, *Inductive limits of Gelfand-Mazur algebras*, Int. J. Pure Appl. Math. **16** (2004), no. 3, 363–378. MR2105545

[5] M. Abel and M. Abtahi, *Description of closed maximal ideals in topological algebras of continuous vector-valued functions*, Mediterr. J. Math. **11** (2014), no. 4, 1185–1193, DOI 10.1007/s00009-013-0366-x. MR3268815

[6] G. R. Allan, *On a class of locally convex algebras*, Proc. London Math. Soc. (3) **17** (1967), 91–114, DOI 10.1112/plms/s3-17.1.91. MR0205102

[7] B. Aupetit, *A primer on spectral theory*, Universitext, Springer-Verlag, New York, 1991. MR1083349

[8] E. Beckenstein, L. Narici, and C. Suffel, *Topological algebras*, North-Holland Publishing Co., Amsterdam-New York-Oxford, 1977. North-Holland Mathematics Studies, Vol. 24; Notas de Matemática, No. 60. [Mathematical Notes, No. 60]. MR0473835

[9] F. F. Bonsall and J. Duncan, *Complete normed algebras*, Springer-Verlag, New York-Heidelberg, 1973. Ergebnisse der Mathematik und ihrer Grenzgebiete, Band 80. MR0423029

[10] P. G. Dixon, *Generalized B^*-algebras*, Proc. London Math. Soc. (3) **21** (1970), 693–715, DOI 10.1112/plms/s3-21.4.693. MR0278079

[11] J. Dugundji, *Topology*, Allyn and Bacon, Inc., Boston, Mass., 1966. MR0193606

[12] M. Fragoulopoulou, *Topological algebras with involution*, North-Holland Mathematics Studies, vol. 200, Elsevier Science B.V., Amsterdam, 2005. MR2172581

[13] M. Fragoulopoulou, A. Inoue, and M. Weigt, *Tensor products of generalized B^*-algebras*, J. Math. Anal. Appl. **420** (2014), no. 2, 1787–1802, DOI 10.1016/j.jmaa.2014.06.046. MR3240107

[14] L. Gillman, M. Henriksen, and M. Jerison, *On a theorem of Gelfand and Kolmogoroff concerning maximal ideals in rings of continuous functions*, Proc. Amer. Math. Soc. **5** (1954), 447–455, DOI 10.2307/2031957. MR66627

[15] A. Inoue and K.-D. Kürsten, *On C^*-like locally convex $*$-algebras*, Math. Nachr. **235** (2002), 51–58, DOI 10.1002/1522-2616(200202)235:1⟨51::AID-MANA51⟩3.3.CO;2-2. MR1889277

[16] A. El Kinani, A. Najmi, and M. Oudadess, *Advertibly complete locally m-convex two-sided algebras*, Rend. Circ. Mat. Palermo (2) **56** (2007), no. 2, 185–197, DOI 10.1007/BF03031438. MR2356470

[17] E. Kissin and V. S. Shulman, *Dense Q-subalgebras of Banach and C^*-algebras and unbounded derivations of Banach and C^*-algebras*, Proc. Edinburgh Math. Soc. (2) **36** (1993), no. 2, 261–276, DOI 10.1017/S0013091500018368. MR1221047

[18] S.-b. Ng and S. Warner, *Continuity of positive and multiplicative functionals*, Duke Math. J. **39** (1972), 281–284. MR291808

[19] Y. Tsertos, *A characterization of Q-algebras*, Functional analysis, approximation theory and numerical analysis, World Sci. Publ., River Edge, NJ, 1994, pp. 277–280. MR1298668

[20] M. Weigt and I. Zarakas, *On domains of unbounded derivations of generalized B^*-algebras*, Banach J. Math. Anal. **12** (2018), no. 4, 873–908, DOI 10.1215/17358787-2017-0060. MR3858753

DEPARTMENT OF MATHEMATICS AND APPLIED MATHEMATICS, NELSON MANDELA UNIVERSITY, SUMMERSTRAND CAMPUS (SOUTH), PORT ELIZABETH, 6031, SOUTH AFRICA
  Email address: Martin.Weigt@mandela.ac.za, weigt.martin@gmail.com

DEPARTMENT OF MATHEMATICS, HELLENIC MILITARY ACADEMY, ATHENS, 19400, GREECE
  Email address: gzarak@hotmail.gr

Polar decomposition, Aluthge and mean transforms

Fadil Chabbabi and Mostafa Mbekhta

In memory of our friend Ahmed Intissar.
We wish to honor him for his exemplary dedication
to the development of Mathematics.

ABSTRACT. In this paper, we give a new proof of the existence and uniqueness of the polar decomposition $T = V|T|$ of a bounded linear operator T on a complex Hilbert space H. We show that the polar part of the polar decomposition of T is given by an explicit formula:

$$V = \int_0^\infty T \exp(-sT^*T) |T| ds.$$

On the other hand, we establish new results on the Aluthge and mean transforms of a bounded linear operator T acting on H. Among other things, we show under some conditions that Aluthge transform of T has closed range if and only if T itself has closed range. We also prove that the mean transform preserves the class of compact operators and Schatten p-ideal. We end this paper by asking several open questions.

1. Introduction

Throughout this paper, let H be a complex Hilbert space and $\mathcal{B}(H)$ the algebra of all bounded linear operators on H. For an arbitrary operator $T \in \mathcal{B}(H)$, we denote by $\sigma(T), \sigma_p(T), \sigma_{ap}, \mathcal{R}(T), \mathcal{N}(T)$ and T^* the spectrum, the point spectrum, the approximate point spectrum, the range, the null subspace and the adjoint operator of T, respectively. For any closed subspace M of H, let P_M denote the orthogonal projection onto M.

An operator $T \in \mathcal{B}(H)$ is a *partial isometry* when $TT^*T = T$ (or, equivalently T^*T is an orthogonal projection; in this case $T^*T = P_{\mathcal{N}(T)^\perp}$). In particular T is an *isometry* if $T^*T = I$, and T is *unitary* if it is a surjective isometry.

The well known polar decomposition of any operator $T \in \mathcal{B}(H)$ is a generalization of the polar decomposition of a nonzero complex number $z = |z|\exp(i\theta)$, $\theta \in \mathbb{R}$, and consists of writing T as a product

(1.1) $$T = V|T|,$$

2010 *Mathematics Subject Classification.* 47A05, 47A10, 47B49, 46L40.

Key words and phrases. normal, quasi-normal operators, polar decomposition, Aluthge and mean transforms

This work was supported in part by the Labex CEMPI (ANR-11-LABX-0007-01).

where $|T| := (T^*T)^{\frac{1}{2}}$ is the modulus of T and $V \in \mathcal{B}(H)$ is a partial isometry satisfying $\mathcal{N}(V) = \mathcal{N}(T)$. From such a decomposition, the Aluthge transform of T is defined in [1] by

$$\Delta(T) := |T|^{\frac{1}{2}} V |T|^{\frac{1}{2}},$$

and more generally, for any scalar $\lambda \in [0,1]$, the λ-Aluthge transform of T is defined in [22] by

$$\Delta_\lambda(T) := |T|^\lambda V |T|^{1-\lambda}.$$

Interest in these transforms was enhanced by Jung, Ko and Pearcy proved in [15] where it was shown that T has a nontrivial invariant subspace if and only if its Aluthge transform does. Since then the Aluthge transform has received much more attention and several interesting results on these transforms have been obtained; see [4, 6, 9, 15–18] and the references therein. While, the mean transform of the operator T was recently introduced in [20], and is given by

$$\widehat{T} := \frac{1}{2}\left(V|T| + |T|V\right).$$

Therein a number of questions about the mean transform and its relationship to the Aluthge transform were addressed.

This paper contains six sections and is organized as follows. In Section 2, we give a new proof of the existence and uniqueness of the polar decomposition of a bounded operator acting on a Hilbert space, and we recall some properties of polar decomposition, which plays an important role in this paper. In Section 3, we review certain known permanence properties of the Aluthge transform and then establish new ones. Among other things, we investigate when a Hilbert space bounded linear operator has a normal Aluthge transform, and show that if T is a bounded linear operator on H whose kernel is contained in that of its adjoint then T has a closed range if and only if so has its Aluthge transform. Likewise, in Section 4, we review some known properties of the mean transform and then show that the mean transform of an operator $T \in \mathcal{B}(H)$ is selfadjoint if and only if the partial isometry factor of the polar decomposition of T is selfadjoint. Moreover, we prove that under certain conditions the Aluthge and mean transforms of an operator coincide if and only if such an operator is normal. In Section 5, we show that the mean transform preserves the ideal of compact operators and Schatten p-ideal classes, and provide a new characterization of Hilbert-Schmidt normal operators. Section 6 contains several open questions.

2. Polar decomposition of bounded linear operators on a Hilbert space

We start this section by giving a new proof of the existence and uniqueness of the polar decomposition of a bounded operator on a complex Hilbert space.

THEOREM 2.1. *For each $T \in \mathcal{B}(H)$, there is a unique partial isometrie $V \in B(H)$ such that $T = V|T|$ and $V^*V = P_{\mathcal{N}(T)^\perp}$. Furthermore, we have*

$$V = \int_0^\infty T \exp(-tT^*T)|T|dt,$$

the convergence is in the strong operator topology on $\mathcal{B}(H)$.

PROOF. Let $A(t) = \int_0^t T\exp(-sT^*T)|T|ds$, $t \geq 0$. Then

$$\begin{aligned}
A(t)^*A(t) &= \left(\int_0^t \exp(-sT^*T)|T|T^*ds\right)\left(\int_0^t T\exp(-sT^*T)|T|dt\right) \\
&= \left(\int_0^t \exp(-sT^*T)ds\right)|T|T^*T|T|\left(\int_0^t \exp(-sT^*T)ds\right) \\
&= \left(\int_0^t \exp(-sT^*T)ds\right)(T^*T)^2\left(\int_0^t \exp(-sT^*T)ds\right) \\
&= \left(\int_0^t \exp(-sT^*T)T^*Tds\right)^2 \\
&= \left(\sum_{n\geq 0}(-1)^n(T^*T)^{n+1}\frac{t^{n+1}}{(n+1)!}\right)^2 \\
&= (I - \exp(-tT^*T))^2.
\end{aligned}$$

Therefore,

$$\|A(t)\| \leq \|I - \exp(-tT^*T)\| = \sup_{\lambda \in \sigma(T^*T)}\{1 - \exp(-t\lambda)\} \leq 1$$

for all $t \geq 0$. By the weak compactness of the unit ball of $\mathcal{B}(H)$, there exists an accumulation point $V \in \mathcal{B}(H)$ of $A(t)$. Let $P = P_{\mathcal{N}(T)^\perp}$ be the orthogonal projection onto $\mathcal{N}(T)^\perp$. Then $P = P_{\overline{\mathcal{R}(|T|)}}$ which implies that $P|T| = |T| = |T|P$. Therefore $A(t)P = A(t)$ and hence, $VP = V$.

On the other hand,

$$\begin{aligned}
\|A(t)|T| - T\|^2 &= \left\|T\left(\int_0^t \exp(-sT^*T)T^*Tds - I\right)\right\|^2 \\
&= \left\|T\left(\sum_{n\geq 0}(-1)^n(T^*T)^{n+1}\frac{t^{n+1}}{(n+1)!} - I\right)\right\|^2 \\
&= \|T\exp(-tT^*T)\|^2 \\
&= \|\exp(-2tT^*T)T^*T\| \\
&= \sup_{\lambda \in \sigma(T^*T)} \lambda\exp(-2t\lambda).
\end{aligned}$$

Hence, letting $t \to +\infty$, we obtain $T = V|T|$.

Now, we have $|T|P|T| = T^*T = |T|V^*V|T|$. It follows that $|T|(P - V^*V)|T| = 0$. Since $\mathcal{N}(|T|) = \mathcal{N}(P)$, we get $P(P-V^*V)P = 0$ and $P = PV^*VP$. On the other hand, since $VP = V$, we have $V^*VP = V^*V = PV^*V$. Thus, $P = PV^*VP = PV^*V = V^*VP = V^*V$. Therefore, $V^*V = P$ and hence V is a partial isometry such that $\mathcal{N}(V) = \mathcal{N}(P) = \mathcal{N}(T)$.

Finally, suppose that $T = V|T| = V'|T|$ is another polar decomposition of T. Then $(V' - V)|T| = 0$. Hence, $V' = V$ on $\mathcal{R}(|T|)$ and since $\mathcal{N}(V') = \mathcal{N}(|T|) = \mathcal{N}(V)$, we conclude that $V' = V$. □

REMARK 2.1. (1) Clearly, Theorem 2.1 remains true in the more general context of von Neumann algebras.

(2) Theorem 2.1 implies that in the polar decomposition of an operator $T \in \mathcal{B}(H)$, the partial isometry part is a function of T and T^*.

The next corollary gives a characterization of the polar factor of the polar decomposition of any operator $T \in \mathcal{B}(H)$ in terms of solution of the equation $XT^*X - T = 0$.

COROLLARY 2.1. *Let $T \in \mathcal{B}(H)$ be an operator and $V \in \mathcal{B}(H)$ a partial isometry. Then V is a solution of the equation $XT^*X - T = 0$ and $V^*T \geq 0$ if and only if $T = V|T|$ is the polar decomposition of T.*

PROOF. Suppose that $T = V|T|$ is the polar decomposition of T. Then by the previous theorem, we have

$$\begin{aligned} VT^*V - T &= \int_0^\infty T\exp(-tT^*T)|T|dt\, T^* \int_0^\infty T\exp(-tT^*T)|T|dt - T \\ &= T\int_0^\infty \exp(-tT^*T)dt|T|T^*T|T|\int_0^\infty \exp(-tT^*T)dt - T \\ &= T\left(\int_0^\infty T^*T\exp(-tT^*T)dt\right)^2 - T \\ &= T\left([-\exp(-tT^*T)]_0^{+\infty}\right)^2 - T = T - T = 0. \end{aligned}$$

Therefore, V is a solution of the equation $XT^*X - T = 0$. On the other hand, since $|T|$ commutes with $\exp(-tT^*T)$, we have

$$V^*T = |T|\left(\int_0^\infty \exp(-tT^*T)|T|dt\right)|T| \geq 0,$$

and the "*only if*" part is proved.

Conversely, suppose that V is a solution of the equation $XT^*X - T = 0$ and $V^*T \geq 0$. Then $\mathcal{N}(V) \subseteq \mathcal{N}(T)$ and $\mathcal{R}(T) \subseteq \mathcal{R}(V)$. Since V^*V and VV^* are projections, we obtain
$$T = TV^*V \text{ and } T = VV^*T.$$
Thus
$$T^* = V^*VT^* \text{ and } T^* = T^*VV^*.$$
As the operator $A := V^*T$ is positive, we have $A^2 = A^*A = T^*VV^*T = T^*T$ and $A = |T|$. We also have $VA = VV^*T = T$, and then $T = VA = V|T|$. The proof is therefore complete. □

We recall several properties of the polar decomposition, useful in the sequel. If $T = V|T|$ is the polar decomposition of an operator $T \in \mathcal{B}(H)$, then

$$V^*V = P_{\mathcal{N}(T)^\perp} = P_{\overline{\mathcal{R}(|T|)}} = P_{\overline{\mathcal{R}(T^*)}} \quad \text{and} \quad VV^* = P_{\mathcal{N}(T^*)^\perp} = P_{\overline{\mathcal{R}(|T^*|)}} = P_{\overline{\mathcal{R}(T)}}.$$

Therefore,
$$V^*VT^* = T^*, \quad V^*V|T| = |T|, \quad VV^*T = T \text{ and } VV^*|T^*| = |T^*|.$$

Also, we have

$$\begin{aligned} T \text{ is injective} &\iff V \text{ is isometry (i.e. } V^*V = I). \\ \mathcal{N}(T) = \mathcal{N}(T^*) &\iff V \text{ is normal (i.e. } VV^* = V^*V). \\ T \text{ and } T^* \text{ are injective} &\iff V \text{ is unitary.} \end{aligned}$$

Moreover, the following important properties hold as well.
(1) If $\alpha = |\alpha|\exp(i\theta) \in \mathbb{C}$, then $\alpha T = (\exp(i\theta)V)(|\alpha||T|) = (\exp(i\theta)V)|\alpha T|$ is the polar decomposition of αT.
(2) $T = V|T| = |T^*|V = VT^*V$.
(3) (i) $T^* = V^*|T^*|$ is the polar decomposition of T^*, and
(ii) $T^* = |T|V^* = V^*TV^*$.
(4) $|T| = V^*T = T^*V = V^*|T^*|V$.
(5) $|T^*| = VT^* = TV^* = V|T|V^*$.
(6) Moreover, according to (1), we have $|T^*|V = V|T|$. Hence, for any polynomial $P \in \mathbb{C}[X]$, we have $P(|T^*|)V = VP(|T|)$. Using the Stone-Weierstrass' theorem, we deduce that

$$T = |T^*|^\lambda V|T|^{1-\lambda} = |T^*|^{1-\lambda}V^*|T|^\lambda \text{ and } T^* = |T|^\lambda V|T^*|^{1-\lambda} = |T|^{1-\lambda}V^*|T^*|^\lambda$$

for all $\lambda \in [0,1]$.
(7) $T = T^*$ if and only if $V = V^*$ and $V|T| = |T|V$.
(8) T is normal if and only if V is normal and $V|T| = |T|V$.

3. Aluthge transform

In this section, we first review certain known results showing that the Aluthge transform preserves some hyponormality properties and spectral sets such as the spectrum and its parts. Second, we establish some new results and provide necessary and sufficient conditions so that the Aluthge transform of an injective operator is normal. Finally, we investigate when an operator and its Aluthge transform both have closed ranges.

We first point out that if $T \in \mathcal{B}(H)$ is an operator, then, without the condition that $\mathcal{N}(V) = \mathcal{N}(T)$, the factorization of $T = V|T|$ as the product of a partial isometry $V \in \mathcal{B}(H)$ and the module of T need not be unique. We also note that the transformation of Aluthge of any operator $T \in \mathcal{B}(H)$ does not depend on the choice of the polar factors of such a factorization. Indeed, suppose that V_1 and V_2 are two partial isometries such that $T = V_1|T| = V_2|T|$. Then $V_1|T|^{\frac{1}{2}} = V_2|T|^{\frac{1}{2}}$ on $\mathcal{R}(|T|^{\frac{1}{2}})$, and thus $V_1|T|^{\frac{1}{2}} = V_2|T|^{\frac{1}{2}} = 0$ on $\mathcal{N}(|T|^{\frac{1}{2}})$. Accordingly, $V_1|T|^{\frac{1}{2}} = V_2|T|^{\frac{1}{2}}$ on H. Therefore, by multiplying this relation on the left by $|T|^{\frac{1}{2}}$, we obtain that

$$\Delta(T) = |T|^{\frac{1}{2}}V_1|T|^{\frac{1}{2}} = |T|^{\frac{1}{2}}V_2|T|^{\frac{1}{2}}.$$

Now, we recall some basic notions that will be used in the sequel. An operator $T \in \mathcal{B}(H)$ is normal if $T^*T = TT^*$, and is said to be quasi-normal if it commutes with T^*T (i.e., $TT^*T = T^*T^2$). Note that if $T = V|T|$ is the polar decomposition of T then T is quasi-normal if and only if $|T|$ and V commute. The operator T is called hyponormal if $T^*T \geq TT^*$, and is said to be a p–hyponormal operator if $(T^*T)^p \geq (TT^*)^p$ for a positive scalar p. It is well known that if $p \geq q > 0$, then every p–hyponormal operator is a q–hyponormal operator. In particular, if $p \geq 1$ then every p–hyponormal operator is hyponormal. Obviously, every normal operator is quasi-normal and hyponormal but the converse is not true in general. However, in finite dimensional spaces, every quasi-normal or hyponormal operator is normal. It is also well known that there exists a hyponormal operator $T \in \mathcal{B}(H)$ such that T^2 is not a hyponormal operator. On the other hand, it is easy to see that if T is quasi-normal, then T^2 is also quasi-normal, but the converse is false as it is shown by nonzero nilpotent operators. Finally, we point out that quasi-normal operators are exactly the fixed points of the aluthge and mean transform (see [9, 15]). That is

(3.1) $\quad T$ quasi-normal $\iff V|T| = |T|V \iff \Delta_\lambda(T) = T \iff \widehat{T} = T$.

In [1], Aluthge introduced the Aluthge transform to study p–hyponormal and log–hyponormal operators. Among other things, he obtained the following result that shows that such a transform preserves some hyponormality properties.

THEOREM 3.1. [1] Let $T \in \mathcal{B}(H)$ be a p-hyponormal operator $(p > 0)$. Then the following assertions hold.
  (1) If $\frac{1}{2} \leq p < 1$, then $\Delta(T)$ is hyponormal.
  (2) If $0 < p < \frac{1}{2}$, then $\Delta(T)$ is $(p + \frac{1}{2})$-hyponormal.

Let $T = V|T|$ be the polar decomposition of an operator $T \in \mathcal{B}(H)$ and $\lambda \in [0, 1]$. Let $A = |T|^\lambda$ and $B = V|T|^{1-\lambda}$, and note that

$$\Delta_\lambda(T) = AB \quad \text{and} \quad T = BA.$$

So, applying Jacobson's lemma, one can show that the Aluthge transform preserves the spectrum and its parts of original operator. Precisely, the following result hold.

THEOREM 3.2. [15] If $\lambda \in [0, 1]$ and $T \in \mathcal{B}(H)$, then

$$\sigma(T) = \sigma(\Delta_\lambda(T)), \quad \sigma_p(T) = \sigma_p(\Delta_\lambda(T)), \quad \text{and} \quad \sigma_{ap}(T) = \sigma_{ap}(\Delta_\lambda(T)).$$

Let $T = V|T|$ be the polar decomposition of an operator $T \in \mathcal{B}(H)$ and $\lambda \in [0, 1]$. Since $\Delta_\lambda(T)|T|^\lambda = |T|^\lambda T$ and $V|T|^{1-\lambda}\Delta_\lambda(T) = TV|T|^{1-\lambda}$, we deduce that

$$P(\Delta_\lambda(T))|T|^\lambda = |T|^\lambda P(T) \quad \text{and} \quad V|T|^{1-\lambda}P(\Delta_\lambda(T)) = P(T)V|T|^{1-\lambda}$$

for all polynomials $P \in \mathbb{C}[X]$. Using the Stone-Weierstrass' theorem, we obtain that

$$f(\Delta_\lambda(T))|T|^\lambda = |T|^\lambda f(T) \quad \text{and} \quad V|T|^{1-\lambda}f(\Delta_\lambda(T)) = f(T)V|T|^{1-\lambda},$$

and

$$\|f(\Delta_\lambda(T))\| \leq \|f(\Delta_1(T))\|^\lambda \|f(\Delta_0(T))\|^{1-\lambda} \leq \|f(T)\|$$

for all analytic functions f defined in an open neighborhood of $\sigma(T) = \sigma(\Delta_\lambda(T))$. We refer the reader to [2, 11], for more details and their applications concerning these inequalities.

Let $T = V|T|$ be the polar decomposition of an operator $T \in \mathcal{B}(H)$. If $T \in \mathcal{B}(H)$ is normal operator, then V and $|T|$ commute, and so are V and $|T|^{\frac{1}{2}}$. In this case, on has $\Delta(T) = |T|^{\frac{1}{2}}V|T|^{\frac{1}{2}} = T$ and $\Delta(T)$ is normal. However, in general, if $\Delta(T)$ is normal then T need not be normal. Indeed, let $0 \neq T \in \mathcal{B}(H)$ be an operator such that $T^2 = 0$, and note that T is not normal but $\Delta(T) = 0$ is normal. The next result is new and characterizes when the Aluthge transform of an injective operator is normal.

THEOREM 3.3. Let $T \in \mathcal{B}(H)$ be an injective operator and $T = V|T|$ be its polar decomposition. Then the following statements are equivalent.
  (1) $\Delta(T)$ is normal.
  (2) $V|T|V^* = V^*|T|V$.
  (3) $V^2|T| = |T|V^2$ and V is unitary.

(4) $V^2T = TV^2$ and V is unitary.
(5) $VTV^* = V^*TV$.

PROOF. We organize the proof as follows: (1) $\Longleftrightarrow$ (2) $\Rightarrow$ (3) $\Rightarrow$ (4) $\Rightarrow$ (5) $\Rightarrow$ (2).

(1) $\Longleftrightarrow$ (2). We have

$$\begin{aligned}
\Delta(T)^*\Delta(T) = \Delta(T)\Delta(T)^* &\Longleftrightarrow |T|^{\frac{1}{2}}V^*|T|V|T|^{\frac{1}{2}} = |T|^{\frac{1}{2}}V|T|V^*|T|^{\frac{1}{2}} \\
&\Longleftrightarrow |T|^{\frac{1}{2}}\{V^*|T|V - V|T|V^*\}|T|^{\frac{1}{2}} = 0 \\
&\Longleftrightarrow V^*|T|V - V|T|V^* = 0 \quad \text{(since } T \text{ is injective)} \\
&\Longleftrightarrow V^*|T|V = V|T|V^*.
\end{aligned}$$

Hence (1) $\Longleftrightarrow$ (2).

(2) $\Rightarrow$ (3). Assume that $V|T|V^* = V^*|T|V$. It is easy to see that $\||T|^{\frac{1}{2}}V^*x\| = \||T|^{\frac{1}{2}}Vx\|$ for all $x \in H$. Since T is injective, we have $\mathcal{N}(V^*) = \mathcal{N}(V) = \{0\}$. Therefore, V is unitary.

On the other hand, multiply the equality $V|T|V^* = V^*|T|V$ on both sides by V to see that $V^2|T| = |T|V^2$, and (3) is proved.

(3) $\Rightarrow$ (4). We have

$$V^2|T| = |T|V^2 \Rightarrow VT = |T|V^2 \Rightarrow V^2T = V|T|V^2 = TV^2.$$

Therefore, $V^2T = TV^2$.

(4) $\Rightarrow$ (5). Multiply the equality $V^2T = TV^2$ on both sides by V^*, we obtain $VTV^* = V^*TV$.

(5) $\Rightarrow$ (2). Since T is injective, we have $V^*V = I$. Then

$$VTV^* = V^*TV \Rightarrow VV|T|V^* = V^*V|T|V \Rightarrow V^2|T|V^* = |T|V \Rightarrow V|T|V^* = V^*|T|V.$$

The proof is complete. $\square$

Theorem 3.2 tells us that the Aluthge transform share many spectral properties with the original operator. However, the Aluthge transform of an operator $T \in \mathcal{B}(H)$ may have a closed range without T having a closed range as shown in Example 3.1. In the rest of this section, we investigate when an operator and its Aluthge transform both have closed ranges. To do so, we need to introduce the reduced minimum modulus that measures the closedness of the range of an operator, and then provide some auxiliary lemmas. Recall that the reduced minimum modulus of an operator $T \in \mathcal{B}(H)$ is defined by

$$\gamma(T) := \begin{cases} \inf\{\|Tx\|; \|x\| = 1, x \in \mathcal{N}(T)^\perp\} & \text{if } T \neq 0 \\ +\infty & \text{if } T = 0. \end{cases}$$

The reduced minimum modulus of an operator $T \in \mathcal{B}(H)$ has the following properties.

(1) T has a closed range if and only if $\gamma(T) > 0$.
(2) $\gamma(T) = \inf \sigma^*(|T|)$, (here $\sigma^*(S) = \sigma(S) \setminus \{0\}$).
(3) The following equalities are valid

$$\gamma(T)^2 = \gamma(T^*T) = \gamma(TT^*) = \gamma(T^*)^2.$$

For further information, the interested reader may consult [**14, 19, 21**].

The following lemma provides a useful formula for the reduced minimum modulus of the Aluthge transform of any operator in $\mathcal{B}(H)$.

LEMMA 3.1. *If* $T = V|T| \in \mathcal{B}(H)$, *then*
$$\gamma(\Delta(T)) = \gamma(|T|^{\frac{1}{2}}|T^*|^{\frac{1}{2}}) = \gamma(|T^*|^{\frac{1}{2}}|T|^{\frac{1}{2}}).$$

PROOF. Using [**21**, Corollaire 1.6], we get
$$\begin{aligned}
\gamma(\Delta(T))^2 &= \inf \sigma^*(\Delta(T)\Delta(T)^*) \\
&= \inf \sigma^*(|T|^{\frac{1}{2}}V|T|V^*|T|^{\frac{1}{2}}) \\
&= \inf \sigma^*(|T|^{\frac{1}{2}}|T^*||T|^{\frac{1}{2}}) \\
&= \inf \sigma^*(|T|^{\frac{1}{2}}|T^*|^{\frac{1}{2}}(|T|^{\frac{1}{2}}|T^*|^{\frac{1}{2}})^*) \\
&= \gamma(|T|^{\frac{1}{2}}|T^*|^{\frac{1}{2}})^2 = \gamma(|T^*|^{\frac{1}{2}}|T|^{\frac{1}{2}})^2.
\end{aligned}$$

Hence, the result holds. □

The next lemma tells us when an operator and its Aluthge transform have the same null subspace.

LEMMA 3.2. *For any operator* $T = V|T| \in \mathcal{B}(H)$, *the following statements hold.*
  (1) *If* $\mathcal{N}(T) \subseteq \mathcal{N}(T^*)$, *then* $\mathcal{N}(\Delta(T)) = \mathcal{N}(T)$.
  (2) *If* $\mathcal{N}(T^*) \subseteq \mathcal{N}(T)$, *then* $\mathcal{N}(\Delta(T)^*) = \mathcal{N}(T)$.

PROOF. The inclusion $\mathcal{N}(T) \subseteq \mathcal{N}(\Delta(T))$ is obvious. To show the other inclusion, suppose that $\mathcal{N}(T) \subseteq \mathcal{N}(T^*)$, and let $x \in \mathcal{N}(\Delta(T))$. Then $|T|^{\frac{1}{2}}V|T|^{\frac{1}{2}}x = 0$. It follows that
$$V|T|^{\frac{1}{2}}x \in \mathcal{N}(|T|^{\frac{1}{2}}) \cap \mathcal{R}(V) = \mathcal{N}(T) \cap \mathcal{R}(V) \subseteq \mathcal{N}(T^*) \cap \mathcal{R}(V) \subseteq \mathcal{R}(T)^\perp \cap \mathcal{R}(V) = \{0\}.$$
Then $V|T|^{\frac{1}{2}}x = 0$ and thus $|T|^{\frac{1}{2}}x = V^*V|T|^{\frac{1}{2}}x = 0$. Therefore $x \in \mathcal{N}(|T|^{\frac{1}{2}}) = \mathcal{N}(T)$. Hence, $\mathcal{N}(\Delta(T)) \subseteq \mathcal{N}(T)$. This achieves the proof of the first statement.

With similar arguments, one proves the second statement. □

The following result shows that if $T \in \mathcal{B}(H)$ is an operator such that its null subspace is contained in that of its adjoint then the ranges of T and its Aluthge transform are both either closed or not closed.

THEOREM 3.4. *Let* $T \in \mathcal{B}(H)$. *If* $\mathcal{N}(T) \subseteq \mathcal{N}(T^*)$, *then*
$$\mathcal{R}(T) \text{ is closed if and only if } \mathcal{R}(\Delta(T)) \text{ is closed}.$$
In particular, if T *is hyponormal (i.e.* $TT^* \leq T^*T$*), then the previous equivalence remains true.*

PROOF. Suppose that $\mathcal{R}(T)$ is closed and $\mathcal{R}(\Delta(T))$ is not closed. By Lemma 3.1, we have $\gamma(|T|^{\frac{1}{2}}|T^*|^{\frac{1}{2}}) = \gamma(\Delta(T)) = 0$. So, we can choose a sequence of unit vectors $x_n \in \mathcal{N}(|T|^{\frac{1}{2}}|T^*|^{\frac{1}{2}})^\perp \subseteq \mathcal{N}(|T^*|^{\frac{1}{2}})^\perp$ such that $|T|^{\frac{1}{2}}|T^*|^{\frac{1}{2}}x_n \longrightarrow 0$. Since $\mathcal{R}(|T^*|^{\frac{1}{2}})$ is closed and each $x_n \in \mathcal{N}(|T^*|^{\frac{1}{2}})^\perp$, there exists $\eta > 0$ such that $\||T^*|^{\frac{1}{2}}x_n\| \geq \eta$ for all n. Put $y_n := \dfrac{|T^*|^{\frac{1}{2}}x_n}{\| |T^*|^{\frac{1}{2}}x_n\|}$ and note that
$$y_n \in \mathcal{R}(|T^*|^{\frac{1}{2}}) = \mathcal{R}(T) \subseteq \mathcal{R}(T^*) = \mathcal{R}(|T|^{\frac{1}{2}}) = \mathcal{N}(|T|^{\frac{1}{2}})^\perp$$

for all n. It follows that

$$\| |T|^{\frac{1}{2}} y_n \| \leq \frac{1}{\eta} \| |T|^{\frac{1}{2}} |T^*|^{\frac{1}{2}} x_n \|$$

for all n, and thus $|T|^{\frac{1}{2}} y_n \longrightarrow 0$. This contradicts the fact that $\mathcal{R}(|T|^{\frac{1}{2}}) = \mathcal{R}(T)$ is closed.

Reciprocally, suppose that $\mathcal{R}(\Delta(T))$ is closed and $\mathcal{R}(T)$ is not closed. Then, there exists a sequence of unit vectors $x_n \in \mathcal{N}(T)^\perp$ such that $Tx_n \longrightarrow 0$. By Cauchy-schawrtz inequality, we have

$$\| |T|^{\frac{1}{2}} x_n \|^2 = \langle |T| x_n, x_n \rangle \leq \|Tx_n\|$$

for all n. Hence, $|T|^{\frac{1}{2}} x_n \longrightarrow 0$. Now, by Lemma 3.2, we have $\mathcal{N}(\Delta(T)) = \mathcal{N}(T)$. Therefore, each $x_n \in \mathcal{N}(\Delta(T))^\perp$ and $\Delta(T) x_n \longrightarrow 0$, which is a contradiction with the fact that $\mathcal{R}(\Delta(T))$ is closed.

It is easy to see that if T is hyponormal, then the condition $\mathcal{N}(T) \subseteq \mathcal{N}(T^*)$ is verified. So we can apply the previous equivalence. The proof of the theorem is therefore complete. $\square$

As a consequence of the previous theorem, we have the following result.

COROLLARY 3.1. *Let $T \in \mathcal{B}(H)$ such that $\mathcal{N}(T) = \mathcal{N}(T^*)$. Then the following assertions are equivalent.*

(1) $\mathcal{R}(T)$ *is closed.*
(2) $\mathcal{R}(\Delta(T))$ *is closed.*
(3) $\mathcal{R}(\Delta(T^*))$ *is closed.*

The following example shows that if $T \in \mathcal{B}(H)$ such that $\mathcal{R}(\Delta(T))$ is closed, then $\mathcal{R}(T)$ need not be closed. It also tells us that the condition that $\mathcal{N}(T) \subseteq \mathcal{N}(T^*)$ in the above result is unavoidable.

EXAMPLE 3.1. Let $T : \ell^2(\mathbb{N}) \to \ell^2(\mathbb{N})$ be the unilateral weighted shift defined by $Te_n = \alpha_n e_{n+1}$ for all $n \in \mathbb{N}$, where

$$\alpha_n = \begin{cases} 0 & \text{if } n \text{ even} \\ \frac{1}{n} & \text{if } n \text{ odd} . \end{cases}$$

Then $T^2 = 0$. Hence $\Delta(T) = 0$ has closed range, but $\mathcal{R}(T)$ is not closed.

4. Mean transform

In this section, we first provide some properties of the mean transform, recently obtained in [9]. Then we establish new results and discuss when the mean transform of an operator in $\mathcal{B}(H)$ is selfadjoint or idempotent or normal. Among other results, we show that the mean transform of an operator in $\mathcal{B}(H)$ is selfadjoint if and only if the polar part of its polar decomposition is too selfadjoint. We also prove that an operator in $\mathcal{B}(H)$ is normal if and only if its Aluthge and mean transforms coincide and the null subspace of its adjoint is included in its own null subspace.

Contrary to what happens with the Aluthge transform, the mean transform does depend on the polar factors of the polar decomposition of the given operator.

For example, consider $T = \begin{pmatrix} 0 & 1 \\ 0 & 0 \end{pmatrix}$ acting on $\mathbb{C}^2$. The canonical polar decomposition of T is $T = V|T|$, where $|T| = \sqrt{T^*T} = \begin{pmatrix} 0 & 0 \\ 0 & 1 \end{pmatrix}$ and $V = \begin{pmatrix} 0 & 1 \\ 0 & 0 \end{pmatrix}$. On the other hand, we can also write $T = U_{max}|T|$, where $U_{max} = \begin{pmatrix} 0 & 1 \\ 1 & 0 \end{pmatrix}$ is unitary. This is the so-called *maximal* polar decomposition of T, since the partial isometry is unitary. In this case,

$$U_{max}|T| + |T|U_{max} = \begin{pmatrix} 0 & 1 \\ 1 & 0 \end{pmatrix} \neq V|T| + |T|V = \begin{pmatrix} 0 & 1 \\ 0 & 0 \end{pmatrix}.$$

This shows that the mean transform depends on the polar decomposition.

In what follows, we will always use the canonical polar decomposition when dealing with the mean transform.

Next three results are quoted from [9] and provide some properties of the mean transform. The first one shows that the mean transform preserves the null subspace of operators. The second one tells us that such a transform is homogenous and maps the identity operator into itself. The last result shows that an operator in $\mathcal{B}(H)$ is invertible if and only if its range is close and its mean transform is too invertible.

PROPOSITION 4.1. ([9]) For any operator $T \in \mathcal{B}(H)$, we have

$$\mathcal{N}(\widehat{T}) = \mathcal{N}(T).$$

In particular, $\widehat{T} = 0$ if and only if $T = 0$.

PROPOSITION 4.2. ([9]) Let $T \in \mathcal{B}(H)$. Then the following properties hold.
  (i) For every $\alpha \in \mathbb{C}$, we have $\widehat{\alpha T} = \alpha \widehat{T}$.
  (ii) For every unitary or anti-unitary operator $U : H \to H$, we have

$$\widehat{UTU^*} = U\widehat{T}U^*.$$

  (iii) $\widehat{T} = I \iff T = I$.

THEOREM 4.1. ([9]) Let $T \in \mathcal{B}(H)$. Then the following statements are equivalent.
  (i) T is invertible.
  (ii) $\widehat{T}$ is invertible and $\mathcal{R}(T)$ is closed.

REMARK 4.1. In Theorem 4.1 (ii), the condition that "$\mathcal{R}(T)$ is closed" is required. Without it, the reverse implication is false, as shown by the following example.

EXAMPLE 4.1. Let $(e_n)_{n\in\mathbb{Z}}$ be the canonical basis of $\ell^2(\mathbb{Z})$, and set

$$\alpha_n := \begin{cases} 1 & \text{if } n \text{ even} \\ \frac{1}{n^2} & \text{if } n \text{ odd} \end{cases}$$

for all $n \in \mathbb{Z}$. Let $T : \ell^2(\mathbb{Z}) \to \ell^2(\mathbb{Z})$ be the weighted bilateral shift defined by $Te_n = \alpha_n e_{n+1}$ for all $n \in \mathbb{Z}$. Clearly, $Te_{2n+1} \xrightarrow[n\to\infty]{} 0$, and therefore T is not invertible.

Now, set
$$\widehat{\alpha}_n := \frac{\alpha_n + \alpha_{n+1}}{2} = \begin{cases} \dfrac{1 + \frac{1}{(n+1)^2}}{2} & \text{if } n \text{ even} \\ \dfrac{1 + \frac{1}{n^2}}{2} & \text{if } n \text{ odd} \end{cases}$$

for $n \in \mathbb{Z}$. Note that $\widehat{T}e_n = \widehat{\alpha}_n e_{n+1}$ for $n \in \mathbb{Z}$, and $\widehat{T}$ is also a bilateral weighted shift. On the other hand, we have $1 \geq \widehat{\alpha}_n \geq \frac{1}{2}$ for all $n \in \mathbb{Z}$, and then $\widehat{T}$ is invertible. $\square$

REMARK 4.2. In general, we have
(1) $\sigma(T) \neq \sigma(\widehat{T})$ (see [20]);
(2) $\left(\widehat{T}\right)^{-1} \neq \widehat{T^{-1}}$ (see [18]).

The following result shows that the mean transform of an operator in $\mathcal{B}(H)$ has a closed range provided that the range of such an operator is closed.

THEOREM 4.2. Let $T \in \mathcal{B}(H)$. Then
$$\mathcal{R}(T) \text{ is closed} \Rightarrow \mathcal{R}(\widehat{T}) \text{ is also closed.}$$

PROOF. Assume that $\mathcal{R}(T)$ is closed and $\mathcal{R}(\widehat{T})$ is not closed. Then there exists a sequence of unit vectors $x_n \in \mathcal{N}(\widehat{T})^\perp = \mathcal{N}(T)^\perp$ such that $\widehat{T}x_n \longrightarrow 0$. Hence,
$$2V^*\widehat{T}x_n = (|T| + V^*|T|V)x_n \longrightarrow 0.$$
Since $|T|$ and $V^*|T|V$ are positive operators, we have
$$|T|x_n \longrightarrow 0 \quad \text{and} \quad V^*|T|Vx_n \longrightarrow 0.$$
Therefore, $Tx_n \longrightarrow 0$, which is a contradiction with the fact that $\mathcal{R}(T)$ is closed. $\square$

REMARK 4.3. In Theorem 4.2, the reverse implication is false as shown by Example 4.1.

The next result characterizes when the mean transform of an operator in $\mathcal{B}(H)$ is selfadjoint. Precisely, it states that the mean transform of an operator in $\mathcal{B}(H)$ is selfadjoint if and only if the polar part of its polar decomposition is too selfadjoint.

PROPOSITION 4.3. Let $T \in \mathcal{B}(H)$ and $T = V|T|$ be its polar decomposition. Then
$$\widehat{T} = \widehat{T}^* \iff V = V^*.$$

PROOF. If $V = V^*$, then $\widehat{T}^* = \frac{1}{2}(|T|V^* + V^*|T|) = \widehat{T}$.
Now, suppose that $\widehat{T}$ is self-adjoint. Then
(4.1) $$|T|V^* + V^*|T| = V|T| + |T|V.$$
First, we show that $\mathcal{N}(V) \subseteq \mathcal{N}(V^*)$. Let $x \in \mathcal{N}(V)$, and note that $Vx = |T|x = 0$. By (4.1), we get $|T|V^*x = 0$. It follows that $VV^*x = 0$ and thus $V^*x = 0$.
By using again (4.1), we obtain
$$(V^* - V)|T| = |T|(V - V^*).$$
Therefore,
$$(V^* - V)|T|(V - V^*) = |T|(V - V^*)^2 = (V - V^*)^2|T|.$$

In particular, $(V - V^*)^2|T| = -4\mathcal{I}m(V)^2|T|$ is a positive operator. It follows that the operators $\mathcal{I}m(V)^2$ and $|T|$ commute. Hence, $\mathcal{I}m(V)^2|T| \geq 0$ is a product of two positive commuting operators. Consequently, we deduce that $\mathcal{I}m(V)^2|T| = 0$. Thus, $\mathcal{I}m(V)^2 = 0$ on $\mathcal{R}(|T|)$. Since $\mathcal{N}(|T|) = \mathcal{N}(V) \subseteq \mathcal{N}(V^*)$, we have $V - V^* = 0$ on $\mathcal{N}(|T|)$. Hence, $\mathcal{I}m(V)^2 = 0$ on H and $V = V^*$. $\square$

REMARK 4.4. In general, "$\widehat{T}$ is self-adjoint" dose not implies that "T is self-adjoint". For example, consider $T = \begin{pmatrix} 2 & 1 \\ -1 & -2 \end{pmatrix}$ acting on $\mathbb{C}^2$. The polar decomposition of T is $T = V|T|$, where $|T| = \begin{pmatrix} 2 & 1 \\ 1 & 2 \end{pmatrix}$ and the partial isometry $V = \begin{pmatrix} 1 & 0 \\ 0 & -1 \end{pmatrix}$. Since, V is self-adjoint, by the preceding theorem, $\widehat{T}$ is self-adjoint, contrary of T which is not self-adjoint.

The following result states that the mean transform of an idempotent operator in $\mathcal{B}(H)$ is too idempotent.

PROPOSITION 4.4. *If $T \in \mathcal{B}(H)$ is an idempotent operator, then $\widehat{T}$ is also an idempotent.*

PROOF. Let $T = V|T|$ be the polar decomposition of T. Since $\mathcal{N}(|T|) = \mathcal{N}(T)$, we have

$$\begin{aligned}
|T|[I - T] = 0 &\Rightarrow |T|[I - T]V = 0 \\
&\Rightarrow |T|V - |T|V|T|V = 0 \\
&\Rightarrow |T|V = (|T|V)^2.
\end{aligned}$$

Therefore, $|T|V$ is an idempotent.

We show that $\mathcal{N}(|T|V) = \mathcal{N}(V) = \mathcal{N}(T)$. Indeed, obviously, $\mathcal{N}(V) \subseteq \mathcal{N}(|T|V)$. For the other inclusion, suppose that $|T|Vx = 0$. Then $Vx \in \mathcal{N}(|T|) \cap \mathcal{R}(V) = \mathcal{N}(T) \cap \mathcal{R}(T) = \{0\}$ (since $T^2 = T$). On the other hand, $\mathcal{R}(|T|V) \subseteq \mathcal{R}(|T|) = \mathcal{R}(|T|)^\perp = \mathcal{N}(T)^\perp$. Therefore, $|T|V = P_{\mathcal{N}(T)^\perp} = P_{\mathcal{R}(T^*)}$. It follows that

$$T^* P_{\mathcal{R}(T^*)} = P_{\mathcal{R}(T^*)} \text{ and } P_{\mathcal{R}(T^*)} T^* = T^*, \text{ hence } |T|V = |T|VT \text{ and } T = T|T|V.$$

Now, we have

$$\begin{aligned}
(\widehat{T})^2 &= \frac{1}{4}[T + |T|V]^2 \\
&= \frac{1}{4}[T + |T|V + T|T|V + |T|VT] \\
&= \frac{1}{4}[2T + 2|T|V] \\
&= \frac{1}{2}[T + |T|V] \\
&= \widehat{T}.
\end{aligned}$$

Therefore, $\widehat{T}$ is an idempotent. $\square$

We close this section with the following result that gives a new characterization of normal operators.

THEOREM 4.3. *An operator $T \in \mathcal{B}(H)$ is normal if and only if $\mathcal{N}(T^*) \subseteq \mathcal{N}(T)$ and $\Delta(T) = \widehat{T}$.*

PROOF. ($\Leftarrow$). This implication holds without any condition on T since all normal operators are quasi-normal and thus are the fixed points of both the Aluthge and the mean transforms.

($\Rightarrow$). Let $T = V|T|$ the polar decomposition of T. By the assumption, we have $\Delta(T) = \widehat{T}$, and then

(4.2) $$|T|^{\frac{1}{2}} V |T|^{\frac{1}{2}} = \frac{1}{2}(T + |T|V).$$

Thus, $T = 2|T|^{\frac{1}{2}} V |T|^{\frac{1}{2}} - |T|V$. Since $\overline{\mathcal{R}(|T|)} = \overline{\mathcal{R}(|T|^{\frac{1}{2}}} \subseteq \overline{\mathcal{R}(T^*)}$, we have
$$\overline{\mathcal{R}(T)} = \overline{\mathcal{R}(2|T|^{\frac{1}{2}} V |T|^{\frac{1}{2}} - |T|V)} \subseteq \overline{\mathcal{R}(T^*)}.$$
By the condition that the "rank of T is finite or $\mathcal{N}(T^*) \subseteq \mathcal{N}(T)$", it follows that

(4.3) $$\overline{\mathcal{R}(T)} = \overline{\mathcal{R}(|T|)} = \overline{\mathcal{R}(T^*)}.$$

Consequently, $V^*V = VV^*$ and $VV^*|T| = |T|$.

Now, multiplying the equality in (4.2) on the left side by V^*, we get
$$V^*|T|^{\frac{1}{2}} V |T|^{\frac{1}{2}} = \frac{1}{2}(|T| + V^*|T|V).$$
We put $A = V^*|T|^{\frac{1}{2}} V$ and $B = |T|^{\frac{1}{2}}$, and obviously note that A and B are positive operators. Since $VV^*|T|^{\frac{1}{2}} = |T|^{\frac{1}{2}}$, we infer that
$$A^2 = V^*|T|^{\frac{1}{2}} VV^*|T|^{\frac{1}{2}} V = V^*|T|V \quad \text{and} \quad B^2 = |T|.$$
By (4), it follows that $AB = \frac{1}{2}(A^2 + B^2)$. In particular, $AB = BA$ and $(A-B)^2 = A^2 + B^2 - 2AB = 0$. Since $A - B$ is a self-adjoint operator, then $A - B = 0$ and thus $V^*|T|V = |T|$. Hence, $|T|V = VV^*|T|V = V|T|$ and $V^*|T| = |T|V^*$. Accordingly, $TT^* = V|T|^2 V^* = VV^*|T|^2 = |T|^2 = T^*T$, which completes the proof. $\square$

5. Mean transform and Schatten p-ideals

In this section, H is a separable Hilbert space. Let $\mathcal{K}(H)$ be the closed ideal of all compact operators on H. The singular values of any compact operator $T \in \mathcal{K}(H)$ are denoted by $(s_n(T))_{n \in \mathbb{N}}$ and are the eigenvalues of $|T|$, arranged in non-increasing order. Given $1 \leq p < \infty$, the Schatten p-class of H, denoted by $\mathcal{S}_p(H)$, is defined as the space of all compact operators $T \in \mathcal{K}(H)$ with p-summable singular values. That is
$$\mathcal{S}_p(H) = \left\{ T \in \mathcal{K}(H); \, \|T\|_p := \left(\sum_{n \geq 1} s_n(T)^p \right)^{\frac{1}{p}} < \infty \right\}.$$
It is well known that $\mathcal{S}_p(H)$ is a two-sided ideal in the full algebra $\mathcal{B}(H)$, and that
$$\|T\|_p = \left(\sum_{n \geq 1} s_n(T)^p \right)^{\frac{1}{p}} = (tr(|T|^p))^{\frac{1}{p}}$$
for all $T \in \mathcal{S}_p(H)$, where $tr(.)$ denotes the canonical trace in $\mathcal{B}(H)$. For general theory of Schatten p-ideals, the interested reader is referred to [2, 13, 23].

For every nonzero vectors $x, y \in H$, let $x \otimes y$ stand for the rank one operator defined by
$$(x \otimes y)u = \langle u, y \rangle x, \quad (u \in H).$$

Note that $x \otimes y$ is an orthogonal projection if and only if $x = y$ and $\|x\| = 1$, and that every rank one operator has the previous form. The following proposition gives the mean transform of rank one operators, and can be found in [9].

PROPOSITION 5.1. ([9]) Let $x, y \in H$ be two nonzero vectors. If $T = x \otimes y$, then
$$\widehat{T} = \frac{1}{2}\left(x + \frac{\langle x, y \rangle}{\|y\|^2} y\right) \otimes y.$$

From the preceding result, we remark that the mean transform of any rank one operator is also a rank one operator. In the next result, we show that the mean transform preserves in both directions the ideal of compact operators and Schatten p-ideals.

THEOREM 5.1. Let $T \in \mathcal{B}(H)$ and $p \geq 1$. Then
$$\widehat{T} \in \mathcal{K}(H) \iff T \in \mathcal{K}(H),$$
and,
$$\widehat{T} \in \mathcal{S}_p(H) \iff T \in \mathcal{S}_p(H).$$
In the later case, we have
$$\|\widehat{T}\|_p \leq \|T\|_p.$$

We start by the following lemma, and then give the proof of Theorem 5.1.

LEMMA 5.1. Let $A, B \in \mathcal{B}(H)$ be two positive operators. Then
$$A + B \in \mathcal{K}(H) \iff A, B \in \mathcal{K}(H),$$
and
$$A + B \in \mathcal{S}_p(H) \iff A, B \in \mathcal{S}_p(H).$$

PROOF. If $A, B \in \mathcal{K}(H)$, then $A + B \in \mathcal{K}(H)$ since $\mathcal{K}(H)$ is an ideal. Conversely, assume that $A + B \in \mathcal{K}(H)$, and let us first prove that A is compact. Let $u_n \rightharpoonup 0$ be a weakly convergent sequence in H (i.e. $\lim_n \langle u_n, x \rangle = 0$ for every $x \in H$). We shall show that $(Au_n)_n$ strongly converges to 0. We have
$$\begin{aligned} \|A^{\frac{1}{2}} u_n\| &= \langle Au_n, u_n \rangle \\ &\leq \langle (A+B)u_n, u_n \rangle \\ &\leq \|(A+B)u_n\| \|u_n\| \end{aligned}$$
for all n. Since $A + B$ is compact and $\|u_n\|$ is bounded, we see that $A^{\frac{1}{2}} u_n \to 0$ as $n \to \infty$. Therefore, $Au_n \to 0$ strongly, and A is compact. It then follows that $B = (A + B) - A$ is also compact since it is a sum of two compact operators.

Now, assume that $A + B \in \mathcal{S}_p(H)$, and let $(e_n)_n$ be an orthonormal basis of H. Then
$$(\langle Ae_n, e_n \rangle)^p \leq (\langle (A+B)e_n, e_n \rangle)^p$$
for all $n \in \mathbb{N}$. Thus,
$$\sum_n (\langle Ae_n, e_n \rangle)^p \leq \sum_n \left(\langle (A+B)e_n, e_n \rangle\right)^p < \infty.$$

Therefore, $A \in \mathcal{S}_p(H)$ and also $B = A + B - A \in \mathcal{S}_p(H)$. The reverse implication is trivial since $\mathcal{S}_p(H)$ is an ideal, and the proof of this lemma is then complete. □

Now, we are in a position to prove Theorem 5.1.

PROOF. (Theorem 5.1) Let $T = V|T|$ be the polar decomposition of T. Suppose that T is compact, and note that, since $\mathcal{K}(H)$ is an ideal, $\widehat{T}$ is too compact.

Conversely, suppose that $\widehat{T}$ is compact and note that $V^*\widehat{T} = \frac{1}{2}(|T| + V^*|T|V)$ is also compact. Since $|T|$ and $V^*|T|V$ are positive operators, Lemma 5.1 implies that both $|T|$ and $V^*|T|V$ are compact. Therefore $T = V|T|$ is compact.

Now, suppose that $T \in \mathcal{S}_p(H)$. Then also $|T| \in \mathcal{S}_p$, with the fact that $\mathcal{S}_p$ is an ideal of $\mathcal{B}(H)$. Hence $|T|V \in \mathcal{S}_p$. Therefore $\widehat{T} = \frac{1}{2}(T + |T|V) \in \mathcal{S}_p$.

Conversely, assume that $\widehat{T} \in \mathcal{S}_p(H)$. Then
$$V^*\widehat{T} = \frac{1}{2}(|T| + V^*|T|V) \in \mathcal{S}_p(H).$$

By using Lemma 5.1, we have
$$|T| \in \mathcal{S}_p(H).$$

It follows that $T = V|T| \in \mathcal{S}_p(H)$.

On the other hand, we have
$$\begin{aligned} \|\widehat{T}\|_p &= \frac{1}{2}\|V|T| + |T|V\|_p \\ &\leq \frac{1}{2}[\|V|T|\|_p + \||T|V\|_p] \\ &\leq \frac{1}{2}[\||T|_p\| + \||T|\|_p] \\ &= \|T\|_p. \end{aligned}$$

$\square$

In the case $p = 2$, we have the following result which gives a new characterization of normal operators.

THEOREM 5.2. If $T \in \mathcal{S}_2(H)$, then
$$\|\widehat{T}\|_2 = \|T\|_2 \iff T \text{ is normal}.$$

PROOF. If T is normal, then $\widehat{T} = T$ and thus $\|\widehat{T}\|_2 = \|T\|_2$.

For the other implication, let $T = V|T|$ be the canonical polar decomposition, and note that
$$\begin{aligned} \|\widehat{T}\|_2^2 &= tr(\widehat{T}\widehat{T}^*) \\ &= \frac{1}{4}tr\left((T + |T|V)(T^* + V^*|T|)\right) \\ &= \frac{1}{4}tr\left(TT^* + TV^*|T| + |T|VT^* + |T|VV^*|T|\right) \\ &= \frac{1}{4}tr\left(TT^* + V|T|V^*|T| + |T|V|T|V^* + |T|VV^*|T|\right) \\ &= \frac{1}{4}tr\left(TT^* + |T^*||T| + |T||T^*| + |T|VV^*|T|\right) \\ &= \frac{1}{4}\left(\|T\|_2^2 + 2tr(|T^*||T|) + tr(|T|VV^*|T|)\right). \end{aligned}$$

It follows that
(5.1) $$4\|\widehat{T}\|_2^2 = \|T\|_2^2 + 2tr(|T^*||T|) + tr(|T|VV^*|T|).$$

Hence, if $\|\widehat{T}\|_2 = \|T\|_2$, then we get
$$4\|T\|_2^2 = \|T\|_2^2 + 2tr(|T^*||T|) + tr(|T|VV^*|T|).$$
Therefore,
$$2(\|T\|_2^2 - tr(|T^*||T|)) = tr(|T|VV^*|T|) - \|T\|_2^2,$$
and thus
$$\begin{aligned} 2(\|T\|_2^2 - tr(|T^*||T|)) &= tr(|T|VV^*|T|) - \|T\|_2^2 \\ &= tr(|T|VV^*|T|) - tr(|T|^2) \\ &= tr(|T|VV^*|T| - |T|^2). \end{aligned}$$

Hence,
$$2(\|T\|_2^2 - tr(|T^*||T|)) = tr(|T|(VV^* - I)|T|).$$
Since $VV^* - I \leq 0$, we obtain that
(5.2) $$\|T\|_2^2 - tr(|T^*||T|) \leq 0.$$
On the other hand, we also have $\widehat{T} \in \mathcal{S}_2(H)$ and
$$\begin{aligned} \|\widehat{T} - T\|_2^2 &= \|\widehat{T}\|_2^2 - 2tr(\widehat{T}T^*) + \|T\|_2^2 \\ &= 2\|T\|_2^2 - tr(TT^* + |T|VT^*) \\ &= \|T\|_2^2 - tr(|T|VT^*) \\ &= \|T\|_2^2 - tr(|T||T^*|) \\ &= \|T\|_2^2 - tr(|T^*||T|) \leq 0 \quad \text{(by (5.2))}. \end{aligned}$$

This implies that $\|\widehat{T} - T\|_2^2 = 0$, and $\widehat{T} = T$. Consequently, T is a quasi-normal operator. Now, since every quasi-normal operator is hyponormal, then $T^*T - TT^* \geq 0$ and also $T^*T - TT^* \in \mathcal{S}_1(H)$ (is the fact that $T \in \mathcal{S}_2(H)$). Hence, $\|T^*T - TT^*\|_1 = Tr(T^*T - TT^*) = 0$, and thus $T^*T - TT^* = 0$, Clearly, T is normal and the proof is therefore complete. $\square$

6. Open questions

In this section, we make some remarks and comments, and cite some questions for possible further study. As pointed earlier, recall that $\Delta\left(\widehat{T}\right) = \widehat{\Delta(T)} = T$ for all quasi-normal operators $T \in \mathcal{B}(H)$. So, it would be certainly interesting to know which operators in $\mathcal{B}(H)$ satisfy any of these identities. This is therefore left as an open problem.

QUESTION 6.1. What is the class of all operators $T \in \mathcal{B}(H)$ satisfying $\widehat{\Delta(T)} = \Delta(\widehat{T})$?

Consider the mean-Aluthge transform, $\widehat{\Delta} : \mathcal{B}(H) \to \mathcal{B}(H)$, defined by
$$\widehat{\Delta}(T) := \widehat{\Delta(T)}, \; T \in \mathcal{B}(H).$$
Beside quasi-normal operators, one may wonder if there are any other fixed points of this transform. On the other hand, one may ask whether some spectral properties of $\widehat{\Delta}(T)$ are inherited from of those of T. These are accumulated in the following problem.

QUESTION 6.2. (1) Determine the fixed points of this transform (i.e., those operators $T \in \mathcal{B}(H)$ for which $\widehat{\Delta}(T) = T$).
(2) What are the spectral properties of $\widehat{\Delta}(T)$ as a function of those of T?

Let $\lambda \in [0, \frac{1}{2}]$, and recall that the λ-Aluthge transform of an operator $T \in \mathcal{B}(H)$ with the polar decomposition $T = V|T|$ is defined by $\Delta_\lambda(T) = |T|^\lambda V |T|^{1-\lambda}$. Now, consider the transform, $\mathcal{M}_\lambda : \mathcal{B}(H) \to \mathcal{B}(H)$, defined by

$$\mathcal{M}_\lambda(T) := \frac{1}{2}\left(\Delta_\lambda(T) + \Delta_{1-\lambda}(T)\right), \ T \in \mathcal{B}(H).$$

Observe that
$$\mathcal{M}_\lambda(T) = \mathcal{M}_{1-\lambda}(T),$$
and
$$\mathcal{M}_0(T) = \widehat{T}$$

for all $\mathcal{B}(H)$. We also note that quasi-normal operators are fixed points of this transform. So, the following question suggests itself, and considers a similar question as the above one but when the transform $\widehat{\Delta}$ is replaced by the transform $\mathcal{M}_\lambda$.

QUESTION 6.3. (1) Determine the fixed points of the mean-Aluthge transform (i.e., those operators $T \in \mathcal{B}(H)$ for which $\mathcal{M}_\lambda(T) = T$).
(2) What are the spectral properties of $\mathcal{M}_\lambda(T)$ as a function of those of T?

Recently in [3], F. Botelho, L. Molnár and G. Nagy studied the linear bijective mapping on von Neumann algebras which commutes with the λ-Aluthge transforms. Given $\lambda \in (0,1)$, they mainly focussed on bijective linear maps Φ on $\mathcal{B}(H)$ satisfying

(6.1) $\qquad \Delta_\lambda(\Phi(T)) = \Phi(\Delta_\lambda(T)) \quad$ for all $T \in \mathcal{B}(H)$.

In [4–8], the current authors characterized bijective maps Φ on $\mathcal{B}(H)$ (not assumed to be linear or continuous) satisfying

(6.2) $\qquad \Delta_\lambda(\Phi(A) \star \Phi(B)) = \Phi(\Delta_\lambda(A \star B)) \ $ for all $\ A, B \in \mathcal{B}(H)$,

where the operation $A \star B$ means one of the following :
(1) $A \star B = AB$, the standard product.
(2) $A \star B = A \circ B = \frac{1}{2}(AB + BA)$, the jordan product.
(3) $A \star B = A \circ B^* = \frac{1}{2}(AB^* + B^*A)$, the star Jordan product .
(4) $A \star B = A^n B A^m$ with $n, m \in \mathbb{N}$, such that $n + m \geq 1$, the (n,m)-Jordan triple product.
(5) $A \star B = A + \omega B, \ 0 \neq \omega \in \mathbb{C}$.

One may wonder if the main results of [3, 4, 6–8] remain valid when the λ-Aluthge transformation is replaced by any of the transforms $\widehat{\Delta}$ and $\mathcal{M}_\lambda$. More precisely, if we denote the transform

$$\Lambda \in \{\widehat{\Delta}, \mathcal{M}_\lambda \ \lambda \in [0, \frac{1}{2}]\},$$

we have the following questions:

QUESTION 6.4. Describe all bijective linear maps $\Phi : \mathcal{B}(H) \to \mathcal{B}(H)$ that commute with the transform Λ (i.e., $\Phi(\Lambda(T)) = \Lambda(\Phi(T))$ for all $T \in \mathcal{B}(H)$).

For a nonlinear version of this question, we state the following one.

QUESTION 6.5. Describe all bijective maps Φ on $\mathcal{B}(H)$ satisfying
$$\Lambda(\Phi(A) \star \Phi(B)) = \Phi(\Lambda(A \star B)) \text{ for all } A, B \in \mathcal{B}(H), \tag{6.3}$$
where $A \star B$ is one of the operations defined above.

In [5], the current authors characterized bijective maps Φ on $\mathcal{B}(H)$ (not assumed to be linear or continuous) satisfying
$$\Delta_\lambda(\Phi(T_1) \star \cdots \star \Phi(T_n)) = \Phi(\Delta_\lambda(T_1 \star \cdots \star T_n)) \text{ for all } T_1, \ldots, T_n \in \mathcal{B}(H), \tag{6.4}$$
where $T_1 \star \cdots \star T_n$ stands for different products of operators that include the usual product and the general Jordan product.

QUESTION 6.6. Describe all bijective maps Φ on $\mathcal{B}(H)$ satisfying
$$\Lambda\left(\Phi(T_1) \star \cdots \star \Phi(T_n)\right) = \Phi\left(\Lambda(T_1 \star \cdots \star T_n)\right) \tag{6.5}$$
for all $T_1, \ldots, T_n \in \mathcal{B}(H)$.

Finally, we close this section with a remark that if $U : H \to H$ is a unitary or anti-unitary operator, then the map
$$T \mapsto \Phi(T) := UTU^*, T \in \mathcal{B}(H)$$
commutes with any of the above transforms.

References

[1] A. Aluthge, *On p-hyponormal operators for $0 < p < 1$*, Integral Equations Operator Theory **13** (1990), no. 3, 307–315, DOI 10.1007/BF01199886. MR1047771

[2] J. Antezana, P. Massey, and D. Stojanoff, *λ-Aluthge transforms and Schatten ideals*, Linear Algebra Appl. **405** (2005), 177–199, DOI 10.1016/j.laa.2005.03.016. MR2148169

[3] F. Botelho, L. Molnár, and G. Nagy, *Linear bijections on von Neumann factors commuting with λ-Aluthge transform*, Bull. Lond. Math. Soc. **48** (2016), no. 1, 74–84, DOI 10.1112/blms/bdv092. MR3455750

[4] F. Chabbabi, *Product commuting maps with the λ-Aluthge transform*, J. Math. Anal. Appl. **449** (2017), no. 1, 589–600, DOI 10.1016/j.jmaa.2016.12.027. MR3595221

[5] F. Chabbabi and M. Mbekhta, *General product nonlinear maps commuting with the λ-Aluthge transform*, Mediterr. J. Math. **14** (2017), no. 2, Art. 42, 10, DOI 10.1007/s00009-017-0860-7. MR3619403

[6] F. Chabbabi and M. Mbekhta, *Jordan product maps commuting with the λ-Aluthge transform*, J. Math. Anal. Appl. **450** (2017), no. 1, 293–313, DOI 10.1016/j.jmaa.2017.01.036. MR3606169

[7] F. Chabbabi and M. Mbekhta, *Nonlinear maps commuting with the λ-Aluthge transform under (n,m)–Jordan-triple product*, Linear Multilinear Algebra **67** (2019), no. 12, 2382–2398, DOI 10.1080/03081087.2018.1492514. MR4017720

[8] F. Chabbabi and M. Mbekhta, *Maps commuting with the λ-Aluthtge transform under ω-addition*, Operator theory: themes and variations, Theta Ser. Adv. Math., vol. 20, Theta, Bucharest, 2018, pp. 119–129. MR3839550

[9] F. Chabbabi, R. E. Curto, and M. Mbekhta, *The mean transform and the mean limit of an operator*, Proc. Amer. Math. Soc. **147** (2019), no. 3, 1119–1133, DOI 10.1090/proc/14277. MR3896061

[10] R. G. Douglas, *On majorization, factorization, and range inclusion of operators on Hilbert space*, Proc. Amer. Math. Soc. **17** (1966), 413–415, DOI 10.2307/2035178. MR203464

[11] C. Foiaş, I. B. Jung, E. Ko, and C. Pearcy, *Complete contractivity of maps associated with the Aluthge and Duggal transforms*, Pacific J. Math. **209** (2003), no. 2, 249–259, DOI 10.2140/pjm.2003.209.249. MR1978370

[12] T. Furuta, *Invitation to linear operators*, Taylor & Francis, Ltd., London, 2001. From matrices to bounded linear operators on a Hilbert space. MR1978629

[13] I. C. Gohberg and M. G. Kreĭn, *Introduction to the theory of linear nonselfadjoint operators*, Translated from the Russian by A. Feinstein. Translations of Mathematical Monographs, Vol. 18, American Mathematical Society, Providence, R.I., 1969. MR0246142

[14] R. Harte and M. Mbekhta, *Generalized inverses in C^*-algebras. II*, Studia Math. **106** (1993), no. 2, 129–138, DOI 10.4064/sm-106-2-129-138. MR1240309

[15] I. B. Jung, E. Ko, and C. Pearcy, *Aluthge transforms of operators*, Integral Equations Operator Theory **37** (2000), no. 4, 437–448, DOI 10.1007/BF01192831. MR1780122

[16] I. B. Jung, E. Ko, and C. Pearcy, *Spectral pictures of Aluthge transforms of operators*, Integral Equations Operator Theory **40** (2001), no. 1, 52–60, DOI 10.1007/BF01202954. MR1829514

[17] I. B. Jung, E. Ko, and C. Pearcy, *The iterated Aluthge transform of an operator*, Integral Equations Operator Theory **45** (2003), no. 4, 375–387, DOI 10.1007/s000200300012. MR1971744

[18] S. Jung, E. Ko, and S. Park, *Subscalarity of operator transforms*, Math. Nachr. **288** (2015), no. 17-18, 2042–2056, DOI 10.1002/mana.201500037. MR3434298

[19] T. Kato, *Perturbation theory for linear operators*, Classics in Mathematics, Springer-Verlag, Berlin, 1995. Reprint of the 1980 edition. MR1335452

[20] S. H. Lee, W. Y. Lee, and J. Yoon, *The mean transform of bounded linear operators*, J. Math. Anal. Appl. **410** (2014), no. 1, 70–81, DOI 10.1016/j.jmaa.2013.08.003. MR3109820

[21] M. Mbekhta, *Conorme et inverse généralisé dans les C^*-algèbres* (French, with French summary), Canad. Math. Bull. **35** (1992), no. 4, 515–522, DOI 10.4153/CMB-1992-068-8. MR1191512

[22] K. Okubo, *On weakly unitarily invariant norm and the Aluthge transformation*, Linear Algebra Appl. **371** (2003), 369–375, DOI 10.1016/S0024-3795(03)00485-3. MR1997382

[23] B. Simon, *Trace ideals and their applications*, London Mathematical Society Lecture Note Series, vol. 35, Cambridge University Press, Cambridge-New York, 1979. MR541149

Université de Lille, Département de Mathématiques, Laboratoire CNRS-UMR 8524 P. Painlevé, 59655 Villeneuve d'Ascq Cedex, France
Email address: fadil.chabbabi@univ-lille.fr

Université de Lille, Département de Mathématiques, Laboratoire CNRS-UMR 8524 P. Painlevé, 59655 Villeneuve d'Ascq Cedex, France
Email address: mostafa.mbekhta@univ-lille.fr

Recent progress on local spectrum-preserving maps

Abdellatif Bourhim and Javad Mashreghi

ABSTRACT. Let X and Y be two infinite-dimensional complex Banach spaces, and $\mathcal{B}(X)$ (resp. $\mathcal{B}(Y)$) be the algebra of all bounded linear operators on X (resp. on Y). Fix two nonzero vectors $x_0 \in X$ and $y_0 \in Y$, and let $\mathcal{B}^2_{x_0}(X)$ (resp. $\mathcal{B}^2_{y_0}(Y)$) be the collection of all operators in $\mathcal{B}(X)$ (resp. in $\mathcal{B}(Y)$) vanishing at x_0 (resp. at y_0) or having zero square. On one hand, we review and discuss recent development of maps from $\mathcal{B}(X)$ onto $\mathcal{B}(Y)$ preserving the local spectrum of different products of operators and matrices. On the other hand, we establish new related results and show that if two maps φ_1 and φ_2 from $\mathcal{B}(X)$ onto $\mathcal{B}(Y)$ satisfy

$$\sigma_{\varphi_2(T)\varphi_1(S)\varphi_2(T)}(y_0) = \sigma_{TST}(x_0), \qquad (T,\ S \in \mathcal{B}(X)),$$

then φ_2 maps $\mathcal{B}_{x_0}(X)$ onto $\mathcal{B}_{y_0}(Y)$ and there exist two bijective linear mappings $A: X \to Y$ and $B: Y \to X$, and a function $\alpha: \mathcal{B}(X) \setminus \mathcal{B}_{x_0}(X) \to \mathbb{C} \setminus \{0\}$) such that $\varphi_1(T) = ATB$ for all $T \in \mathcal{B}(X)$, and $\varphi_2(T)^2 = B^{-1}T^2 A^{-1}$ and $\varphi_2(T)y_0 = \alpha(T)B^{-1}Tx_0$ for all $T \notin \mathcal{B}^2_{x_0}(X)$. When $X = Y = \mathbb{C}^n$, we show that the surjectivity condition on φ_1 and φ_2 is redundant. Furthermore, some known results are obtained as immediate consequences of our main results.

1. Introduction

In our article survey [34], we mainly discussed *linear preserver problems* that ask for the characterization of linear maps between algebras that leave invariant certain properties or subsets or relations. We mainly focussed on the famous Kaplansky's conjecture and its partial results and presented a survey on linear preserver problems of spectra and local spectra. This conjecture was motivated by two classical results. The first result is the well-known theorem of Gleason-Kahane-Żelazko in the theory of Banach algebra [86, 103] that states that every unital invertibility preserving linear map from a Banach algebra to a semisimple commutative Banach algebra is multiplicative. While, the second one is due to Marcus and Moyls [106] and states that every linear map on the algebra $\mathcal{M}_n(\mathbb{C})$ of all complex $n \times n$-matrices preserving eigenvalues is either an automorphism or an anti-automorphism. Such a conjecture asserts that every surjective unital invertibility preserving linear map between two semisimple Banach algebras is a Jordan homomorphism. It has not yet fully solved and remains open even for general C^*-algebras, but it has been confirmed for von Neumann algebras [13] and for the algebra of all bounded linear

---

2010 *Mathematics Subject Classification.* Primary 47B49; Secondary 47A10, 47A11.

Key words and phrases. Nonlinear preservers, Local spectrum, the single-valued extension property, finite rank operators.

©2020 American Mathematical Society

operators on a Banach spaces [51, 99, 141]. Some details and explanations about this conjecture can be found in [34, 48].

In recent years, there has been considerable interest in studying the so-called *nonlinear preserver problems*. These problems involve maps between algebras that leave invariant certain properties or subsets or relations without assuming any algebraic condition like linearity or additivity or multiplicativity but a weak algebraic condition is often imposed through the preserving property. The first result of this kind is due to Kowalski and Słodkowski [105] and dates back to 1980. It states that a complex-valued function f on a Banach algebra $\mathcal{A}$ is multiplicative and linear provided that $f(0) = 0$ and $f(x) - f(y)$ lies in the spectrum of $x - y$ for all $x, y \in \mathcal{A}$, and shows that the theorem of Gleason-Kahane-Żelazko holds without the linearity hypothesis of f. Since then, a number of techniques have been developed to treat nonlinear preserver problems and many results have been obtained; see for instance [16, 27, 70, 94, 109, 112, 124, 137] and the references therein. In [27], Bhatia, Šemrl and Sourour described the form of all surjective maps defined on the algebra $\mathcal{M}_n(\mathbb{C})$ of all complex $n \times n$-matrices and preserving the spectral radius of the difference of matrices. In [112], Molnár studied maps preserving the spectrum of operator or matrix products, and his result has been extended in several direction for uniform algebras and semisimple commutative Banach algebras; see for instance [2, 32, 34, 53, 85, 90, 91, 95–97, 101, 146, 147].

Recently, there has been an upsurge of interest in linear and nonlinear *local spectra preserver problems*, which demand the characterization of maps on matrices or Banach space operators that leave a local spectral set or quantity or property invariant. Bourhim and Ransford were the first ones to consider this type of preserver problem, characterizing in [41] additive maps on, the algebra $\mathcal{B}(X)$ of all linear bounded operators on a complex Banach space X, which preserve the local spectrum of operators at each vector of X. Their results opened the way for several authors to describe maps on matrices or operators that preserve the local spectrum, local spectral radius, and local inner spectral radius; see for instance [34, 59, 60] and the references therein. In recent development of this active research area it has been shown particularly by Costara [58] that surjective linear maps on $\mathcal{B}(X)$ preserving the local spectrum at a nonzero fixed vector are automatically continuous and thus a complete description of such maps is obtained.

In this paper, we continue our discussion on preserver problems of spectra and local spectra that was initiated in [34], to which we refer the reader for motivation and discussions about the famous Kaplansky's conjecture, and related problems and results. Mainly, we intend to review recent progress of nonlinear local spectra preserver problems investigated in [1, 4–7, 19–22, 29–31, 33, 57, 58]. Section 2 provides some definitions and basic properties of the local spectra with the objective of providing a convenient background about local spectral theory for easy references. In Section 3, we first review some basic properties of Jordan homomorphisms of algebras and then discuss Molnár's results and their variants on multiplicatively spectrum-preserving maps on the full algebra of all bounded linear operators on a complex Hilbert space $\mathcal{H}$. Furthermore, recent results on multiplicatively spectrum-preserving between semisimple Banach algebras are also discussed. Section 4 is devoted for Costara's recent result about the automatic continuity of surjective linear maps on $\mathcal{B}(X)$ compressing the local spectrum at a nonzero fixed vector. The main results of Section 5 are quoted from [30] and describe the form of all maps φ_1 and φ_1 on $\mathcal{B}(X)$ such that, for every T and S in $\mathcal{B}(X)$, the local spectra of ST and $\varphi_1(S)\varphi_2(T)$ are the same at a nonzero fixed vector x_0 of X. In Section 6, we obtain the form of all surjective maps φ_1 and φ_2 on $\mathcal{B}(X)$ such that, for every T and S in $\mathcal{B}(X)$, the local spectra of STS and $\varphi_2(S)\varphi_1(T)\varphi_2(S)$ are the same at a nonzero fixed vector x_0 of X. The obtained results are new and extend the main results of several papers including [7] and [35] where maps preserving the local spectrum of triple product and skew-triple product of operators are described. In Section 7 and Section 8, we discuss surjective maps preserving the local spectrum of Jordan and Lie products of operators and of matrices at a nonzero vector. The last section is devoted for final remarks and outlines some open problems for further research.

2. Background from local spectral theory

Throughout this paper by $\mathcal{M}_n(\mathbb{C})$ we mean as usual the algebra of all $n \times n$-complex matrices and by $\mathcal{B}(X,Y)$ the space of all bounded linear maps from a complex Banach space X into another one Y. When $X = Y$, we simply write $\mathcal{B}(X)$ instead of $\mathcal{B}(X,X)$ and denote its identity operator by $\mathbf{1}$. For an operator $T \in \mathcal{B}(X)$ and $\lambda \in \mathbb{C}$, we write $T - \lambda$ instead of $T - \lambda\mathbf{1}$. For any matrix $T \in \mathcal{M}_n(\mathbb{C})$, we denote by T^{tr} its transpose. For x in X and f in the dual space X^* of X, let $x \otimes f$ stand for the operator of rank at most one defined by

$$(2.1) \qquad (x \otimes f)y := f(y)x, \ (y \in X).$$

When x and y are two elements in a complex Hilbert space $\mathcal{H}$, then $x \otimes y$ is defined by $(x \otimes y)z := \langle z, y \rangle x$, $(z \in \mathcal{H})$. Note that every rank one operator in $\mathcal{B}(X)$ can be written in the form (2.1), and that every finite rank operator $T \in \mathcal{B}(X)$ can be written as a finite sum of rank one operators; i.e., $T = \sum_{k=1}^{n} x_k \otimes f_k$ for some $x_k \in X$ and $f_k \in X^*$, $k = 1, 2, \ldots, n$. Denote by $\mathcal{F}_n(X)$ the set of all operators in $\mathcal{B}(X)$ of rank at most n, and by $\mathcal{F}(X)$ the set of all finite rank operators in $\mathcal{B}(X)$. Lastly, denote by $\mathcal{N}_1(X)$ the set of all rank one nilpotent operators on X, and observe that $x \otimes f \in \mathcal{N}_1(X)$ if and only if $f(x) = 0$.

The spectrum of an operator $T \in \mathcal{B}(X)$ is

$$\sigma(T) := \{\lambda \in \mathbb{C} : T - \lambda \text{ is not invertible}\}.$$

It is a nonempty compact subset of $\mathbb{C}$, and its complement in $\mathbb{C}$, denoted by $\rho(T)$, is the usual resolvent set of T. The spectral radius of T is

$$(2.2) \qquad r(T) := \max\{|\lambda| : \lambda \in \sigma(T)\},$$

and satisfies $r(T) = \lim_{n \to \infty} \|T^n\|^{\frac{1}{n}}$, in the sense that the indicated limit always exists and has the indicated value.

For an operator $T \in \mathcal{B}(X)$, the resolvent function $\lambda \mapsto (T - \lambda)^{-1}$ is analytic on $\rho(T)$. If $x \in X$, then the function

$$R_{T,x}(\lambda) := (T - \lambda)^{-1} x$$

is analytic on $\rho(T)$, and obviously satisfies

$$(2.3) \qquad (T - \lambda) R_{T,x}(\lambda) = x$$

for all $\lambda \in \rho(T)$. A X-valued analytic function F is called an analytic extension of $R_{T,x}$ if its domain $D(F)$ contains $\rho(T)$ and $(T - \lambda)F(\lambda) = x$ for for all $\lambda \in D(F)$.

The function $R_{T,x}$ may have analytic extensions to an open set containing $\rho(T)$. An obvious analytic function of $R_{T,x}$ is given by the power series
$$-\sum_{k\geq 0}\frac{T^k x}{\lambda^{k+1}}.$$
The radius of convergence of this series is
(2.4) $$\mathrm{r}_T(x):=\limsup_{k\to\infty}\|T^k x\|^{\frac{1}{k}},$$
and is nothing but the so-called local spectral radius of T at x.

For an operator $T\in\mathcal{B}(X)$ and a vector $x\in X$, two analytic extensions of $R_{T,x}$ may or may not agree on their common domain. If for any vector $x\in X$, any two analytic extensions of $R_{T,x}$ must agree on their common domain, then T is said to have the single valued extension property (henceforth abbreviated to SVEP). Equivalently, since the set of zeros of a nonconstant analytic function is totally disconnected, T has SVEP if and only if for every open subset U of $\mathbb{C}$, the equation $(T-\lambda)F(\lambda)=0$, $(\lambda\in U)$, has no nontrivial analytic solution F. Every operator $T\in\mathcal{B}(X)$ for which the interior of its point spectrum, $\sigma_p(T)$, is empty enjoys this property. The examples of operators without SVEP can be found among those which are surjective but not invertible; see [**121**, Proposition 1.3.2-(f)].

The local resolvent set of an operator $T\in\mathcal{B}(X)$ at a point $x\in X$ is
$$\rho_T(x):=\bigcup\{D(F):F\text{ is an extension of }R_{T,x}\},$$
and is clearly an open subset containing $\rho(T)$. Obviously, it coincides with the union of all open subsets U of $\mathbb{C}$ for which there is an analytic function $F:U\to X$ such that $(T-\lambda)F(\lambda)=x$, $(\lambda\in U)$. The local spectrum of T at x is defined by
$$\sigma_T(x):=\mathbb{C}\backslash\rho_T(x),$$
and is a closed subset of $\sigma(T)$. Unlike the spectrum, this set might be empty even when x is a nonzero vector. If, however, T has SVEP then $\sigma_T(x)\neq\emptyset$ for all nonzero vectors x in X. In this case, for any vector $x\in X$, the function $R_{T,x}$ has a maximal analytic extension $\widetilde{x}_T(.)$ on $\rho_T(x)$, which is called the local resolvent function of T at x. Unlike the usual resolvent function, there are bounded linear operators on Banach spaces which admit nontrivial bounded local resolvent functions; see for instance [**26, 45, 88, 118–120**]. It is worth mentioning that, as demonstrated by weighted shift operators, sometimes the description of the local spectra of an operator is difficult; see [**40, 42, 108**]. However, the local spectra of matrices is well understood and can be found for instance in [**39, 87, 152**]. Further information about the local spectrum can be found in the remarkable books of Aiena [**8**] and of Laursen and Neumann [**121**].

In what follows, we state some elementary results about the local spectrum for easy references. The first lemma summarizes some known basic properties of the local spectrum.

Lemma 2.1. *For an operator $T\in\mathcal{B}(X)$, two vectors $x,\,y\in X$ and a nonzero scalar $\alpha\in\mathbb{C}$, the following statements hold.*

(a) $\sigma_T(\alpha x)=\sigma_T(x)$ and $\sigma_{\alpha T}(x)=\alpha\sigma_T(x)$.
(b) $\sigma_T(x+y)\subset\sigma_T(x)\cup\sigma_T(y)$. *The equality holds if $\sigma_T(x)\cap\sigma_T(y)=\emptyset$.*
(c) *If T has SVEP, $x\neq 0$ and $Tx=\lambda x$ for some $\lambda\in\mathbb{C}$, then $\sigma_T(x)=\{\lambda\}$.*
(d) *If T has SVEP and $Tx=\alpha y$, then $\sigma_T(y)\subset\sigma_T(x)\subset\sigma_T(y)\cup\{0\}$.*

(e) If $R \in \mathcal{B}(X)$ commutes with T, then $\sigma_T(Rx) \subset \sigma_T(x)$.

PROOF. See for instance [8] or [121]. □

Let $x_0 \in X$ and $y_0 \in Y$ be nonzero vectors. It is shown in [35, Lemma 2.3 and Lemma 2.4] that if $A : X \to Y$ and $B : Y \to X$ are bijective linear transformations, then $\sigma_T(x_0) = \sigma_{ATB}(y_0)$ for all rank one operators $T \in \mathcal{B}(X)$ if and only if A is continuous, $A^{-1} = B$ and $Ax_0 = \alpha x_0$ for some nonzero scalar $\alpha \in \mathbb{C}$. If, however, $A : X^* \to Y$ and $B : Y \to X^*$ are bijective linear transformations, then there is a rank one operator $T_0 \in \mathcal{B}(X)$ such that $\sigma_{T_0}(x_0) \neq \sigma_{AT_0^*B}(y_0)$. The second lemma shows a little more than what have been established in [33, Lemma 3.8 and Lemma 3.9].

Lemma 2.2 (Bourhim–Mabrouk [33]). *Let $x_0 \in X$ and $y_0 \in Y$ be nonzero vectors, and let Ω be an open neighborhood of the identity operator in $\mathcal{B}(X)$. Then the following assertions hold.*
  (a) *If $A : X \to Y$ is bijective bounded linear transformation, then $\sigma_T(x_0) = \sigma_{ATA^{-1}}(y_0)$ for all $T \in \Omega$ if and only if $Ax_0 = \alpha x_0$ for some nonzero scalar $\alpha \in \mathbb{C}$.*
  (b) *If $A : X^* \to Y$ and $B : Y \to X^*$ is a bijective bounded linear transformation, then there is $T_0 \in \Omega$ such that $\sigma_{T_0}(x_0) \neq \sigma_{AT_0^*A^{-1}}(y_0)$.*

It is well known that the spectrum function $T \mapsto \sigma(T)$ is always upper semi-continuous on $\mathcal{B}(X)$ but, in general, it is not lower semicontinuous. For a fixed nonzero vector $x_0 \in X$, the local spectrum function $T \mapsto \sigma_T(x_0)$ is, in general, neither upper nor lower semicontinuous; see [75, Example 1 and Example 2]. However, Newburgh's theorem [14, Corollary 3.4.5] tells us that the spectrum function is continuous on $\mathcal{M}_n(\mathbb{C})$ and Dollinger and Oberai in [75, Example 1 and Example 2] showed that if $x_0 \in \mathbb{C}^n$ is a nonzero fixed point then the local spectrum function $T \mapsto \sigma_T(x_0)$ is lower semi-continuous on $\mathcal{M}_n(\mathbb{C})$. Most of nonlinear local spectra preserver problems on $\mathcal{M}_n(\mathbb{C})$ use a density argument together with the continuity of the spectrum function and lower semicontinuity of the local spectrum function on $\mathcal{M}_n(\mathbb{C})$ to transform them into nonlinear global spectra preserver problems.

Lemma 2.3 (Dollinger–Oberai [75]). *If $x_0 \in \mathbb{C}^n$ is a nonzero vector, the local spectrum function $T \mapsto \sigma_T(x_0)$ is lower semi-continuous on $\mathcal{M}_n(\mathbb{C})$. That is, if $(T_k)_{k \geq 1} \subset \mathcal{M}_n(\mathbb{C})$ is a converging sequence to a matrix $T \in \mathcal{M}_n(\mathbb{C})$, then $\sigma_T(x_0) \subset \liminf_{k \to \infty} \sigma_{T_k}(x_0)$.*

PROOF. See [75, Corollary 2.3]. □

For any elements a and b in a ring R with identity $\mathbf{1}$, Jacobson's lemma states that if $\mathbf{1} - ab$ is invertible then so is $\mathbf{1} - ba$. Indeed, it suffices to check that

$$(\mathbf{1} - ba)\left(\mathbf{1} + b(\mathbf{1} - ab)^{-1}a\right) = \left(\mathbf{1} + b(\mathbf{1} - ab)^{-1}a\right)(\mathbf{1} - ba) = \mathbf{1},$$

and thus

$$(\mathbf{1} - ba)^{-1} = \mathbf{1} + b(\mathbf{1} - ab)^{-1}a.$$

Halmos observed in [89] that this can be *formally* checked by writing

$$\begin{aligned}(\mathbf{1} - ba)^{-1} &= \mathbf{1} + ba + baba + bababa + \ldots \\ &= \mathbf{1} + b(\mathbf{1} + ab + abab + \ldots)a \\ &= \mathbf{1} + b(\mathbf{1} - ab)^{-1}a.\end{aligned}$$

If $\sigma(a) := \{\lambda \in \mathbb{C} : \lambda\mathbf{1} - a \text{ is not invertible}\}$ denotes the spectrum of any element a in a unital complex algebra $\mathcal{A}$, then it is immediate from Jacobson's lemma that

$$\sigma(ab)\backslash\{0\} = \sigma(ba)\backslash\{0\}$$

for all $a, b \in \mathcal{A}$. This identity extends to a large number of different parts of the spectrum, such as the left/right spectrum, the essential spectrum, etc. Over the years, several authors established Jacobson's lemma for various kinds of invertibility and studied the common spectral and local spectral properties of product of elements in Banach algebras and of product of Banach space operators RS and SR; see for instance [17, 25, 52, 65, 116, 126, 143, 144, 148–150].

The following result gives the relation between local spectra of product of Banach space operators RS and SR and is quoted from [25].

Lemma 2.4 (Benhida–Zerouali [25]). *If $S, T \in \mathcal{B}(X)$, then $\sigma_{ST}(Sx) \cup \{0\} = \sigma_{TS}(x) \cup \{0\}$ for all $x \in X$.*

PROOF. See [25, Proposition 3.1] or [149, Theorem 2.3]. □

To conclude this section, the so-called nonzero local spectrum is introduced and related lemmas are appended. Recall that the nonzero local spectrum of $T \in \mathcal{B}(X)$ at any $x \in X$ is defined by

$$\sigma_T^*(x) := \begin{cases} \{0\} & \text{if } \sigma_T(x) \subseteq \{0\}; \\ \sigma_T(x)\backslash\{0\} & \text{if } \sigma_T(x) \nsubseteq \{0\}. \end{cases}$$

The last lemma is an elementary observation that gives the nonzero local spectrum of any rank one operator, and its proof will be omitted.

Lemma 2.5. *Let x_0 be a nonzero vector in X. For any $x \in X$ and $f \in X^*$, we have*

$$\sigma_{x \otimes f}^*(x_0) = \begin{cases} \{0\} & \text{if } f(x_0) = 0, \\ \{f(x)\} & \text{otherwise.} \end{cases}$$

3. Multiplicatively spectrum preserving maps

Let $\mathcal{A}$ and $\mathcal{B}$ be unital complex Banach algebras, and denote their identities by $\mathbf{1}_\mathcal{A}$ and $\mathbf{1}_\mathcal{B}$ or simply by $\mathbf{1}$ when there is no confusion. For any $x \in \mathcal{A}$ and $\lambda \in \mathbb{C}$, write $x - \lambda$ instead of $x - \lambda\mathbf{1}$, and denote the spectrum of any element $x \in \mathcal{A}$ by $\sigma(x) := \{\lambda \in \mathbb{C} : x - \lambda \text{ is not invertible}\}$ and the spectral radius by $\mathrm{r}(x) = \max\{|\lambda| : \lambda \in \sigma(x)\}$. A map φ between $\mathcal{A}$ and $\mathcal{B}$ is said to be spectrum-preserving if $\sigma(\varphi(x)) = \sigma(x)$ for all $x \in \mathcal{A}$. Recall also that $\mathcal{A}$ is said to be semisimple if its Jacobson radical is trivial. This means that there are no nonzero elements $x_0 \in \mathcal{A}$ such that $\sigma(x_0 x) = \{0\}$ for all $x \in \mathcal{A}$.

In [9, 46], the question whether for given two elements $a, b \in \mathcal{A}$ the condition $\sigma(ax) = \sigma(bx)$ for all $x \in \mathcal{A}$ entails $a = b$ was studied, and an affirmative answer has been obtained in [9, 46] for various special cases and for some classes of algebras, including C^*–algebras. While in [43], Braatvedt and Brits showed that the answer is always affirmative if $\mathcal{A}$ is a semismple Banach algebra. Among other results and problems, a spectrum-preserving problem has been discussed in [46] but omitting the condition of linearity, and studying maps φ from $\mathcal{A}$ onto $\mathcal{B}$ satisfying

(3.5) $\qquad \sigma\bigl(\varphi(a)\varphi(b)\bigr) = \sigma(ab), \qquad (a,b \in \mathcal{A}).$

These maps are called multiplicatively spectrum-preserving, and have been first investigated by Molnár in [**112**]. He obtained some characterizations of multiplicatively spectrum-preserving maps between some known Banach algebras.

In this section, we first review some basic properties of Jordan homomorphisms of algebras and then discuss Molnár's results on multiplicatively spectrum-preserving maps on the full algebra of all bounded linear operators on a complex Hilbert space $\mathcal{H}$. Lastly, we present some recent results on multiplicatively spectrum-preserving between semisimple Banach algebras.

3.1. Jordan homomorphisms of algebras.
A linear map φ between two complex algebras $\mathcal{A}$ and $\mathcal{B}$ is said to be *homomorphism* if $\varphi(ab) = \varphi(a)\varphi(b)$ for all $a, b \in \mathcal{A}$, and is called *anti-homomorphism* if $\varphi(ab) = \varphi(b)\varphi(a)$ for all $a, b \in \mathcal{A}$. It is called *Jordan homomorphism* if it preserves squares, i.e., $\varphi(a^2) = \varphi(a)^2$ for all $a \in \mathcal{A}$. Equivalently, the map φ is a Jordan homomorphism if and only if it preserves the Jordan structure. More explicitly, this means that $\varphi(ab+ba) = \varphi(a)\varphi(b)+\varphi(b)\varphi(a)$ for all $a, b \in \mathcal{A}$. A bijective homomorphism (*resp. anti-homomorphism*) is called an *isomorphism* (*resp. anti-isomorphism*) and a bijective Jordan homomorphism is called a *Jordan isomorphism*. If $\mathcal{A}$ is a C^*-algebra, then a linear map φ from $\mathcal{A}$ into $\mathcal{B}$ is a Jordan homomorphism if and only if $\varphi(h^2) = \varphi(h)^2$ for all self-adjoint elements $h \in \mathcal{A}$.

Finally, recall that an isomorphism φ between two C^*-algebras $\mathcal{A}$ and $\mathcal{B}$ is called a *∗-isomorphism* if it is an isomorphism and $\varphi(x^*) = \varphi(x)^*$ for all $x \in \mathcal{A}$. In a similar way, *∗-anti-isomorphisms* and *Jordan ∗-isomorphisms* are defined. Kadison, in his celebrated paper [**102**], proved that a surjective linear map between two C^*-algebras $\mathcal{A}$ and $\mathcal{B}$ is an isometry if and only it is a Jordan ∗-isomorphism multiplied by a unitary element in $\mathcal{B}$. It is also worth mentioning that a surjective linear map φ on $\mathcal{B}(\mathcal{H})$ is an isometry if and only if there are two unitary operators $U, V \in \mathcal{B}(\mathcal{H})$ such that either $\varphi(T) = UTV$ for all $T \in \mathcal{B}(\mathcal{H})$, or $\varphi(T) = UT^tV$ for all $T \in \mathcal{B}(\mathcal{H})$. Here, T^t denotes the transpose of T relative to an arbitrary but fixed orthonormal basis of $\mathcal{H}$.

The following result summarizes some well known properties of Jordan homomorphisms. Their proofs can be found in [**92**].

Proposition 3.1. *If $\varphi : \mathcal{A} \to \mathcal{B}$ is a Jordan homomorphism between two unital complex algebras $\mathcal{A}$ and $\mathcal{B}$, then the following statements hold.*

(a) *For every $a, b \in \mathcal{A}$, we have*

(3.6) $$\varphi(aba) = \varphi(a)\varphi(b)\varphi(a).$$

(b) *For every positive integer n, we have*

(3.7) $$\varphi(a^n) = (\varphi(a))^n$$

for all $a \in \mathcal{A}$.

(c) *For every $a, b, c \in \mathcal{A}$, we have*

(3.8) $$\varphi(abc + cba) = \varphi(a)\varphi(b)\varphi(c) + \varphi(c)\varphi(b)\varphi(a).$$

(d) *If there exists $a_0 \in \mathcal{A}$ such that $\varphi(a_0) = \mathbf{1}$, then $\varphi(\mathbf{1}) = \mathbf{1}$ and $\varphi(a^{-1}) = \varphi(a)^{-1}$ for all invertible elements $a \in \mathcal{A}$.*

PROOF. See for instance [**92**]. □

Obviously, every homomorphism and every anti-homomorphism is a Jordan homomorphism. Conversely, in [98], Jacobson and Rickart proved that a Jordan homomorphism from an arbitrary ring into a domain is either a homomorphism or an antihomomorphism. The same conclusion holds for Jordan homomorphisms onto prime rings, as shown by Herstein [92] and Smiley [142]. It is also worth mentioning that the problem of whether Jordan homomorphisms can be expressed through homomorphisms and antihomomorphisms was considered by several authors; see for instance [18, 47] and the reference therein.

Theorem 3.2 (Herstein [92]). *If $\mathcal{B}$ is prime and φ is a Jordan homomorphism from $\mathcal{A}$ onto $\mathcal{B}$, then either φ is a homomorphism or an anti-homomorphism.*

PROOF. See [92, Theorem H]. □

In the above discussion, we assumed that Jordan homomorphisms are linear. It should be clear that the above results remain valid when the linearity is replaced by additivity. We also would like to mention that a number of authors studies maps φ satisfying either (3.6) or (3.7) or (3.8). It is known that if $\mathcal{R}$ is a prime ring and contains a nontrivial idempotent then every multiplicative bijective mapping of $\mathcal{R}$ onto an arbitrary ring is additive; see [107]. In [138], Šemrl proved that if $\epsilon > 0$, $\mathcal{A}$ and $\mathcal{B}$ are standard operator algebras on Banach space X and Y with $\dim X = \infty$ and $\varphi : \mathcal{A} \to \mathcal{B}$ is a bijective mapping satisfying

$$\|\varphi(TS) - \varphi(T)\varphi(S)\| \leq \epsilon$$

for all S, $T \in \mathcal{A}$ then there is either a bounded linear bijective operator or a bounded conjugate linear bijective operator $A : X \to Y$ such that $\varphi(T) = ATA^{-1}$ for all $T \in \mathcal{A}$. In [113], Molnár $\varphi : \mathcal{A} \to \mathcal{B}$ is bijective mappings $\varphi : \mathcal{A} \to \mathcal{B}$ satisfying (3.6); i.e.,

$$\varphi(TST) = \varphi(T)\varphi(S)\varphi(T)$$

for all S, $T \in \mathcal{A}$. He showed that such a map must be a ring isomorphism or a ring anti-isomorphism multiplied by 1 or -1. For further related results, we refer the interested reader to [10–12, 28, 77, 81, 84, 104, 110, 111, 122, 127–129, 131, 136, 145] and the references therein.

3.2. Molnár's results on spectrum preservers. A variant problem of Kaplansky's conjecture asks if any surjective spectrum-preserving linear map between semisimple Banach algebras is a Jordan isomorphism? Motivated by this problem, Molnár studied in [112], not necessarily linear, multiplicatively spectrum-preserving self-maps of certain algebras $\mathcal{A}$. In particular, he obtained the following result.

Theorem 3.3 (Molnár [112]). *If $\mathcal{H}$ is an infinite-dimensional complex Hilbert space, then a map φ from $\mathcal{B}(\mathcal{H})$ onto itself satisfies*

(3.9) $$\sigma(\varphi(S)\varphi(T)) = \sigma(ST)$$

for all S, $T \in \mathcal{B}(\mathcal{H})$ if and only if there is an invertible operator $A \in \mathcal{B}(\mathcal{H})$ such that $\varphi(T) = ATA^{-1}$ for all $T \in \mathcal{B}(\mathcal{H})$, or $\varphi(T) = -ATA^{-1}$ for all $T \in \mathcal{B}(\mathcal{H})$.

Some remarks concerning this theorem are in order. Note that no linearity of maps φ satisfying (3.9) is not assumed, just surjectivity suffices. Note that if φ is an algebra anti-automorphism of $\mathcal{B}(\mathcal{H})$, then $\sigma(ST) = \sigma(\varphi(ST)) = \sigma(\varphi(T)\varphi(S))$ for all S, $T \in \mathcal{B}(\mathcal{H})$. Therefore,

(3.10) $$\sigma(ST) \cup \{0\} = \sigma(\varphi(S)\varphi(T)) \cup \{0\}$$

for all $S, T \in \mathcal{B}(\mathcal{H})$. If, however, $\mathcal{H} = \mathbb{C}^n$ is finite dimensional, then a map φ on $\mathcal{M}_n(\mathbb{C})$ satisfies (3.9) if and only if there exists an invertible matrix $A \in \mathcal{M}_n(\mathbb{C})$ such that either $\varphi(T) = \pm ATA^{-1}$ for all $T \in \mathcal{M}_n(\mathbb{C})$, or $\varphi(T) = \pm AT^{tr}A^{-1}$ for all $T \in \mathcal{M}_n(\mathbb{C})$. This result has been first proved by Molnár in [112] under the surjectivity condition on φ, and then it was shown in [53] that such a condition is redundant.

Molnár also obtained in [112] a similar result when the usual product in (3.9) is replaced by the skew product ST^*. He showed that if $\mathcal{H}$ is an infinite-dimensional complex Hilbert space, then a map φ from $\mathcal{B}(\mathcal{H})$ onto itself satisfies

$$(3.11) \qquad \sigma\big(\varphi(S)\varphi(T)^*\big) = \sigma\big(ST^*\big)$$

for all $S, T \in \mathcal{B}(\mathcal{H})$ if and only if there are two unitary operators $U, V \in \mathcal{B}(\mathcal{H})$ such that $\varphi(T) = UTV$ for all $T \in \mathcal{B}(\mathcal{H})$. If, however, $\mathcal{H} = \mathbb{C}^n$ is finite dimensional, then a map φ on $\mathcal{M}_n(\mathbb{C})$ satisfies (3.11) if and only if there exist two unitary matrices $U, V \in \mathcal{M}_n(\mathbb{C})$ such that either $\varphi(T) = UTV$ for all $T \in \mathcal{M}_n(\mathbb{C})$, or $\varphi(T) = UT^{tr}V$ for all $T \in \mathcal{M}_n(\mathbb{C})$.

3.3. Molnár type results on Banach algebras.
Molnár's results [112] have been extended in different directions for some special algebras $\mathcal{A}$ on which a number of authors considered surjective, a priori nonlinear maps which preserve spectral sets or properties of the usual products or Jordan triple-products or certain so-called skew products of elements of $\mathcal{A}$; see for instance [53, 54, 70, 85, 90, 91, 94–96, 101, 117, 124, 132, 133, 146, 147]. In [53], Chan, Li and Sze characterized all maps preserving spectra of generalized products and generalized Jordan products of matrices. Recently, Abdelali in [2] gave a complete characterezations of all maps preserving spectra of polynomial products of matrices. His results extend and unify several results obtained earlier on maps preserving the spectrum of generalized product or generalized Jordan product of matrices.

In what follows, we shall only discuss certain recent results that are connected with Molnár's paper and give a complete description of maps between Banach algebras preserving the spectrum of product or triple product of elements. Before doing so, we review some concepts needed in the sequel. If $\mathcal{A}$ has a minimal left (or right) ideal, then its *socle*, denoted by $\text{soc}(\mathcal{A})$, is the sum of all minimal left (or right) ideals of $\mathcal{A}$. If $\mathcal{A}$ has no minimal one-sided ideals, the socle is then defined to be trivial; i.e., $\text{soc}(\mathcal{A}) = \{0\}$. Note that $\text{soc}(\mathcal{A})$ is an ideal of $\mathcal{A}$ consisting of all elements $a \in \mathcal{A}$ for which $\sigma(xa)$ is finite for all $x \in \mathcal{A}$, and that all its elements are algebraic. A nonzero element $u \in \mathcal{A}$ is said to have *rank one* if for every $x \in \mathcal{A}$, the spectrum $\sigma(ux)$ contains at most one nonzero scalar. The set $\mathcal{F}_1(\mathcal{A})$ of all rank one elements of $\mathcal{A}$ is contained in $\text{soc}(\mathcal{A})$ and in turn $\text{soc}(\mathcal{A})$ is equal to the set of all finite sums of rank one elements of $\mathcal{A}$. An ideal P of $\mathcal{A}$ is *essential* if 0 is the only element a of $\mathcal{A}$ for which $a.P = 0$. We refer the reader to [14] for more details. It should be noted that a semisimple Banach algebra is finite dimensional if and only if it coincides with its socle; see for instance [14, Theorem 5.4.2].

Theorem 3.4 (Bourhim–Mashreghi–Stepanyan [32]). *Assume that $\mathcal{A}$ is semisimple and $\mathcal{B}$ has an essential socle. If maps φ_1 and φ_2 from $\mathcal{A}$ onto $\mathcal{B}$ satisfy*

$$(3.12) \qquad \sigma(\varphi_1(a)\varphi_2(b)) = \sigma(ab), \quad (a, b \in \mathcal{A}),$$

then the maps $\varphi_1\varphi_2(\mathbf{1})$ and $\varphi_1(\mathbf{1})\varphi_2$ coincide and are Jordan isomorphisms.

If φ_1 and φ_2 are two maps from $\mathcal{A}$ onto $\mathcal{B}$ satisfying (3.12), then it was shown in [32] that $\mathcal{A}$ is semisimple if and only if so is $\mathcal{B}$, and that $\mathcal{A}$ has an essential socle if and only if $\mathcal{B}$ has an essential socle. So, Theorem 3.4 remains valid regardless which algebra is semisimple and which one has an essential socle. Note that in the statement of the above theorem the only restriction on the maps φ_1 and φ_2 is the surjectivity; no linearity or additivity or continuity is assumed. However, the conclusion of such theorem shows that these maps φ_1 and φ_2 are automatically linear.

As a consequence of Theorem 3.4, we obtain a complete description of maps on a semisimple Banach algebra $\mathcal{A}$ of large socle that preserve the spectrum of skew product of elements of $\mathcal{A}$.

Corollary 3.5. *Assume that $\mathcal{A}$ and $\mathcal{B}$ are C^* algebras such that one of them has an essential socle. If a map φ from $\mathcal{A}$ onto $\mathcal{B}$ satisfies*

$$(3.13) \qquad \sigma\left(\varphi(a)\varphi(b)^*\right) = \sigma\left(ab^*\right), \quad (a, b \in \mathcal{A}),$$

then φ is a Jordan $*$–isomorphism multiplied by a unitary element of $\mathcal{B}$.

PROOF. Since φ is surjective, there is $a_0 \in \mathcal{A}$ such that $\varphi(a_0) = 0$ and then $\{0\} = \sigma\left(\varphi(a_0)\varphi(a_0)^*\right) = \sigma(a_0 a_0^*)$. This implies that $0 = \mathrm{r}(a_0 a_0^*) = \|a_0 a_0^*\| = \|a_0\|^2$, and $a_0 = 0$. Clearly, $\varphi(0) = 0$.

By (3.13), we have $\sigma\left(\varphi(a)\varphi(b^*)^*\right) = \sigma(ab)$ for all $a, b \in \mathcal{A}$. Then Theorem 3.4 tells us that the map $\psi: a \mapsto \varphi(a)\varphi(\mathbf{1})^*$ is a Jordan isomorphism and $\psi(a) = \varphi(\mathbf{1})\varphi(a^*)^*$ for all $a \in \mathcal{A}$. In particular, one has $\varphi(\mathbf{1})\varphi(\mathbf{1})^* = \mathbf{1}$. Note that, since φ is surjective, there is $a_0 \in \mathcal{A}$ such that $\varphi(a_0^*) = \mathbf{1} - \varphi(\mathbf{1})^*\varphi(\mathbf{1})$. We have

$$\psi(a_0) = \varphi(\mathbf{1})\varphi(a_0^*)^* = \varphi(\mathbf{1})\left(\mathbf{1} - \varphi(\mathbf{1})^*\varphi(\mathbf{1})\right) = \varphi(\mathbf{1}) - \varphi(\mathbf{1})\varphi(\mathbf{1})^*\varphi(\mathbf{1}) = \varphi(\mathbf{1}) - \varphi(\mathbf{1}) = 0,$$

and then $a_0 = 0$. This implies that $0 = \varphi(0) = \mathbf{1} - \varphi(\mathbf{1})^*\varphi(\mathbf{1})$ and $\varphi(\mathbf{1})$ is a unitary element of $\mathcal{B}$.

Moreover, for every self-adjoint element $h \in \mathcal{A}$, we have

$$\psi(h) = \varphi(h)\varphi(\mathbf{1})^* = \varphi(\mathbf{1})\varphi(h)^* = \left(\varphi(h)\varphi(\mathbf{1})^*\right)^* = (\psi(h))^*.$$

Then given $a := h + ik \in \mathcal{A}$ where h and k are the real and imaginary parts of a, we have

$$\psi(a^*) = \psi(h - ik) = \psi(h) - i\psi(k) = (\psi(h) + i\psi(k))^* = \psi(a)^*.$$

This implies that ψ is a self-adjoint map and completes the proof. □

The second corollary is an immediate consequence of Theorem 3.4 when φ_1 and φ_2 are the same.

Corollary 3.6. *Assume that $\mathcal{A}$ is semisimple and $\mathcal{B}$ has an essential socle. If a map φ from $\mathcal{A}$ onto $\mathcal{B}$ satisfies (3.5), then $\varphi(\mathbf{1})$ is a central invertible element of $\mathcal{B}$ for which $\varphi(\mathbf{1})^2 = \mathbf{1}$ and $\varphi(\mathbf{1})\varphi$ is a Jordan isomorphism.*

The final result describes the form of all maps φ from $\mathcal{A}$ onto $\mathcal{B}$ satisfying

$$(3.14) \qquad \sigma\left(\varphi(a)\varphi(b)\varphi(a)\right) = \sigma(aba), \quad (a, b \in \mathcal{A})$$

when the socle of $\mathcal{A}$ is an essential ideal of $\mathcal{A}$.

Theorem 3.7 (Bourhim–Mashreghi–Stepanyan [32]). *Assume that $\mathcal{A}$ is semisimple and $\mathcal{B}$ has an essential socle. A surjective map $\varphi: \mathcal{A} \to \mathcal{B}$ satisfies (3.14) if and only if $\varphi(\mathbf{1})$ is a central invertible element of $\mathcal{B}$ for which $\varphi(\mathbf{1})^3 = \mathbf{1}$ and $\varphi(\mathbf{1})^2\varphi$ is a Jordan isomorphism.*

This theorem too remains valid regardless which algebra is semisimple and which one has an essential socle. The proof of the above theorems uses spectral and algebraic methods and shows that the hypotheses of Theorem 3.4 and Theorem 3.7 allow us to exploit the main result of [49] and deduce that the corresponding mappings are Jordan isomorphisms.

Theorem 3.8 (Brešar–Fošner–Šemrl [49]). *If $\mathcal{A}$ and $\mathcal{B}$ are semisimple Banach algebras such that $\mathrm{soc}(\mathcal{A})$ is essential, and $\varphi : \mathcal{A} \to \mathcal{B}$ is a bijective linear map preserving the invertibility, then φ is a Jordan isomorphism.*

4. Linear maps preserving the local spectrum

Throughout this section, let X and Y be two complex Banach spaces. The local spectral radius of an operator $T \in \mathcal{B}(X)$ at a vector $x \in X$ is somehow not entirely optimal because the maximum modulus of the local spectrum of T at x could justifiably also be called the local spectral radius of T at x. This is denoted by
$$\Gamma_T(x) := \max\{|\lambda| : \lambda \in \sigma_T(x)\}$$
with the convention that $\Gamma_T(x) = -\infty$ if $\sigma_T(x) = \emptyset$. Note that $\Gamma_T(x) \leq \mathrm{r}_T(x)$ for all $T \in \mathcal{B}(X)$ and all $x \in X$, and that this inequality can be strict. If, however, T has SVEP, then $\Gamma_T(x) = \mathrm{r}_T(x)$ for all $x \in X$; see for example [121]. Motivated by [41], González and Mbekhta characterized in [87] linear maps on $\mathcal{M}_n(\mathbb{C})$ preserving the local spectrum at a nonzero fixed vector $x_0 \in \mathbb{C}^n$. While in [39], Bourhim and Miller characterized linear maps on $\mathcal{M}_n(\mathbb{C})$ that preserve the local spectral radius at x_0. Both results were extended in [44] by Bračič and Müller to infinite dimensional Banach spaces, by imposing a continuity assumption on the preserving map. They showed that if $x_0 \in X$ is a nonzero vector then the set of all operators $T \in \mathcal{B}(X)$ with $\sigma(T) = \sigma_T(x_0)$ (respectively $\mathrm{r}(T) = \mathrm{r}_T(x_0)$) is residual, and then they used this result to characterize continuous surjective linear maps φ on $\mathcal{B}(X)$ satisfying either

(4.15) $$\sigma_{\varphi(T)}(x_0) = \sigma_T(x_0)$$

for all $T \in \mathcal{B}(X)$, or

(4.16) $$\mathrm{r}_{\varphi(T)}(x_0) = \mathrm{r}_T(x_0)$$

for all $T \in \mathcal{B}(X)$.

In [62, Theorem 1.2], Costara used subharmonicity arguments and showed that surjective linear maps φ on $\mathcal{B}(X)$ satisfying (4.16) are automatically continuous. But, since $\mathrm{r}_T(x_0)$ might be different than $\Gamma_T(x_0)$ for some operators $T \in \mathcal{B}(X)$ without SVEP, no one of the above equations entails the other one and thus the automatic continuity of surjective linear maps φ on $\mathcal{B}(X)$ satisfying (4.15) won't be deduced from [62, Theorem 1.2]. Recently, he showed in [58, Theorem 2.2] that if $y_0 \in Y$ is a fixed nonzero vector and φ is a linear surjective map from $\mathcal{B}(X)$ onto $\mathcal{B}(Y)$ such that

(4.17) $$\Gamma_{\varphi(T)}(y_0) \leq \mathrm{r}(T)$$

for all $T \in \mathcal{B}(X)$, then φ is automatically continuous. As an immediate consequence of this result, one concludes that $x_0 \in X$ and $y_0 \in Y$ are nonzero vectors and φ is a linear surjective map from $\mathcal{B}(X)$ onto $\mathcal{B}(Y)$ such that

(4.18) $$\sigma_{\varphi(T)}(y_0) \subset \sigma_T(x_0)$$

for all $T \in \mathcal{B}(X)$, then φ is continuous. We also would like to point out that the proof of [**58**, Theorem 2.2] is completely different from that of [**62**, Theorem 1.2], and doesn't use subharmonicity arguments. Therefore, it also gives a new and simple proof of [**62**, Theorem 1.2] since (4.16) implies (4.17).

Based on the above discussion and [**34**, Section 7.2], the following theorem summarizes the main results of the aforementioned papers.

Theorem 4.1 (Bračič–Müller [**44**] and Costara [**58**]). *Let $x_0 \in X$ and $y_0 \in Y$ be fixed nonzero vectors, and φ be a linear map φ from $\mathcal{B}(X)$ onto $\mathcal{B}(Y)$. Then*

$$(4.19) \qquad \Gamma_{\varphi(T)}(y_0) = \Gamma_T(x_0), \qquad (T \in \mathcal{B}(X))$$

if and only if there are a bijection $A \in \mathcal{B}(X,Y)$ and a scalar α of modulus one such that $Ax_0 = y_0$ and $\varphi(T) = \alpha A T A^{-1}$ for all $T \in \mathcal{B}(X)$.

5. Product and nonlinear local spectra preservers

Throughout this section, let X and Y be two infinite-dimensional complex Banach spaces. Motivated by Molnár results on multiplicatively spectrum-preserving, Bourhim and Mashreghi obtained in [**35**] a complete description of surjective maps φ (not assumed to be linear or continuous) from $\mathcal{B}(X)$ onto $\mathcal{B}(Y)$ preserving the local spectrum of the product of operators. They showed that if $x_0 \in X$ and $y_0 \in Y$ are nonzero vectors then a map φ from $\mathcal{B}(X)$ onto $\mathcal{B}(Y)$ satisfies

$$(5.20) \qquad \sigma_{\varphi(S)\varphi(T)}(y_0) = \sigma_{ST}(x_0), \qquad (S, T \in \mathcal{B}(X)),$$

if and only if there exists a bijective linear operator $A \in \mathcal{B}(X,Y)$ such that $Ax_0 = y_0$ and either $\varphi(T) = ATA^{-1}$ for all $T \in \mathcal{B}(X)$ or $\varphi(T) = -ATA^{-1}$ for all $T \in \mathcal{B}(X)$. This result opened the way for a number of authors to consider nonlinear maps on $\mathcal{B}(X)$ preserving the local spectrum of different products of operators at a fixed vector. Given two fixed nonzero vectors $h_0 \in \mathcal{H}$ and $k_0 \in \mathcal{K}$ in two infinite-dimensional complex Hilbert spaces $\mathcal{H}$ and $\mathcal{K}$, maps φ from $\mathcal{B}(\mathcal{H})$ onto $\mathcal{B}(\mathcal{K})$ satisfying

$$(5.21) \qquad \sigma_{TS^*}(h_0) = \sigma_{\varphi(T)\varphi(S)^*}(k_0), \qquad (T, S \in \mathcal{B}(X))$$

are described in [**7**]. It is shown that (5.21) is satisfied if and only if there exist two unitary operators $U, V \in \mathcal{B}(\mathcal{H}, \mathcal{K})$ and a nonzero scalar α such that $Uh_0 = \alpha k_0$ and $\varphi(T) = UTV^*$ for all $T \in \mathcal{B}(\mathcal{H})$. When $X = Y = \mathbb{C}^n$ is a finite-dimensional space, it was shown in [**22**] that the aforementioned results remain valid without the surjectivity condition on the map φ. A simplified proof of the main result of [**22**] was suggested in [**33**].

The main results of this section describe the form of all maps φ_1 and φ_2 on $\mathcal{B}(X)$ such that, for every T and S in $\mathcal{B}(X)$, the local spectra of ST and $\varphi_1(S)\varphi_2(T)$ are the same at a nonzero fixed vector x_0 of X. These results extends the main results of the aforementioned papers where maps preserving the local spectrum of product and skew-product of operators are described.

5.1. The infinite-dimensional case. The following theorem extends and unifies the main results of several papers, including [**7, 35**].

Theorem 5.1 (Bourhim–Lee [**30**]). *Let $x_0 \in X$ and $y_0 \in Y$ be nonzero vectors. If two surjective maps φ_1 and φ_2 from $\mathcal{B}(X)$ onto $\mathcal{B}(Y)$ satisfy*

$$(5.22) \qquad \sigma_{\varphi_1(S)\varphi_2(T)}(y_0) = \sigma_{ST}(x_0), \qquad (S, T \in \mathcal{B}(X)),$$

then φ_2 maps $\mathcal{B}_{x_0}(X)$ onto $\mathcal{B}_{y_0}(Y)$ and there exist two bijective linear mappings $A: X \to Y$ and $B: Y \to X$ such that $Ax_0 = y_0$ and

(5.23) $$\varphi_1(T) = ATB, \qquad (T \in \mathcal{B}(X)),$$

and

(5.24) $$\varphi_2(T) = B^{-1}TA^{-1}, \qquad (T \notin \mathcal{B}_{x_0}(X)).$$

Note that the only restriction on the maps φ_1 and φ_2 is the *surjectivity*; no linearity or additivity or continuity is assumed. In fact, since φ_2 arbitrarily maps $\mathcal{B}_{x_0}(X)$ onto $\mathcal{B}_{y_0}(Y)$, this map needs not be continuous or additive. We also would like to point out that, unlike when $\varphi_1 = \varphi_2$, the converse of Theorem 5.1 may not hold in general. Indeed, let T_1 and T_2 be two operators in $\mathcal{B}_{x_0}(X)$ such that T_1 has SVEP and $\sigma_{T_2}(x_0) = \emptyset$. Let φ_1 and φ_2 be two maps on $\mathcal{B}(X)$ such that $\varphi_1(T) = T$ for all $T \in \mathcal{B}(X)$ and

$$\varphi_2(T) = \begin{cases} T & \text{if } T \notin \{T_1, T_2\} \\ T_2 & \text{if } T = T_1 \\ T_1 & \text{if } T = T_2 \end{cases}$$

Clearly, we have $\sigma_{\varphi_1(S)\varphi_2(T)}(x_0) = \sigma_{ST}(x_0)$ for all $S \in \mathcal{B}(X)$ and $T \notin \{T_1, T_2\}$. However, for $S = \mathbf{1}$ and $T = T_1$, we have

$$\sigma_{\varphi_1(S)\varphi_2(T)}(x_0) = \sigma_{T_2}(x_0) = \emptyset \text{ and } \sigma_{ST}(x_0) = \sigma_{T_1}(x_0) = \{0\}.$$

Clearly, $\sigma_{\varphi_1(S)\varphi_2(T)}(x_0) \neq \sigma_{ST}(x_0)$ and (5.22) is not satisfied.

The proof of this theorem relies on several lemmas quoted from [35] to show that if (5.22) is satisfied then φ_1 is a bijective linear map preserving rank one operators in both directions. Therefore, with the help of [34, Theorem 4.12], it is shown that there are bijective linear mappings $P: X \to Y$ and $Q: X^* \to Y^*$ such that $\varphi_1(x \otimes f) = Px \otimes Qf$ for all $x \in X$ and $f \in X^*$. After showing that φ_2 has the form (5.24), this was an important step towards proving that φ_1 has the form (5.23). However, to show that φ_2 has the form (5.24) several new results were established in [30] and used in the proof of Theorem 5.1.

The following lemma shows that if (5.22) is satisfied then $\varphi_2(\mathcal{B}_{x_0}(X)) = \mathcal{B}_{y_0}(Y)$.

Lemma 5.2 (Bourhim–Lee [30]). *For a nonzero vector x_0 in X and an operator $A \in \mathcal{B}(X)$, the following statements are equivalent.*
  (a) $Ax_0 = 0$.
  (b) $\sigma^*_{TA}(x_0) \subseteq \{0\}$ for all operators $T \in \mathcal{B}(X)$.
  (c) $\sigma^*_{TA}(x_0) = \{0\}$ for all rank one operators $T \in \mathcal{B}(X)$.

In term of the local spectrum at a nonzero fixed vector, next lemma gives necessary and sufficient conditions for two operators to be the same. It is used to conclude that (5.24) is satisfied once it is shown that (5.23) holds for all rank one operators.

Lemma 5.3 (Bourhim–Lee [30]). *For a nonzero vector x_0 in X and two operators A and B in $\mathcal{B}(X)$, the following statements are equivalent.*
  (a) *Either $Ax_0 = Bx_0 = 0$ or $A = B$.*
  (b) $\sigma_{TA}(x_0) \cup \{0\} = \sigma_{TB}(x_0) \cup \{0\}$ *for all operators $T \in \mathcal{B}(X)$.*

(c) $\sigma^*_{TA}(x_0) = \sigma^*_{TB}(x_0)$ for all rank one operators $T \in \mathcal{B}(X)$.

The last lemma gives a spectral characterization of rank one operators in term of the local spectrum. It is needed to show that φ_2 maps all rank one operators in $\mathcal{B}(X)\backslash\mathcal{B}_{x_0}(X)$ onto all rank one operators in $\mathcal{B}(Y)\backslash\mathcal{B}_{y_0}(Y)$.

Lemma 5.4 (Bourhim–Lee [30]). *For a nonzero vector x_0 of X and an operator $R \in \mathcal{B}(X)$ for which $Rx_0 \neq 0$, the following statements are equivalent.*

  (a) *R has rank one.*
  (b) *$\sigma^*_{TR}(x_0)$ is a singleton for all $T \in \mathcal{B}(X)$.*
  (c) *$\sigma^*_{TR}(x_0)$ is a singleton for all $T \in \mathcal{F}_2(X)$.*

5.2. The finite-dimensional case. For any nonzero vector $x_0 \in \mathbb{C}^n$, let $\mathcal{M}_{n,x_0}(\mathbb{C})$ be the collection of all matrices in $\mathcal{M}_n(\mathbb{C})$ vanishing at x_0.

Theorem 5.5 (Bourhim–Lee [30]). *Let x_0 be a nonzero vector in $\mathbb{C}^n$. Two maps φ_1 and φ_2 on $\mathcal{M}_n(\mathbb{C})$ satisfy*

$$\sigma_{\varphi_1(S)\varphi_2(T)}(x_0) = \sigma_{ST}(x_0), \quad (S, T \in \mathcal{M}_n(\mathbb{C})), \tag{5.25}$$

if and only if φ_2 maps $\mathcal{M}_{n,x_0}(\mathbb{C})$ into itself and there are two invertible matrices A and B in $\mathcal{M}_n(\mathbb{C})$ such that $Ax_0 = x_0$ and

$$\varphi_1(T) = ATB, \quad (T \in \mathcal{M}_n(\mathbb{C})), \tag{5.26}$$

and

$$\varphi_2(T) = B^{-1}TA^{-1}, \quad (T \notin \mathcal{M}_{n,x_0}(\mathbb{C})). \tag{5.27}$$

The proof of this result uses a density argument. Using some arguments borrowed from [53], it was shown that if φ_1 and φ_2 are two maps on $\mathcal{M}_n(\mathbb{C})$ satisfying (5.25) then there are two linear maps L_1 and L_2 on $\mathcal{M}_n(\mathbb{C})$ such that each φ_i coincides with L_i on GL_n, the group of all invertible matrices in $\mathcal{M}_n(\mathbb{C})$. Roughly speaking, this and the density of the set $\{T \in \mathcal{M}_n(\mathbb{C}) : \sigma_T(x_0) = \sigma(T)\}$ together with the continuity of each L_i and the spectrum allow one to prove that

$$\sigma(L_1(S)L_2(T)) = \sigma(ST)$$

for all $S, T \in \mathcal{M}_n(\mathbb{C})$. Therefore [117, Theorem 1.1] entails that there are two invertible matrices A and B in $\mathcal{M}_n(\mathbb{C})$ such that either

$$L_1(T) = ATB \text{ and } L_2(T) = B^{-1}TA^{-1}, \ (T \in \mathcal{M}_n(\mathbb{C})), \tag{5.28}$$

or

$$L_1(T) = AT^{tr}B \text{ and } L_2(T) = B^{-1}T^{tr}A^{-1}, \ (T \in \mathcal{M}_n(\mathbb{C})). \tag{5.29}$$

Since

$$\sigma_T(x_0) = \sigma_{\varphi_1(T)\varphi_2(I_n)}(x_0) = \sigma_{L_1(T)L_2(I_n)}(x_0) \tag{5.30}$$

for all $T \in GL_n$, Lemma 2.2 implies that (5.29) cannot occur, and thus L_1 and L_2 are of the form (5.28). In this case, (5.30) becomes

$$\sigma_T(x_0) = \sigma_{ATA^{-1}}(x_0) \tag{5.31}$$

for all $T \in GL_n$, and Lemma 2.2 implies that there is a nonzero scalar α such that $Ax_0 = \alpha x_0$.

Since $\varphi_2(T) = L_2(T) = B^{-1}TA^{-1}$ for all $T \in GL_n$, for every $S \in \mathcal{M}_n(\mathbb{C})$ and $T \in GL_n$, we have

$$\begin{aligned}\sigma_{ASBT}(x_0) &= \sigma_{A(SBTA)A^{-1}}(x_0) \\ &= \sigma_{SBTA}(x_0) \\ &= \sigma_{\varphi_1(S)\varphi_2(BTA)}(x_0) \\ &= \sigma_{\varphi_1(S)T}(x_0).\end{aligned}$$

By [33, Lemma 5.2], we conclude that $\varphi_1(S) = ASB$ for all $S \in \mathcal{M}_n(\mathbb{C})$. In a similar way, one shows that

$$\sigma_{SB^{-1}TA^{-1}}(x_0) = \sigma_{S\varphi_2(T)}(x_0)$$

for all $S \in \mathcal{M}_n(\mathbb{C})$ and $T \notin \mathcal{M}_{n,x_0}(\mathbb{C})$ and then conclude that $\varphi_2(T) = B^{-1}TA^{-1}$ for all $T \notin \mathcal{M}_{n,x_0}(\mathbb{C})$. Moreover,

$$\{0\} = \sigma_{A^{-1}SB^{-1}T}(x_0) = \sigma_{\varphi_1(A^{-1}SB^{-1})\varphi_2(T)}(x_0) = \sigma_{S\varphi_2(T)}(x_0)$$

for all $S \in \mathcal{M}_n(\mathbb{C})$ and $T \in \mathcal{M}_{n,x_0}(\mathbb{C})$, and thus Lemma 5.2 tells us that φ_2 maps $\mathcal{M}_{n,x_0}(\mathbb{C})$ into itself.

6. Triple product and nonlinear local spectra preservers

Throughout this section, let X and Y be two infinite-dimensional complex Banach spaces and $x_0 \in X$ and $y_0 \in Y$ be nonzero vectors. It is shown in [36] that a map φ from $\mathcal{B}(X)$ onto $\mathcal{B}(Y)$ satisfies

(6.32) $\quad \sigma_{\varphi(T)\varphi(S)\varphi(T)}(y_0) = \sigma_{TST}(x_0), \quad (T, S \in \mathcal{B}(X)),$

if and only if there exists a bijective mapping $A \in \mathcal{B}(X,Y)$ such that $Ax_0 = y_0$ and $\varphi(T) = \lambda ATA^{-1}$ for all $T \in \mathcal{B}(X)$, where λ is a third root of unity, i.e., $\lambda^3 = 1$. Motivated by this result, Abdelali, Achchi, and Marzouki obtained a similar result by characterizing maps φ preserving the local spectrum at fixed nonzero vector of the skew-triple product TS^*T of Hilbert spaces operators. When $X = Y = \mathbb{C}^n$ is a finite-dimensional space, it was shown in [6, 20, 21] that the main results of [7, 36] remain valid without the surjectivity condition on the map φ. See also [33] for a simple proof.

In this section, we consider a more general local spectrum preserver problem by characterizing all pairs of maps φ_1 and φ_2 from $\mathcal{B}(X)$ onto $\mathcal{B}(Y)$ satisfying

(6.33) $\quad \sigma_{\varphi_2(S)\varphi_1(T)\varphi_2(S)}(y_0) = \sigma_{STS}(x_0), \quad (T, S \in \mathcal{B}(X)).$

The results obtained are new and generalize and unify the main results of several papers, including [6, 7, 20, 21, 36]. Their proofs are based upon a number of known and also new lemmas, and will be presented in entirely self-contained and readable forms.

6.1. The infinite-dimensional case. Let $\mathcal{B}_{x_0}^2(X)$ be the collection of all operators T in $\mathcal{B}(X)$ either vanishing at x_0 or having zero square; i.e.,

$$\mathcal{B}_{x_0}^2(X) := \{T \in \mathcal{B}(X) : \text{ either } T^2 = 0 \text{ or } Tx_0 = 0\}.$$

This set plays an important role in the statement and proof of the following result.

Theorem 6.1. *Two maps φ_1 and φ_2 from $\mathcal{B}(X)$ onto $\mathcal{B}(Y)$ satisfy*

(6.34) $\quad\sigma_{\varphi_2(S)\varphi_1(T)\varphi_2(S)}(y_0) \cup \{0\} = \sigma_{STS}(x_0) \cup \{0\}, \qquad (T, S \in \mathcal{B}(X)),$

if and only if φ_2 maps $\mathcal{B}^2_{x_0}(X)$ onto $\mathcal{B}^2_{y_0}(Y)$ and there exist two bijective linear mappings $A : X \to Y$ and $B : Y \to X$ such that

(6.35) $\qquad\qquad\qquad \varphi_1(T) = ATB, \qquad (T \in \mathcal{B}(X)),$

and for every $S \notin \mathcal{B}^2_{x_0}(X)$ we have

(6.36) $\qquad\qquad\qquad \varphi_2(S)^2 = B^{-1}S^2A^{-1},$

and $\varphi_2(S)y_0 = \alpha_S B^{-1}Sx_0$ for some nonzero scalar α_S depending on S.

Note that if (6.34) is satisfied then one may further choose A so that $\varphi_2(\mathbf{1})Ax_0 = y_0$. Indeed, (6.34), we have

$$\sigma_{\varphi_2(\mathbf{1})ATB\varphi_2(\mathbf{1})}(y_0) \cup \{0\} = \sigma_T(x_0) \cup \{0\}$$

for all $T \in \mathcal{B}(X)$ and then Lemma 2.2 tells us that $\varphi_2(\mathbf{1})Ax_0 = (\varphi_2(\mathbf{1}))^{-1}B^{-1}x_0 = \frac{1}{\alpha_1}y_0$. Multiplying A by $\alpha_\mathbf{1}$ and dividing B by $\alpha_\mathbf{1}$, we can choose such an operator A so that $\varphi_2(\mathbf{1})Ax_0 = y_0$.

The proof of this theorem uses several preliminary results and lemmas. Some of them is quoted from [36] and the rest of them is new and shall be proved. The first lemma will be used to establish the additivity of the surjective map φ_1 when (6.34) is satisfied. Its proof is simple and will be given for the sake of completeness.

Lemma 6.2. *For every $R \in \mathcal{B}(X)$ such that R^2 is a rank one operator, we have*

$$\sigma^*_{R(T+S)R}(x_0) = \sigma^*_{RTR}(x_0) + \sigma^*_{RSR}(x_0)$$

for all operators $T, S \in \mathcal{B}(X)$.

PROOF. Let $R \in \mathcal{B}(X)$ such that R^2 is a rank one operator. By Lemma 2.4 and [35, Lemma 2.2], we have

$$\begin{aligned}\sigma^*_{R(T+S)R}(x_0) &= \sigma^*_{R^2(T+S)}(Rx_0) \\ &= \sigma^*_{R^2T}(Rx_0) + \sigma^*_{R^2S}(Rx_0) \\ &= \sigma^*_{RTR}(Rx_0) + \sigma^*_{RSR}(x_0)\end{aligned}$$

for all $S, T \in \mathcal{B}(X)$. $\square$

Next lemma characterizes in term of the local spectrum when two operators are the same. It will be used, in particular, to show that φ_1 is automatically an injective linear map when (6.34) is satisfied.

Lemma 6.3. *If A, $B \in \mathcal{B}(X)$, then $A = B$ if and only if $\sigma^*_{TAT}(x_0) = \sigma^*_{TBT}(x_0)$ for all rank one operators $T \in \mathcal{B}(X) \setminus \mathcal{B}^2_{x_0}(X)$.*

PROOF. The proof is essentially the same as the one of [36, Theorem 3.1], and is therefore left for the reader. $\square$

The following lemma is quoted from [36] and gives a spectral characterization of rank one operators in term of the local spectrum. We use it to show that φ_1 preserves rank one operators in both directions.

Lemma 6.4 (Bourhim–Mashreghi [36])**.** *For a nonzero operator $R \in \mathcal{B}(X)$, the following statements are equivalent.*

(a) R has rank one.
(b) $\sigma_{TRT}^*(x_0)$ has at most one element for all $T \in \mathcal{B}(X)$.
(c) $\sigma_{TRT}^*(x_0)$ is a singleton for all $T \in \mathcal{F}_2(X)$.

The next lemma tells us that φ_2 arbitrarily maps $\mathcal{B}_{x_0}^2(X)$ onto $\mathcal{B}_{y_0}^2(Y)$ provided that (6.34) is satisfied.

Lemma 6.5. *For an operator $A \in \mathcal{B}(X)$, the following statements are equivalent.*
  (a) $A \in \mathcal{B}_{x_0}^2(X)$; *i.e., $Ax_0 = 0$ or $A^2 = 0$.*
  (b) $\sigma_{ATA}^*(x_0) \subseteq \{0\}$ *for all operators $T \in \mathcal{B}(X)$.*
  (c) $\sigma_{ATA}^*(x_0) = \{0\}$ *for all rank one operators $T \in \mathcal{B}(X)$.*

PROOF. Observe that if $Ax_0 = 0$ or $A^2 = 0$ then $(ATA)^2 x_0 = 0$ for all $T \in \mathcal{B}(X)$. This in its turn implies that $\sigma_{ATA}^*(x_0) \subseteq \{0\}$ for all operators $T \in \mathcal{B}(X)$ and the implication $(a) \Rightarrow (b)$ always holds. Since (b) is obviously entails (c), we only need to establish the implication $(c) \Rightarrow (a)$. So, assume that $\sigma_{TA}^*(x_0) = \{0\}$ for all rank one operators $T \in \mathcal{B}(X)$, and suppose for the sake of contradiction that $Ax_0 \neq 0$ and $A^2 \neq 0$. Take $x \in X$ such that $A^2 x \neq 0$ and a linear functional $f \in X^*$ such that $f(A^2 x) \neq 0$ and $f(x_0) \neq 0$. Then $\sigma_{A(x \otimes f)A}^*(x_0) = \sigma_{(Ax) \otimes (A^* f)}^*(x_0) = \{f(A^2 x)\} \neq \{0\}$. This contradiction shows that the implication $(c) \Rightarrow (a)$ always holds and finishes the proof. □

In term of the local spectrum at a nonzero fixed vector, next lemma gives necessary and sufficient conditions for two operators to have the same squares. It tells us in particular that if A and B are two operators not in $\mathcal{B}_{x_0}^2(X)$, then their squares are equal and Ax_0 and Bx_0 are linearly dependent precisely when $\sigma_{ATA}^*(x_0) = \sigma_{BTB}^*(x_0)$ for all rank one operators $T \in \mathcal{B}(X)$.

Lemma 6.6. *For three operators A, B and C in $\mathcal{B}(X)$ such that $Ax_0 \neq 0$ and $A^2 \neq 0$, the following statements are equivalent.*
  (a) $A^2 = CB$ *and* $Ax_0 = \alpha C x_0$ *for some nonzero scalar α.*
  (b) $\sigma_{ATA}(x_0) \cup \{0\} = \sigma_{BTC}(x_0) \cup \{0\}$ *for all operators $T \in \mathcal{B}(X)$.*
  (c) $\sigma_{ATA}^*(x_0) = \sigma_{BTC}^*(x_0)$ *for all rank one operators $T \in \mathcal{B}(X)$.*

PROOF. Assume that $A^2 = CB$ and $Ax_0 = \alpha C x_0$ for some nonzero scalar α. By Lemma 2.4, we have

$$\sigma_{ATA}(x_0) \cup \{0\} = \sigma_{A^2 T}(Ax_0) \cup \{0\} = \sigma_{CBT}(\alpha C x_0) \cup \{0\} = \sigma_{BTC}(x_0) \cup \{0\}$$

for all $T \in \mathcal{B}(X)$. Therefore, the implication $(a) \Longrightarrow (b)$ holds.

Since the implication $(b) \Longrightarrow (c)$ obviously holds, we need to establish the implication $(c) \Rightarrow (a)$. Assume that $\sigma_{ATA}^*(x_0) = \sigma_{BTC}^*(x_0)$ for all rank one operators $T \in \mathcal{B}(X)$, and note that, since $A \notin \mathcal{B}_{x_0}^2(X)$, we must have $Cx_0 \neq 0$ and $CB \neq 0$. Now, let $x \in X$ and $f \in X^*$ such that $f(Ax_0) \neq 0$ and $f(A^2 x) \neq 0$, and note that Lemma 2.5 entails that $f(Cx_0) \neq 0$ and

$$\{f(CBx)\} = \sigma_{B(x \otimes f)C}^*(x_0) = \sigma_{A(x \otimes f)A}^*(x_0) = \{f(A^2 x)\}.$$

Hence, $f(CBx) = f(A^2 x)$ for all $x \in X$ and all $f \in X^*$ such that $f(Ax_0) \neq 0$ and $f(A^2 x) \neq 0$. From this, we infer that $f(CBx) = f(A^2 x)$ for all $x \in X$ and all $f \in X^*$ and thus $CB = A^2$. Now, assume for the sake of contradiction that Ax_0 and Cx_0 are linearly independent and choose $x \in X$ such that $A^2 x \neq 0$. Then either

A^2x and Ax_0 are linearly independent or A^2x and Cx_0 are linearly independent. If A^2x and Cx_0 are linearly independent, choose $f \in X^*$ such that $f(Cx_0) = 0$, $f(Ax_0) \neq 0$ and $f(A^2x) = 1$. Then, by Lemma 2.5, we have

$$\{0\} = \sigma^*_{B(x \otimes f)C}(x_0) = \sigma^*_{A(x \otimes f)A}(x_0) = \{f(A^2x)\} = \{1\},$$

which is a contradiction. In a similar way, one shows that A^2x and Ax_0 are linearly dependent and thus Ax_0 and Cx_0 are too linearly dependent. This completes the proof of lemma. $\square$

The last lemma gives a spectral characterization of rank one operators in term of the local spectrum. It is a simple consequence of Lemma 2.4 and [**35**, Bourhim-Mashreghi]. We need it to show that if (6.34) is satisfied then φ_2 maps all rank one operators in $\mathcal{B}(X) \backslash \mathcal{B}^2_{x_0}(X)$ onto all rank one operators in $\mathcal{B}(Y) \backslash \mathcal{B}^2_{y_0}(Y)$.

Lemma 6.7. *For an operator $R \in \mathcal{B}(X)$ for which $R^2 \neq 0$ and $Rx_0 \neq 0$, the following statements are equivalent.*

(a) R^2 has rank one.
(b) $\sigma^*_{RTR}(x_0)$ has at most one element for all $T \in \mathcal{B}(X)$.
(c) $\sigma^*_{RTR}(x_0)$ is a singleton for all $T \in \mathcal{F}_2(X)$.

PROOF. Obviously, if R has rank one and $T \in \mathcal{B}(X)$ is an arbitrary operator, then RTR has rank at most one and thus $\sigma^*_{RTR}(x_0)$ is a singleton. This shows that the implication $(a) \Rightarrow (b)$ always holds.

Now, assume that $\sigma^*_{RTR}(x_0)$ is a singleton for all $T \in \mathcal{F}_2(X)$. By Lemma 2.4, we see that $\sigma^*_{R^2T}(Rx_0)$ is a singleton for all $T \in \mathcal{F}_2(X)$ and thus [**35**, Theorem 4.1] tells us that R^2 has rank one. This establishes the implication $(c) \Rightarrow (a)$ and completes the proof of the lemma. $\square$

All the ingredients are collected and we are therefore in a position to prove that Theorem 6.1.

PROOF OF THEOREM 6.1. Assume that φ_1 and φ_2 are two maps $\mathcal{B}(X)$ onto $\mathcal{B}(Y)$ such that φ_2 maps $\mathcal{B}^2_{x_0}(X)$ onto $\mathcal{B}^2_{y_0}(Y)$ and φ_1 and φ_2 take the forms (6.35) and (6.36). Moreover, assume that for every $S \notin \mathcal{B}_{x_0}(X)$ there exists a nonzero scalar α_S such that $\varphi_2(S)y_0 = \alpha_S B^{-1} S x_0$. For every $T \in \mathcal{B}(X)$ and $S \in \mathcal{B}^2_{x_0}(X)$, we have $\varphi_2(S) \in \mathcal{B}^2_{y_0}(Y)$ and

$$(STS)^2 x_0 = 0 \text{ and } (\varphi_2(S)\varphi_1(T)\varphi_2(S))^2 y_0 = 0.$$

Then $\sigma_{STS}(x_0) \subset \{0\}$ and $\sigma_{\varphi_2(S)\varphi_1(T)\varphi_2(S)}(y_0) \subset \{0\}$ for all $T \in \mathcal{B}(X)$ and $S \in \mathcal{B}^2_{x_0}(X)$. If, however, $T \in \mathcal{B}(X)$ and $S \notin \mathcal{B}^2_{x_0}(X)$, then Lemma 2.4 implies that

$$\begin{aligned}
\sigma_{\varphi_2(S)\varphi_1(T)\varphi_2(S)}(y_0) \cup \{0\} &= \sigma_{\varphi_2(S)^2 \varphi_1(T)}(\varphi_2(S)y_0) \cup \{0\} \\
&= \sigma_{B^{-1}S^2TB}(\alpha_S B^{-1} S x_0) \cup \{0\} \\
&= \sigma_{S^2T}(Sx_0) \cup \{0\} \\
&= \sigma_{STS}(x_0) \cup \{0\}.
\end{aligned}$$

Clearly, (6.34) is satisfied.

Conversely, assume that φ_1 and φ_2 surjective maps from $\mathcal{B}(X)$ onto $\mathcal{B}(Y)$ satisfying (6.34), and let us break down the proof of Theorem 6.1 into several steps.

STEP 1. $\varphi_2\left(\mathcal{B}^2_{x_0}(X)\right) = \mathcal{B}^2_{y_0}(Y)$.

This is an immediate consequence of Lemma 6.5 together with (6.34) and the surjectivity of φ_1.

STEP 2. φ_1 is bijective and $\varphi_1(0) = 0$.

If $\varphi_1(A) = \varphi_1(B)$ for some $A, B \in \mathcal{B}(X)$, we get that
$$\sigma_{TAT}(x_0) = \sigma_{\varphi_2(T)\varphi_1(A)\varphi_2(T)}(y_0) = \sigma_{\varphi_2(T)\varphi_1(B)\varphi_2(T)}(y_0) = \sigma_{TBT}(x_0)$$
for all $T \in \mathcal{B}(X)$. By Lemma 6.3, we see that $A = B$ and φ_1 is injective. But since φ_1 is assumed to be surjective, the map φ_1 is, in fact, bijective. For the second part of this step, note that for every $T \in \mathcal{B}(X)$, we have
$$\{0\} = \sigma_{T \cdot 0 \cdot T}(x_0) = \sigma_{\varphi_2(T)\varphi_1(0)\varphi_2(T)}(y_0).$$
Again by Lemma 6.3 and the surjectivity of φ_2, we see that $\varphi_1(0) = 0$.

STEP 3. φ_1 preserves rank one operators in both directions.

Let $R \in \mathcal{B}(X)$ be a rank one operator, and note that, since $\varphi_1(0) = 0$ and φ_1 is bijective, $\varphi_1(R) \neq 0$. Moreover, for every operator $S = \varphi_2(T) \in \mathcal{B}(Y)$, we note that
$$\sigma^*_{TRT}(x_0) = \sigma^*_{\varphi_2(T)\varphi_1(R)\varphi_2(T)}(y_0) = \sigma^*_{S\varphi_1(R)S}(y_0)$$
contains at most one element. By Lemma 6.4, we see that $\varphi_1(R)$ has rank one. The converse holds in a similar way and thus φ_1 preserves the rank one operators in both directions; as desired.

STEP 4. φ_1 is linear.

Let us first show that φ_1 is homogeneous. For every $\lambda \in \mathbb{C}$ and $A, T \in \mathcal{B}(X)$, we have
$$\begin{aligned}
\sigma_{\varphi_2(T)(\lambda\varphi_1(A))\varphi_2(T)}(y_0) &= \lambda\sigma_{\varphi_2(T)\varphi_1(A)\varphi_2(T)}(y_0) \\
&= \lambda\sigma_{TAT}(x_0) \\
&= \sigma_{T(\lambda A)T}(x_0) \\
&= \sigma_{\varphi_2(T)\varphi_1(\lambda A)\varphi_2(T)}(y_0).
\end{aligned}$$
Since φ_2 is surjective, Lemma 6.3 shows that $\varphi_1(\lambda A) = \lambda\varphi_1(A)$ for all $\lambda \in \mathbb{C}$ and $A \in \mathcal{B}(X)$; as desired.

Now, let us show that φ_1 is additive. Let $S \in \mathcal{B}(Y)$ be a rank one operator, by Lemma 6.7 and Step 1, there exists a rank one operator $R \notin \mathcal{B}_{x_0}(X)$ such that $S = \varphi_2(R)$. Therefore, by Lemma 6.2, we have
$$\begin{aligned}
\sigma^*_{S\varphi_1(T_1+T_2)S}(y_0) &= \sigma^*_{\varphi_2(R)\varphi_1(T_1+T_2)\varphi_2(R)}(y_0) \\
&= \sigma^*_{R(T_1+T_2)R}(x_0) \\
&= \sigma^*_{RT_1R}(x_0) + \sigma^*_{RT_2R}(x_0) \\
&= \sigma^*_{\varphi_2(R)\varphi_1(T_1)\varphi_2(R)}(y_0) + \sigma^*_{\varphi_2(R)\varphi_1(T_2)\varphi_2(R)}(y_0) \\
&= \sigma^*_{\varphi_2(R)(\varphi_1(T_1)+\varphi_1(T_2))\varphi_2(R)}(y_0) \\
&= \sigma^*_{S(\varphi_1(T_1)+\varphi_1(T_2))S}(y_0)
\end{aligned}$$
for all $T_1, T_2 \in \mathcal{B}(X)$. By the arbitrariness of the rank one operator $S \in \mathcal{B}(Y)$ for which $Sy_0 \neq 0$ and Lemma 6.3, we deduce that
$$\varphi_1(T_1 + T_2) = \varphi_1(T_1) + \varphi_1(T_2)$$

for all T_1, $T_2 \in \mathcal{B}(X)$, and φ_1 is linear.

STEP 5. There are bijective linear mappings $P : X \to Y$ and $Q : X^* \to Y^*$ such that $\varphi_1(x \otimes f) = Px \otimes Qf$ for all $x \in X$ and $f \in X^*$.

Since φ_1 is a bijective linear map from $\mathcal{F}(X)$ onto $\mathcal{F}(Y)$ and preserves rank one operators in both directions, either there are bijective linear mappings $P : X \to Y$ and $Q : X^* \to Y^*$ such that

(6.37) $$\varphi_1(x \otimes f) = Px \otimes Qf, \quad (x \in X, \ f \in X^*),$$

or there are bijective linear mappings $M : X^* \to Y$ and $N : X \to Y^*$ such that

(6.38) $$\varphi_1(x \otimes f) = Mf \otimes Nx, \quad (x \in X, \ f \in X^*);$$

see for instance [34, Theorem 4.12].

Assume that φ_1 takes the form (6.38), and note that (6.34) implies that

$$\sigma_{\varphi_2(\mathbf{1})(Mf \otimes Nx)\varphi_2(\mathbf{1})}(y_0) = \sigma_{\varphi_2(\mathbf{1})\varphi_1(x \otimes f)\varphi_2(\mathbf{1})}(y_0) = \sigma_{x \otimes f}(x_0)$$

for all $x \in X$ and $f \in X^*$. Choose a nonzero linear functional $g \in Y^*$ such that $g(\varphi_2(\mathbf{1})y_0) = 0$ and set $x = N^{-1}g$. Because x and x_0 are nonzero vectors in X, one can find a linear functional $f \in X^*$ such that $f(x_0) \neq 0$ and $f(x) \neq 0$. Then

$$\{0\} \neq \{f(x)\} = \sigma^*_{x \otimes f}(x_0) = \sigma^*_{\varphi_2(\mathbf{1})(Mf \otimes Nx)\varphi_2(\mathbf{1})}(y_0) = \sigma^*_{\varphi_2(\mathbf{1})(Mf \otimes g)\varphi_2(\mathbf{1})}(y_0) = \{0\}.$$

This contradiction shows that φ only takes the form (6.37); as desired.

STEP 6. For any $x \in X$ and $f \in X^*$, we have

(6.39) $$f(x) = (Qf)\left(\varphi_2(\mathbf{1})^2 Px\right).$$

Given $x \in X$ and $f \in X^*$, the previous step and (6.34) entail that

$$\sigma_{\varphi_2(\mathbf{1})(Px \otimes Qf)\varphi_2(\mathbf{1})}(y_0) = \sigma_{x \otimes f}(x_0).$$

So, assume first that $f(x_0) \neq 0$, and note that
(6.40)
$$\{0\} \neq \{f(x_0)\} = \sigma_{x_0 \otimes f}(x_0) = \sigma_{\varphi_2(\mathbf{1})\varphi_1(x_0 \otimes f)\varphi_2(\mathbf{1})}(y_0) = \sigma_{\varphi_2(\mathbf{1})(Px_0 \otimes Qf)\varphi_2(\mathbf{1})}(y_0).$$

Assume for the sake of contradiction that $Qf(\varphi_2(\mathbf{1})y_0) = 0$, and note that $\varphi_2(\mathbf{1})(Px_0 \otimes Qf)\varphi_2(\mathbf{1})y_0 = 0$ and $\sigma_{\varphi_2(\mathbf{1})(Px_0 \otimes Qf)\varphi_2(\mathbf{1})}(y_0) = \{0\}$. This contradicts (6.40) and shows that $(Qf)(\varphi_2(\mathbf{1})y_0) \neq 0$. Then Lemma 2.5 implies that

$$\{f(x)\} = \sigma^*_{x \otimes f}(x_0) = \sigma^*_{\varphi_2(\mathbf{1})(Px \otimes Qf)\varphi_2(\mathbf{1})}(y_0) = \left\{(Qf)\left(\varphi_2(\mathbf{1})^2 Px\right)\right\},$$

and thus (6.39) holds whenever $f(x_0) \neq 0$. Since any linear functional from X^* can be written as sum of two linear functionals from X^* not vanishing at x_0, we conclude that (6.39) holds for all $x \in X$ and $f \in X^*$; as desired.

STEP 7. P is continuous and $\varphi_2(\mathbf{1})$ is invertible.

First, we show that $\varphi_2(\mathbf{1})$ is injective. If not, there is a nonzero vector $y \in Y$ such that $\varphi_2(\mathbf{1})y = 0$. Take $x = P^{-1}y$, and let $f \in X^*$ be a linear functional such that $f(x) = 1$. By (6.39), we have

$$1 = f(x) = (Qf)\left(\varphi_2(\mathbf{1})^2 Px\right) = (Qf)\left(\varphi_2(\mathbf{1})^2 y\right) = 0.$$

This contradiction tells us that $\varphi_2(\mathbf{1})$ is injective.

Now, let us show that P is continuous and $(\varphi_2(\mathbf{1}))^* Q = P^{*-1}$. Let $(x_n)_n$ be a sequence in X such that $\lim_{n\to\infty} x_n = 0$ and $\lim_{n\to\infty} Px_n = y \in Y$. By (6.39), we have
$$Q(f)(\varphi_2(\mathbf{1})^2 y) = \lim_{n\to\infty} Qf(\varphi_2(\mathbf{1})^2 Px_n) = \lim_{n\to\infty} f(x_n) = 0$$
for all $f \in X^*$. Since Q is bijective, we see that $\varphi_2(\mathbf{1})^2 y = 0$. As $\varphi_2(\mathbf{1})^2$ is injective, we get that $y = 0$ and the closed graph theorem tells us that P is continuous. Moreover, we have
$$f(x) = (Qf)\left(\varphi_2(\mathbf{1})^2 Px\right) = \left(\varphi_2(\mathbf{1})^2 P\right)^* Qf(x)$$
for all $x \in X$ and $f \in X^*$, and thus
(6.41) $$\mathbf{1}_{X^*} = \left(\varphi_2(\mathbf{1})^2 P\right)^* Q = P^* \left(\varphi_2(\mathbf{1})^2\right)^* Q.$$
It then follows that
(6.42) $$P^{*-1} Q^{-1} = \left(\varphi_2(\mathbf{1})^2\right)^*,$$
and therefore $\varphi_2(\mathbf{1})$ is invertible.

STEP 8. $\varphi_2(\mathbf{1}) Px_0 = \alpha y_0$ for some nonzero scalar $\alpha \in \mathbb{C}$.

Assume that there is a linear functional f in X^* such that $f(x_0) = 1$ and $f\left(P^{-1}(\varphi_2(\mathbf{1}))^{-1} y_0\right) = 0$. We have
$$(x_0 \otimes f)x_0 = x_0 \text{ and } \left(P(x_0 \otimes f) P^{-1}(\varphi_2(\mathbf{1}))^{-1}\right)(y_0) = 0.$$
Therefore, by (6.34) and (6.42), we have
$$\begin{aligned} \{1\} = \sigma_{x_0 \otimes f}(x_0) &= \sigma_{\varphi_2(\mathbf{1})(Px_0 \otimes Qf)\varphi_2(\mathbf{1})}(y_0) \\ &= \sigma_{\varphi_2(\mathbf{1})(Px_0 \otimes \varphi_2(\mathbf{1})^* Qf)}(y_0) \\ &= \sigma_{\varphi_2(\mathbf{1}) P(x_0 \otimes f) P^{-1}(\varphi_2(\mathbf{1}))^{-1}}(y_0) \\ &= \{0\}. \end{aligned}$$
This contradiction shows that there is a nonzero scalar $\alpha \in \mathbb{C}$ such that $\varphi_2(\mathbf{1}) Px_0 = \alpha y_0$.

STEP 9. $\varphi_2(\mathbf{1}) Ax_0 = y_0$ and $\varphi_2(T)^2 = B^{-1} T^2 A^{-1}$ for all $T \notin \mathcal{B}_{x_0}(X)$, where $A := \alpha^{-1} P$ and $B := \alpha P^{-1} \varphi_2(\mathbf{1})^{-2}$.

We may and shall assume that $\alpha = 1$ and note that, since $\varphi_2(\mathbf{1}) Ax_0 = y_0$ obviously holds, we only need to prove the second part of this step. For any $x \in X$, $f \in X^*$ and $T \notin \mathcal{B}_{x_0}(X)$, we have
$$\begin{aligned} \sigma_{T(x \otimes f)T}(x_0) &= \sigma_{\varphi_2(T)\varphi_1(x \otimes f)\varphi_2(T)}(y_0) \\ &= \sigma_{\varphi_2(T) P(x \otimes f) Q^* \varphi_2(T)}(y_0) \\ &= \sigma_{\varphi_2(T) P(x \otimes f) P^{-1} \varphi_2(\mathbf{1})^{-2} \varphi_2(T)}(y_0) \\ &= \sigma_{P^{-1} \varphi_2(\mathbf{1})^{-1} \varphi_2(T) P(x \otimes f) P^{-1} \varphi_2(\mathbf{1})^{-2} \varphi_2(T) \varphi_2(\mathbf{1}) P}(x_0). \end{aligned}$$
By Lemma 6.6, we conclude that
$$P^{-1} \varphi_2(\mathbf{1})^{-2} \varphi_2(T) \varphi_2(\mathbf{1}) P P^{-1} \varphi_2(\mathbf{1})^{-1} \varphi_2(T) P = T^2$$
and
$$P^{-1} \varphi_2(\mathbf{1})^{-2} \varphi_2(T) \varphi_2(\mathbf{1}) Px_0 = \alpha_T Tx_0$$

for all $T \notin \mathcal{B}_{x_0}(X)$, where α_T is a nonzero scalar depending on T. Consequently,
$$\varphi_2(T)^2 = \varphi_2(\mathbf{1})^2 PT^2 P^{-1} = B^{-1}T^2 A^{-1},$$
and
$$\varphi_2(T)y_0 = \alpha_T \varphi_2(\mathbf{1})^2 PTx_0 = \alpha_T B^{-1} T x_0$$
for all $T \notin \mathcal{B}_{x_0}(X)$; as desired.

STEP 10. φ_1 has the form (6.35).

Observe that
$$\begin{aligned}\sigma_{STS}(x_0) \cup \{0\} &= \sigma_{\varphi_2(S)\varphi_1(T)\varphi_2(S)}(y_0) \cup \{0\} \\ &= \sigma_{\varphi_2(S)\varphi_1(T)B^{-1}B\varphi_2(S)}(y_0) \cup \{0\} \\ &= \sigma_{B\varphi_2(S)^2 AA^{-1}\varphi_1(T)B^{-1}}(B\varphi_2(S)y_0) \cup \{0\} \\ &= \sigma_{S^2 A^{-1}\varphi_1(T)B^{-1}}(Sx_0) \cup \{0\} \\ &= \sigma_{SA^{-1}\varphi_1(T)B^{-1}S}(x_0) \cup \{0\}.\end{aligned}$$

for all $T \in \mathcal{B}(X)$ and $S \notin \mathcal{B}_{x_0}(X)$. By Lemma 6.3, we have
$$\varphi_1(T) = ATB$$
for all $T \in \mathcal{B}(X)$. $\square$

Let $\mathcal{B}(\mathcal{H})$ (resp. $\mathcal{B}(\mathcal{K})$) be the algebra of all bounded linear operators on an infinite-dimensional complex Hilbert space $\mathcal{H}$ (resp. $\mathcal{K}$). Given two fixed nonzero vectors $h_0 \in \mathcal{H}$ and $k_0 \in \mathcal{K}$, maps φ from $\mathcal{B}(\mathcal{H})$ onto $\mathcal{B}(\mathcal{K})$ satisfying

(6.43) $$\sigma_{ST^*S}(h_0) = \sigma_{\varphi(S)\varphi(T)^*\varphi(S)}(k_0), \qquad (T, S \in \mathcal{B}(\mathcal{H}))$$

are described in [7]. Such a characterization is now a consequence of Theorem 6.1.

Corollary 6.8. *Given two nonzero vectors $h_0 \in \mathcal{H}$ and $k_0 \in \mathcal{K}$, a map φ from $\mathcal{B}(\mathcal{H})$ onto $\mathcal{B}(\mathcal{K})$ satisfies (6.43) if and only if there exist a unitary operator $U \in \mathcal{B}(\mathcal{H}, \mathcal{K})$ and a nonzero scalar α such that $Uh_0 = \alpha k_0$ and $\varphi(T) = UTU^*$ for all $T \in \mathcal{B}(\mathcal{H})$.*

PROOF. The "*if*" part is straightforward. So, assume that (6.43) is satisfied, and note that
$$\sigma_{STS}(h_0) = \sigma_{\varphi(S)\varphi(T^*)^*\varphi(S)}(k_0)$$
for all $T, S \in \mathcal{B}(\mathcal{H})$. By Theorem 6.1, there exist two bijective linear mappings $A: \mathcal{H} \to \mathcal{K}$ and $B: \mathcal{K} \to \mathcal{H}$ such that $\varphi(\mathbf{1})Ah_0 = k_0$ and

(6.44) $$\varphi(T^*)^* = ATB, \qquad (T \in \mathcal{B}(\mathcal{H})),$$

and

(6.45) $$\varphi(T)^2 = B^{-1}T^2A^{-1}, \qquad (T \notin \mathcal{B}_{x_0}^2(\mathcal{H})).$$

Fix a unit vector $x \in \mathcal{H}$, and apply (6.44) and (6.45) to $T = x \otimes x$ to get that
$$B^{-1}(x \otimes x)A^{-1} = \varphi(x \otimes x)^2 = (B^*(x \otimes x)A^*)^2,$$
and thus

(6.46) $$x \otimes x = B\left(B^*(x \otimes x)A^*\right)^2 A = \langle A^*B^*x, x \rangle (BB^*x) \otimes (A^*Ax).$$

From this, we infer that there are two positive scalars λ_A and λ_B such that $A^*A = \lambda_A \mathbf{1}$ and $BB^* = \lambda_B \mathbf{1}$. These together with (6.46) imply that $\lambda_A \lambda_B A^* B^* = \mathbf{1}$, and then
$$\frac{1}{(\lambda_A \lambda_B)^2} \mathbf{1} = (A^* B^*) BA = A^* (B^* B) A = \lambda_B \lambda_A.$$
Clearly, $\lambda_A \lambda_B = 1$ and $Ah_0 = k_0$. Therefore, there is a unitary operator $U \in \mathcal{B}(\mathcal{H}, \mathcal{K})$ and a positive constant α such that $B^{-1} = A = \alpha^{-1} U$ and thus $Uh_0 = \alpha k_0$ and
$$\varphi(T) = ATB = UTU, \quad (T \in \mathcal{B}(\mathcal{H})),$$
and the proof is therefore complete. $\square$

6.2. The finite-dimensional case. We first fix some usual notation and definitions that we shall use in the sequel. For a matrix $T \in \mathcal{M}_n(\mathbb{C})$, let $\operatorname{Tr}(T)$ and T^{tr} denote the trace and the transpose of T, respectively. Let GL_n denote, as usual, the group of all invertible matrices in $\mathcal{M}_n(\mathbb{C})$, and keep in mind that GL_n is a dense open subset of $\mathcal{M}_n(\mathbb{C})$. Recall that a matrix $T \in \mathcal{M}_n(\mathbb{C})$ is said to be *cyclic* with a cyclic vector $x_0 \in \mathbb{C}^n$ if the linear span of $\{T^k x_0 : k \geq 0\}$ is dense in $\mathbb{C}^n$, and consider the set
$$\mathcal{C}_{n,x_0} := \{T \in \mathcal{M}_n(\mathbb{C}) : |\sigma(T)| = n \text{ and } T \text{ is cyclic with cyclic vector } x_0\},$$
where $|\sigma(T)|$ denotes the cardinality of $\sigma(T)$. It is known that $\mathcal{C}_{n,x_0}$ is an open dense subset of $\mathcal{M}_n(\mathbb{C})$ and
(6.47) $$\sigma_T(x_0) = \sigma(T)$$
for all $T \in \mathcal{C}_{n,x_0}$; see for instance [33, 39]. Finally, for any vector $x_0 \in \mathbb{C}^n$, we let $\mathcal{M}^2_{n,x_0}(\mathbb{C})$ be the collection of all matrices in $\mathcal{M}_n(\mathbb{C})$ vanishing at x_0 or of zero square; i.e.,
$$\mathcal{M}^2_{n,x_0}(\mathbb{C}) := \{T \in \mathcal{M}_n(\mathbb{C}) : T^2 = 0 \text{ or } Tx_0 = 0\}.$$
Note that the restriction to infinite-dimensional Banach spaces in the statement of Theorem 6.1 is just for the sake of simplicity. With the surjectivity condition on φ_1 and φ_2, such a result and its proof remain valid for finite-dimensional case. However, if $X = Y = \mathbb{C}^n$ then the surjectivity hypothesis on φ_1 and φ_2 is redundant, as is shown by the following result whose proof is completely different than the one of Theorem 6.1.

Theorem 6.9. *Two maps φ_1 and φ_2 on $\mathcal{M}_n(\mathbb{C})$ satisfy*
(6.48) $$\sigma_{STS}(x_0) = \sigma_{\varphi_2(S)\varphi_1(T)\varphi_2(S)}(x_0), \quad (S, T \in \mathcal{M}_n(\mathbb{C})),$$
if and only if φ_2 maps $\mathcal{M}^2_{n,x_0}(\mathbb{C})$ into itself and there are two invertible matrices $A, B \in \mathcal{M}_n(\mathbb{C})$ such that
(6.49) $$\varphi_1(T) = ATB, \quad (T \in \mathcal{M}_n(\mathbb{C})),$$
and for every $S \notin \mathcal{M}^2_{n,x_0}(\mathbb{C})$ we have
(6.50) $$\varphi_2(S)^2 = B^{-1} S^2 A^{-1},$$
and $\varphi_2(S) x_0 = \alpha_S B^{-1} S x_0$ for some non zero scalar α depending on S

The following lemma is a variant of Lemma 6.6, and is needed for the proof of Theorem 6.9. In its proof, we shall identify $\mathcal{M}_n(\mathbb{C})$ with $\mathcal{B}(X)$, the algebra of all linear operators on a complex Banach space X of dimension n.

Lemma 6.10. *Let O be a dense subset of $\mathcal{M}_n(\mathbb{C})$. For four matrices A, B, C and D in $\mathcal{M}_n(\mathbb{C})$ such that $B \notin \mathcal{M}_{n,x_0}^2(\mathbb{C})$ and $BA \neq 0$, the following statements are equivalent.*

(a) $BA = DC$ and $Bx_0 = \alpha D x_0$ for some nonzero scalar α.
(b) $\sigma_{ATB}^*(x_0) = \sigma_{CTD}^*(x_0)$ for all $T \in O$.

PROOF. The implication (a) $\implies$ (b) follows immediately from Lemma 2.4. Assume that $BA = DC$ and $Bx_0 = \alpha D x_0$ for some nonzero scalar α, and note that Lemma 2.4 implies that

$$\sigma_{ATB}^*(x_0) = \sigma_{BAT}^*(Bx_0) = \sigma_{DCT}^*(\alpha D x_0) = \sigma_{DCT}^*(D x_0) = \sigma_{CTD}^*(x_0)$$

for all $T \in O$.

Conversely, assume that $\sigma_{ATB}^*(x_0) = \sigma_{CTD}^*(x_0)$ for all $T \in O$, and let us first show that Dx_0 and Bx_0 are linearly dependent. If not, choose $x \in X$ such that $BAx \neq 0$ and note that either BAx and Bx_0 are linearly independent or BAx and Dx_0 are linearly independent. We may and shall assume that BAx and Dx_0 are linearly independent and pick up a linear functional $f \in X^*$ such that $f(Dx_0) = 0$, $f(Bx_0) \neq 0$ and $f(BAx) = 1$. We have

$$\{f(BAx)\} = \sigma_{BA(x \otimes f)}^*(Bx_0) = \sigma_{A(x \otimes f)B}^*(x_0) = \sigma_{C(x \otimes f)D}^*(x_0) = \{0\}.$$

This contradiction shows that Dx_0 and Bx_0 are linearly dependent. It is easy to see that Dx_0 must be nonzero and thus $Bx_0 = \alpha D x_0$ for some nonzero scalar α.

By the density of O in $\mathcal{M}_n(\mathbb{C})$ and the continuity of the spectrum and lower semi-continuity of the local spectrum, we obtain that

(6.51) $\sigma_{BAT}^*(Bx_0) \cup \sigma_{DCT}^*(Bx_0) = \sigma_{ATB}^*(x_0) \cup \sigma_{CTD}^*(x_0) \subset \sigma(ATB) \cap \sigma(CTD)$

for all $T \in \mathcal{M}_n(\mathbb{C})$. Now, let $x \in X$ be a nonzero vector and let us show that $f(BAx) = f(DCx)$ for all $f \in X^*$. Pick up a linear functional $f \in X^*$ such that $f(BAx) \neq 0$. If $f(Bx_0) \neq 0$ then apply (6.51) to $T := x \otimes f$ to get that

$$\{0\} \neq \{f(BAx)\} \subseteq \sigma(BAx \otimes f) \cap \sigma(DCx \otimes f) = \{0, f(BAx)\} \cap \{0, f(DCx)\}$$

and $f(BAx) = f(DCx)$. If, however, $f(x_0) = 0$, then BAx and Bx_0 must be linearly independent. In this case, take a linear functional $g \in X^*$ such that $g(Bx_0) = 1$ and $g(BAx) = t$ for some nonzero scalar $t \neq -f(BAx)$, and note that what has been shown previously applied to g and $f + g$ shows that

$$g(BAx) = g(DCx) \text{ and } (f+g)(BAx) = (f+g)(DCx).$$

From this it follows that $f(BAx) = f(DCx)$ in this case too.

Now, if $f \in X^*$ is a linear functional such that $f(BAx) = 0$ then (6.51) gives

$$\{f(DCx)\} \subseteq \sigma(BAx \otimes f) \cap \sigma(DCx \otimes f) = \{0\}$$

and $f(DCx) = 0$. This and what has been shown previously imply that $f(BAx) = f(DCx)$ for all $f \in X^*$ and $BAx = DCx$. Clearly, $BA = DC$ and the proof is complete. $\square$

All the necessary ingredients are collected, and therefore we are in a position to prove Theorem 6.9.

PROOF OF THEOREM 6.9. Checking the "*if*" part is straightforward.

For the "*only if*" part assume that φ satisfies (6.48), and let us first show that the restriction of both φ_1 and φ_2 on GL_n equal to bijective linear maps L_1 and L_2.

Indeed, let $T_0 \in GL_n$, and note that, since $\mathcal{C}_{n,x_0}$ is a nonempty open set and the fact that $(S,T) \mapsto STS$ is continuous and the fact that $\mathcal{C}_{n,x_0} \subseteq \{ST_0S : S \in GL_n\}$, there is an open neighborhood $\mathcal{V}_{T_0}$ of T_0 and a nonempty open set $\mathcal{O}_{T_0} \subseteq GL_n$ such that

(6.52) $$STS \in \mathcal{C}_{n,x_0}$$

for all $S \in \mathcal{O}_{T_0}$ and $T \in \mathcal{V}_{T_0}$. This together with (6.48) and (6.47) imply that

$$\sigma(STS) = \sigma(\varphi_2(S)\varphi_1(T)\varphi_2(S)) \text{ and } |\sigma(STS)| = n$$

for all $S \in \mathcal{O}_{T_0}$ and $T \in \mathcal{V}_{T_0}$. It then follows that

$$\text{Tr}(STS) = \text{Tr}(\varphi_2(S)\varphi_1(T)\varphi_2(S))$$

for all $S \in \mathcal{O}_{T_0}$ and $T \in \mathcal{V}_{T_0}$. Therefore, following the same argument as the one in the proof of Assertion 1 and Assertion 2 of [53, Theorem 2.1], one concludes that there is a bijective linear maps L_1 on $\mathcal{M}_n(\mathbb{C})$ such that $\varphi_1 = L_1$ on GL_n.

Let $T \in \mathcal{M}_n(\mathbb{C})$ and $S \in GL_n$, and let $(S_k)_k \subset GL_n \cap \mathcal{C}_{n,x_0}$ such that $\lim_{k \to \infty} S_k = STS$. Since $|\sigma(S_k)| = n$ and $\sigma(S_k) = \sigma_{S_k}(x_0)$ for all k, we have

$$\begin{aligned}\sigma(S_k) &= \sigma_{S_k}(x_0) = \sigma_{\varphi_2(S)\varphi_1(S^{-1}S_kS^{-1})\varphi_2(S)}(x_0) \\ &= \sigma_{\varphi_2(S)L_1(S^{-1}S_kS^{-1})\varphi_2(S)}(x_0) \\ &= \sigma(\varphi_2(S)L_1(S^{-1}S_kS^{-1})\varphi_2(S))\end{aligned}$$

for all k. Once more the density of GL_n in $\mathcal{M}_n(\mathbb{C})$ and the continuity of both the spectrum and the linear map L_1 imply that

(6.53) $$\sigma(STS) = \sigma(\varphi_2(S)L_1(T)\varphi_2(S))$$

for all $(S,T) \in GL_n \times \mathcal{M}_n(\mathbb{C})$. It then follows that the linear map $T \mapsto \varphi_2(\mathbf{1})L_1(T)\varphi_2(\mathbf{1})$ preserves the spectrum, and thus there is an invertible matrix P in $\mathcal{M}_n(\mathbb{C})$ such that either

(6.54) $$\varphi_2(\mathbf{1})L_1(T)\varphi_2(\mathbf{1}) = PTP^{-1}, \ (T \in \mathcal{M}_n(\mathbb{C})),$$

or

(6.55) $$\varphi_2(\mathbf{1})L_1(T)\varphi_2(\mathbf{1}) = PT^{tr}P^{-1}, \ (T \in \mathcal{M}_n(\mathbb{C}));$$

see [**106**]. On the other hand, from (6.48), we see that

(6.56) $$\sigma_T(x_0) = \sigma_{\varphi_2(\mathbf{1})\varphi_1(T)\varphi_2(\mathbf{1})}(x_0) = \sigma_{\varphi_2(\mathbf{1})L_1(T)\varphi_2(\mathbf{1})}(x_0)$$

for all $T \in GL_n$. This and Lemma 2.2 imply that (6.55) cannot occur, and thus L_1 is of the form (6.54). In this case, (6.56) becomes

(6.57) $$\sigma_T(x_0) = \sigma_{PTP^{-1}}(x_0)$$

for all $T \in GL_n$, and once more Lemma 2.2 implies that there is a nonzero scalar α such that $Px_0 = \alpha x_0$. Dividing P by α, we may and shall assume that $Px_0 = x_0$.

Now, we show that φ_1 and φ_2 have the asserted forms. Keep in mind that $\varphi_1(T) = L_1(T) = \varphi_2(\mathbf{1})^{-1}PTP^{-1}\varphi_2(\mathbf{1})^{-1} = ATB$ for all $T \in GL_n$, where $A = \varphi_2(\mathbf{1})^{-1}P$ and $B = P^{-1}\varphi_2(\mathbf{1})^{-1}$. Clearly, $\varphi_2(\mathbf{1})Ax_0 = x_0$. Moreover, for every $S \in \mathcal{M}_n(\mathbb{C})$ and $T \in GL_n$, we have

$$\begin{aligned}\sigma_{SA^{-1}TB^{-1}S}(x_0) &= \sigma_{\varphi_2(S)\varphi_1(A^{-1}TB^{-1})\varphi_2(S)}(x_0) \\ &= \sigma_{\varphi_2(S)T\varphi_2(S)}(x_0).\end{aligned}$$

By Lemma 6.10, we conclude that, for any $S \in \mathcal{M}_n(\mathbb{C}) \setminus \mathcal{M}_{n,x_0}^2(\mathbb{C})$, we have
$$\varphi_2(S)^2 = B^{-1}S^2 A^{-1}$$
and $\varphi_2(S)x_0 = \alpha_S B^{-1} S x_0$ for some nonzero scalar α_S depending on S; as desired.

Similarly, for every $T \in \mathcal{M}_n(\mathbb{C})$ and $S \notin \mathcal{M}_{n,x_0}^2(\mathbb{C})$, we have
$$\begin{aligned}\sigma_{STS}(x_0) &= \sigma_{\varphi_2(S)\varphi_1(T)\varphi_2(S)}(x_0)\\ &= \sigma_{\varphi_2(S)^2\varphi_1(T)}(\varphi_2(S)x_0)\\ &= \sigma_{B^{-1}S^2 A^{-1}\varphi_1(T)}(B^{-1}Sx_0)\\ &= \sigma_{S^2 A^{-1}\varphi_1(T) B^{-1}}(Sx_0)\\ &= \sigma_{S A^{-1}\varphi_1(T) B^{-1} S}(x_0).\end{aligned}$$

By Lemma 6.3, we conclude that $A^{-1}\varphi_1(T)B^{-1} = T$ for all $T \in \mathcal{M}_n(\mathbb{C})$. Thus
$$\varphi_1(T) = ATB$$
for all $T \in \mathcal{M}_n(\mathbb{C})$. Moreover,
$$\{0\} = \sigma_{SA^{-1}TB^{-1}S}(x_0) = \sigma_{\varphi_2(S)\varphi_1(A^{-1}TB^{-1})\varphi_2(S)}(x_0) = \sigma_{\varphi_2(S)T\varphi_2(S)}(x_0)$$
for all $T \in \mathcal{M}_n(\mathbb{C})$ and $S \in \mathcal{M}_{n,x_0}^2(\mathbb{C})$, and thus Lemma 6.5 tells us that $\varphi_2(S) \in \mathcal{M}_{n,x_0}^2(\mathbb{C})$ and hence φ_2 maps $\mathcal{M}_{n,x_0}^2(\mathbb{C})$ into itself. The proof is therefore complete. $\square$

For a nonzero vector x_0 in $\mathbb{C}^n$ and two fixed scalars μ and ν in $\mathbb{C}$ simultaneously nonzero, Abdelali, Achchi and Marzouki characterized in [6] all maps φ on $\mathcal{M}_n(\mathbb{C})$ satisfying
$$\sigma_{\mu ST^*S + \nu T^*S}(x_0) = \sigma_{\mu\varphi(S)\varphi(T)^*\varphi(S) + \nu\varphi(T)^*\varphi(S)}(x_0)$$
for all $S, T \in \mathcal{M}_n(\mathbb{C})$. This provides, in particular, a complete description of all maps on $\mathcal{M}_n(\mathbb{C})$ preserving the local spectrum of the skew double product "ST^*" or the skew double product "ST^*S". The following corollary is an immediate consequence of Theorem 6.9 and shows that Corollary 6.8 remains valid without the surjectivity condition of φ when $\mathcal{H} = \mathcal{K} = \mathbb{C}^n$ is a finite dimensional space. The proof is similar to that of Corollary 6.8 and will therefore be omitted.

Corollary 6.11. *A map φ on $\mathcal{M}_n(\mathbb{C})$ satisfies*
$$(6.58) \qquad \sigma_{ST^*S}(x_0) = \sigma_{\varphi(S)\varphi(T)^*\varphi(S)}(x_0), \ (S, T \in \mathcal{M}_n(\mathbb{C})),$$
if and only if there exist a unitary matrix U and a nonzero scalar α such that $Ux_0 = \alpha x_0$ and $\varphi(T) = UTU^*$ for all $T \in \mathcal{M}_n(\mathbb{C})$.

7. Jordan product and nonlinear local spectra preservers

In this section, we discuss surjective maps preserving the local spectrum of Jordan product of operators and of matrices at a nonzero vector.

7.1. The infinite-dimensional case. Let X and Y be two infinite-dimensional complex Banach spaces. The following theorem is quoted from [31] and gives a complete description of all surjective maps φ on $\mathcal{B}(X)$ preserving the local spectrum of Jordan product of operators at a nonzero fixed vector. It shows that such a map is either an algebra automorphism or the negative of an algebra automorphism of $\mathcal{B}(X)$ and the fixed vector is an eigenvector of the intertwining operator.

Theorem 7.1 (Bourhim–Mabrouk [31]). *If $x_0 \in X$ and $y_0 \in Y$ are nonzero vectors, then a map φ from $\mathcal{B}(X)$ onto $\mathcal{B}(Y)$ satisfies*

(7.59) $$\sigma_{\varphi(T)\varphi(S)+\varphi(S)\varphi(T)}(y_0) = \sigma_{TS+ST}(x_0), \ (T, \ S \in \mathcal{B}(X)),$$

if and only if there exists a bijective bounded linear mapping A from X into Y such that $Ax_0 = y_0$ and $\varphi(T) = \pm ATA^{-1}$ for all $T \in \mathcal{B}(X)$.

The proof of this result is technical and makes use of a series of results, which seem of independent interest. One of them gives a complete description of the local spectrum at a fixed vector of Jordan product of every rank one operator and arbitrary operator in $\mathcal{B}(X)$. It is quit useful and was repeatedly used to establish several auxiliary results needed for the proof of the above theorem. Another one establishes a *local spectral identity principle*, which states that two operators A and B in $\mathcal{B}(X)$ coincide if and only if $\sigma^*_{AR+RA}(x_0) = \sigma^*_{BR+RB}(x_0)$ for all rank one operators $R \in \mathcal{B}(X)$. With this, one shows that if (7.59) is satisfied then φ is a homogeneous injective map. One more gives a local spectral characterization of rank one nilpotent operators in term of local spectrum of Jordan product of operators. It shows that a nonzero operator $N \in \mathcal{B}(X)$ is a rank one nilpotent operator if and and only if either $\sigma_{TN+NT}(x_0)$ is a singleton for all $T \in \mathcal{B}(X)$ and $\sigma_N(x_0) = \{0\}$, or $\sigma^*_{TN+NT}(x_0)$ is a singleton for all $T \in \mathcal{B}(X)$ and there exists $T_0 \in \mathcal{B}(X)$ such that $\sigma_{T_0N+NT_0}(x_0) = \{0,a\}$ for some nonzero $a \in \mathbb{C}$. Such a characterization allows one to show that if a map φ from $\mathcal{B}(X)$ onto $\mathcal{B}(Y)$ satisfies (7.59), then φ preserves rank one nilpotent operators in both directions. This together with the local spectral identity principles and the following observation (i.e., Lemma 7.2) allow one to show that

(7.60) $$\varphi(N_1 + N_2) = \varphi(N_1) + \varphi(N_2).$$

for all $N_1, N_2 \in \mathcal{N}_1(X)$ for which $N_1 + N_2 \in \mathcal{N}_1(X)$.

Lemma 7.2 (Bourhim–Mabrouk [31]). *For every operator $N \in \mathcal{N}_1(X)$, we have*
$$\sigma^*_{(T+S)N+N(T+S)}(x_0) = \sigma^*_{TN+NT}(x_0) + \sigma^*_{SN+NS}(x_0)$$
for all $T, S \in \mathcal{B}(X)$.

Roughly speaking, once (7.60) is established one can use the following useful result quoted from [76, Lemma 2.2] to show that there exists a bijective transformation $A \in \mathcal{B}(X, Y)$ such that $Ax_0 = y_0$ and

(7.61) $$\varphi(N) = \pm ANA^{-1}$$

for all $N \in \mathcal{N}_1(X)$. Then the last step is to show that (7.61) holds in fact for all operators $T \in \mathcal{B}(X)$.

Lemma 7.3 (Du–Hou–Bai [76]). *If φ is a bijective map from $\mathcal{N}_1(X)$ into $\mathcal{N}_1(Y)$ for which*
$$N_1 + N_2 \in \mathcal{N}_1(X) \iff \varphi(N_1) + \varphi(N_2) \in \mathcal{N}_1(Y)$$
for all $N_1, N_2 \in \mathcal{N}_1(X)$, then one of the following statements holds.

(1) *There exists a bijective bounded linear or conjugate linear transformation $A : X \to Y$ such that*

(7.62) $$\varphi(N) = \tau_N ANA^{-1}$$

for all $N \in \mathcal{N}_1(X)$, where τ_N is a scalar depending on N.

(2) There exists a bijective bounded linear or conjugate linear transformation $A : X^* \to Y$ such that

(7.63) $$\varphi(N) = \tau_N A N^* A^{-1}$$

for all $N \in \mathcal{N}_1(X)$, where τ_N is a scalar depending on N.

7.2. The finite-dimensional case. Given three scalars μ, ν and ξ simultaneously nonzero such that either $\nu = \xi = 0$ or $\nu + \xi \neq 0$, the so-called *quadratic product* on $\mathcal{M}_n(\mathbb{C})$ is defined by

(7.64) $$\mathsf{q}(S, T) := \mu STS + \nu TS + \xi ST, \ (S, T \in \mathcal{M}_n(\mathbb{C})).$$

This product includes the usual product "TS" when $\mu = \xi = 0$ and $\nu = 1$, the triple product "STS" if $\mu = 1$ and $\xi = \nu = 0$ and Jordan product "$ST + TS$" if $\mu = 0$ and $\xi = \nu = 1$. In [33], Bourhim and Mabrouk showed that Theorem 7.1 remains valid when $X = Y = \mathbb{C}^n$ is a finite-dimensional space and without the surjectivity condition of the map φ. The proof given there in is completely different from that of Theorem 7.1. Later, a little bit more has been shown in [4] where Abdelali and Bourhim established the following result that characterizes maps on $\mathcal{M}_n(\mathbb{C})$ that preserve the local spectrum of the quadratic product of matrices at a nonzero fixed point of $\mathbb{C}^n$. It unifies and extends the main results of [21, 22, 33], where nonlinear maps preserving the local spectrum of the usual product "TS", the triple product "STS" and Jordan product "$ST + TS$" of matrices have been characterized.

Theorem 7.4 (Abdelali–Bourhim [4]). *Let x_0 be a nonzero vector in $\mathbb{C}^n$. A map φ on $\mathcal{M}_n(\mathbb{C})$ satisfies*

(7.65) $$\sigma_{\mathsf{q}(S,T)}(x_0) = \sigma_{\mathsf{q}(\varphi(S),\varphi(T))}(x_0), \ (S, T \in \mathcal{M}_n(\mathbb{C}))$$

if and only if there is an invertible matrix A and a complex scalar λ such that $Ax_0 = x_0$ and

(7.66) $$\varphi(T) = \lambda A T A^{-1}, \ (T \in \mathcal{M}_n(\mathbb{C})),$$

with

(7.67) $$\begin{cases} \lambda^3 = 1 & \text{if} \quad (\nu, \xi) = (0, 0); \\ \lambda^2 = 1 & \text{if} \quad \mu = 0; \\ \lambda = 1 & \text{if} \quad \mu \neq 0 \text{ and } (\nu, \xi) \neq (0, 0). \end{cases}$$

The proof of the above result is simpler and shorter than the proofs of the main results of [21, 22, 33]. First, for every x, $y \in \mathbb{C}^n$ and $T \in \mathcal{M}_n(\mathbb{C})$, an elementary lemma that describes the local spectrum of $\mathsf{q}(x \otimes y, T)$ at a nonzero vector $x_0 \in \mathbb{C}^n$ was given. It was then used in the proof of a local spectral identity principle that says that if G is a dense subset of $\mathcal{M}_n(\mathbb{C})$ and is a neighborhood of **1** then two matrices A and B in $\mathcal{M}_n(\mathbb{C})$ are equal if and only if $\sigma_{\mathsf{q}(S,A)}(x_0) = \sigma_{\mathsf{q}(S,B)}(x_0)$ for all $S \in G$. With a density argument, it is then shown that if a map φ on $\mathcal{M}_n(\mathbb{C})$ satisfies (7.65) then there is an open dense and arcwise connected subset G of $\mathcal{M}_n(\mathbb{C})$ and a bijective linear map L on $\mathcal{M}_n(\mathbb{C})$ such that φ and L coincide, and

(7.68) $$\sigma\left(\mathsf{q}(S, T)\right) = \sigma\left(\mathsf{q}(L(S), L(T))\right), \ (S, T \in \mathcal{M}_n(\mathbb{C})).$$

Then the following lemma tells us that there is an invertible matrix $M \in \mathcal{M}_n(\mathbb{C})$ and a scalar λ satisfying (7.67) such that either $L(T) = \lambda M T M^{-1}$ for all $T \in \mathcal{M}_n(\mathbb{C})$ or $L(T) = \lambda M T^{tr} M^{-1}$ for all $T \in \mathcal{M}_n(\mathbb{C})$.

Lemma 7.5 (Abdelali–Bourhim [4]). *Let α and β be two scalars simultaneously nonzero, and let L be a bijective linear map on $\mathcal{M}_n(\mathbb{C})$ such that*

(7.69) $\quad\quad \sigma\left(\alpha T^3 + \beta T^2\right) = \sigma\left(\alpha L(T)^3 + \beta L(T)^2\right), \ (T \in \mathcal{M}_n(\mathbb{C}))$.

Then there is an invertible matrix $M \in \mathcal{M}_n(\mathbb{C})$ and a complex scalar λ such that

(7.70) $\quad\quad\quad\quad L(T) = \lambda M T M^{-1}, \ (T \in \mathcal{M}_n(\mathbb{C}))$,

or

(7.71) $\quad\quad\quad\quad L(T) = \lambda M T^{tr} M^{-1}, \ (T \in \mathcal{M}_n(\mathbb{C}))$,

with

(7.72) $\quad\quad \begin{cases} \lambda^3 = 1 & \text{if} \quad \beta = 0; \\ \lambda^2 = 1 & \text{if} \quad \alpha = 0; \\ \lambda = 1 & \text{if} \quad \alpha\beta \neq 0. \end{cases}$

Once the above lemma is applied, choose a nonzero scalar t such that $\mu t + \nu + \xi \neq 0$ and $t\mathbf{1} \in G$, and note that (7.65) implies that

$$\sigma_{t(\mu t + \nu + \xi)T}(x_0) = \sigma_{\mathsf{q}(t\mathbf{1}, T)}(x_0)$$
$$= \sigma_{\mathsf{q}(\varphi(t\mathbf{1}), \varphi(T))}(x_0)$$
$$= \sigma_{\mathsf{q}(L(t\mathbf{1}), L(T))}(x_0)$$
$$= \sigma_{t(\mu t + \nu + \xi)\lambda^{-1} L(T)}(x_0)$$

for all $T \in G$. This implies that $\sigma_{\lambda^{-1} L(T)}(x_0) = \sigma_T(x_0)$ for all $T \in G$ and thus it was concluded that L takes only the form $T \mapsto \lambda M T M^{-1}$ on $\mathcal{M}_n(\mathbb{C})$ with λ satisfying (7.67) and that there is a nonzero scalar α such that $M x_0 = \alpha x_0$. As $\varphi(S) = L(S) = \lambda M S M^{-1}$ for all $S \in G$, it was shown that

$$\sigma_{\mathsf{q}(S, \varphi(T))}(x_0) = \sigma_{\mathsf{q}(S, L(T))}(x_0)$$

for all $S \in G$ and $T \in \mathcal{M}_n(\mathbb{C})$ and then the local spectral identity principle permits to conclude that $\varphi(T) = L(T) = \lambda M T M^{-1}$ for all $T \in \mathcal{M}_n(\mathbb{C})$. If necessary, let $A = M/\alpha$ and observe that $A x_0 = x_0$ and $\varphi(T) = \lambda A T A^{-1}$ for all $T \in \mathcal{M}_n(\mathbb{C})$.

8. Lie product and nonlinear local spectra preservers

On any ring $\mathcal{R}$, the Lie product is defined by

$$[x, y] := xy - yx, \ (x, y \in \mathcal{R}).$$

If $\mathcal{R}$ is a $*$-ring, then the skew Lie product on $\mathcal{R}$ is defined by

$$[x, y]_* := xy - yx^*, \ (x, y \in \mathcal{R}).$$

A number of authors classified maps on matrix algebras or operator algebras or general Banach algebras preserving several algebraic and spectral quantities of the Lie and skew Lie products; see [**15, 50, 53, 55, 67, 68, 73, 74, 76, 82, 83, 85, 90, 91, 93, 95, 114, 115, 125, 130, 137, 139, 140, 151, 153**] and their references. In [**67**], Cui, Li and Poon studied preservers of a unitary similarity function f on Lie products of matrices and then obtain results on special cases when the function f is the pseudo spectrum and the pseudo spectral radius. In [**93**], Hou and He gave a complete classification of nonlinear surjective maps on $\mathcal{B}_s(H)$, the space of all self-adjoint operators on a complex separable Hilbert space, preserving, respectively, numerical radius and numerical range of Lie product. Maps on factor Von Neumann algebras preserving skew Lie product, strong skew Lie product and zero skew Lie product are consider by several authors; see [56, 68, 69, 71, 82, 151]. In [71], Cui and Li proved that, if $\mathcal{A}$ and $\mathcal{B}$ are factor Von Neumann algebras and $\Phi : \mathcal{A} \to \mathcal{B}$ is a bijective map preserving skew Lie products (i.e., $\Phi([S,T]_*) = [\Phi(S), \Phi(T)]_*$ for all $S, T \in \mathcal{A}$), then Φ is a $*$-ring isomorphism. In [72], Cui and Hou characterized, in particular, all linear bijective maps preserving zero skew Lie product of operators. In [56], the authors characterized all maps on $\mathcal{B}(\mathcal{H})$ preserving numerical range and the maps preserving pseudo-spectrum of skew Lie product of operators.

In this section, we discuss surjective maps preserving the local spectrum of Lie and skew-Lie products of operators and of matrices at a nonzero vector.

8.1. The infinite-dimensional case.
Let X and Y be two infinite-dimensional complex Banach spaces. For any nonzero vector $x_0 \in X$, let

$$\mathcal{E}_{x_0}(X) := \{T \in \mathcal{B}(X) : Tx_0 \in \mathbb{C}x_0 \text{ and } T^2 + \alpha T + \beta \mathbf{1} = 0 \text{ for some } \alpha, \beta \in \mathbb{C} \text{ with } \alpha^2 - 4\beta \neq 0\}.$$

Note that this set is nothing but the collection of all operators $T \in \mathcal{B}(X)$ for which x_0 is an eigenvector and $(T - r\mathbf{1}_X)^2$ is a nonzero scalar operator for some scalar $r \in \mathbb{C}$.

The following theorem characterizes all maps from $\mathcal{B}(X)$ onto $\mathcal{B}(Y)$ preserving the local spectrum of Lie product of operators at fixed vectors.

Theorem 8.1 (Abdelali–Bourhim–Mabrouk [5]). *Let $x_0 \in X$ and $y_0 \in Y$ be two nonzero vectors. A map φ from $\mathcal{B}(X)$ onto $\mathcal{B}(Y)$ satisfies*

(8.73) $$\sigma_{[\varphi(T),\ \varphi(S)]}(y_0) = \sigma_{[T,\ S]}(x_0), \quad (T, S \in \mathcal{B}(X)),$$

if and only if there are two functions $\eta : \mathcal{B}(X) \to \mathbb{C}$ and $\xi : \mathcal{B}(X) \to \{-1, 1\}$, and a bijective map $A \in \mathcal{B}(X, Y)$ such that $Ax_0 = y_0$, the function ξ is constant on $\mathcal{B}(X) \backslash \mathcal{E}_{x_0}(X)$ and

(8.74) $$\varphi(T) = \xi(T) A T A^{-1} + \eta(T) \mathbf{1}_Y$$

for all $T \in \mathcal{B}(X)$.

Note that the function η freely maps $\mathcal{B}(X)$ into $\mathbb{C}$ and need not be linear or continuous. We also note that ξ may map $\mathcal{E}_{x_0}(X)$ arbitrarily into $\{-1, 1\}$ and thus it is not linear and may not be too continuous on $\mathcal{B}(X)$. This tells us that if φ satisfies (8.73) then φ may or may not be linear or continuous. It also shows that no one of Theorem 7.1 and Theorem 8.1 can be deduced from each other. We also point out that the proof of this theorem is long and uses several technical lemmas. One of them describes the local spectra of the Lie product of any rank operator in $\mathcal{B}(X)$ by an arbitrary operator in $\mathcal{B}(X)$, and was repeatedly used to establish other intermediate results. Another one characterizes rank one operators modulo scalars in term of the local spectrum at a nonzero vector $x_0 \in X$ of their Lie product by operators in $\mathcal{B}(X)$. It shows that an operator $A \in \mathcal{B}(X)$ is rank one modulo a scalar operator if and only if for every $T \in \mathcal{B}(X)$, the nonzero local spectrum $\sigma^*_{[A,T]}(x_0)$ is either a singleton or has two elements that are symmetric about the origin. Then those rank one operators $A \in \mathcal{B}(X)$ for which $\sigma^*_{[A,T]}(x_0)$ is a singleton for all $T \in \mathcal{B}(X)$ are also characterized and their description is also used in the proof of the above theorem.

Theorem 8.2 (Abdelali–Bourhim–Mabrouk [5]). *A nonzero operator $A \in \mathcal{B}(X)$ is rank one modulo a scalar operator if and only if for every operator $T \in \mathcal{B}(X)$ there exists a scalar α such that $\sigma^*_{[A,T]}(x_0) \subseteq \{-\alpha, \alpha\}$.*

For any subset $\mathcal{F}$ of $\mathcal{B}(X)$, let
$$\widetilde{\mathcal{F}} := \{T + \lambda \mathbf{1}_X \ : \ T \in \mathcal{F} \text{ and } \lambda \in \mathbb{C}\},$$
and set
$$\mathcal{N}_1^{x_0}(X) := \{N \in \mathcal{N}_1(X) : Nx_0 = 0\}.$$
The following result shows that operators $A \in \mathcal{B}(X)$ for which $\sigma^*_{[A,T]}(x_0)$ is a singleton for all $T \in \mathcal{B}(X)$ are nothing but those belonging to $\widetilde{\mathcal{N}_1^{x_0}}(X)$.

Theorem 8.3 (Abdelali–Bourhim–Mabrouk [5]). *A nonzero operator A belongs to $\widetilde{\mathcal{N}_1^{x_0}}(X)$ if and only if $\sigma^*_{[A,T]}(x_0)$ is a singleton for all operators $T \in \mathcal{B}(X)$.*

Besides these results, there are other lemmas that were used in the proof of Theorem 8.1. Here, we cite one of them which plays an important role in the proof of Theorem 8.1 and is of interest in its own right. It describes all bijective maps ψ from $\mathcal{N}_1^{x_0}(X)$ onto $\mathcal{N}_1^{y_0}(Y)$ for which

(8.75) $$\psi(\lambda N_1 + N_2) = \lambda \psi(N_1) + \psi(N_2)$$

for all $\lambda \in \mathbb{C}$ and $N_1, N_2 \in \mathcal{N}_1^{x_0}(X)$ for which $\lambda N_1 + N_2 \in \mathcal{N}_1^{x_0}(X)$. Here, $\mathcal{N}_1^{x_0}(X)$ is the set of all $N \in \mathcal{N}_1(X)$ vanishing at x_0, and $\mathcal{N}_1^{y_0}(Y)$ is the set of all $N \in \mathcal{N}_1(Y)$ vanishing at y_0.

Lemma 8.4 (Abdelali–Bourhim–Mabrouk [5]). *Let $\psi : \mathcal{N}_1^{x_0}(X) \cup \{0\} \to \mathcal{N}_1^{y_0}(Y) \cup \{0\}$ be a bijective map such that $\psi(0) = 0$ and $\psi(\lambda N_1 + N_2) = \lambda \psi(N_1) + \psi(N_2)$ for all $\lambda \in \mathbb{C}$ and $N_1, N_2 \in \mathcal{N}_1^{x_0}(X) \cup \{0\}$ for which $\lambda N_1 + N_2 \in \mathcal{N}_1^{x_0}(X) \cup \{0\}$. Then there exist a bijective linear mapping $A : X \to Y$ and two nonzero scalars λ_0 and β_0 such that $Ax_0 = \beta_0 y_0$ and*

(8.76) $$\psi(N) = \lambda_0 A N A^{-1}$$

for all $N \in \mathcal{N}_1^{x_0}(X)$.

8.2. The finite-dimensional case. Note that if $X = Y = \mathbb{C}^n$ are finite dimensional spaces and $x_0 = y_0 \in \mathbb{C}^n$ is a nonzero vector, then a map φ on $\mathcal{M}_n(\mathbb{C})$ satisfies (8.73) provided it is of the form (8.74). Indeed, assume that there are two functions $\eta : \mathcal{M}_n(\mathbb{C}) \to \mathbb{C}$ and $\xi : \mathcal{M}_n(\mathbb{C}) \to \{-1, 1\}$, and an invertible matrix $A \in \mathcal{M}_n(\mathbb{C})$ such that ξ is a constant function on $\mathcal{B}(X) \backslash \mathcal{E}_{x_0}(\mathbb{C}^n)$, $Ax_0 = x_0$ and

$$\varphi(T) = \xi(T) A T A^{-1} + \eta(T)$$

for all $T \in \mathcal{M}_n(\mathbb{C})$. Let $S, T \in \mathcal{M}_n(\mathbb{C})$, and note that

(8.77) $$\sigma_{[\varphi(T), \varphi(S)]}(x_0) = \sigma_{\xi(T)\xi(S) A [T,S] A^{-1}}(x_0) = \xi(T)\xi(S) \sigma_{[T,S]}(x_0).$$

If both S and T are not in $\mathcal{E}_{x_0}(\mathbb{C}^n)$, then $\xi(T)\xi(S) = 1$ and $\sigma_{[\varphi(T),\varphi(S)]}(x_0) = \sigma_{[T,S]}(x_0)$. If, however, $Tx_0 \in \mathbb{C}x_0$ and $(T-r)^2 = s\mathbf{1}$ for some scalars $r, s \in \mathbb{C}$ such that $s \neq 0$, we note that, since

$$[T, S] = (T-r)^{-1} [S, (T-r)] (T-r) = (T-r)^{-1} [S, T] (T-r),$$

we have
$$\begin{aligned}\sigma_{[T,S]}(x_0) &= \sigma_{(T-r)^{-1}[S,T](T-r)}(x_0) \\ &= \sigma_{[S,T]}((T-r)x_0) \\ &= \sigma_{[S,T]}(x_0) \\ &= -\sigma_{[T,S]}(x_0).\end{aligned}$$

Since $\xi(T)\xi(S) = \pm 1$, this identity and (8.77) entail that $\sigma_{[\varphi(T),\varphi(S)]}(y_0) = \sigma_{[T,S]}(x_0)$ whenever T or S belongs to $\mathcal{E}_{x_0}(\mathbb{C}^n)$. Therefore, (8.73) is satisfied for all $S, T \in \mathcal{M}_n(\mathbb{C})$ provided that φ is a map on $\mathcal{M}_n(\mathbb{C})$ of the form (8.74).

However, it is unknown whether or not the converse remains valid. So, it is natural to wonder if maps φ on $\mathcal{M}_n(\mathbb{C})$ satisfying (8.73) are of the form (8.74). This is therefore left as an open problem.

PROBLEM 1. Let $x_0 \in \mathbb{C}^n$ be a nonzero vector. Which maps φ on $\mathcal{M}_n(\mathbb{C})$ (not assumed to be surjective) satisfy

(8.78) $$\sigma_{[\varphi(T),\varphi(S)]}(x_0) = \sigma_{[T,S]}(x_0) \quad (T, S \in \mathcal{M}_n(\mathbb{C}))?$$

Motivated by this problem and the above cited papers on preservers and skew Lie product, Abdelali, Bourhim and Mabrouk characterized in [1] all maps φ on $\mathcal{M}_n(\mathbb{C})$ preserving the local spectrum at a fixed nonzero vector $x_0 \in \mathbb{C}^n$ of the skew Lie product of matrices.

Theorem 8.5 (Abdelali–Bourhim–Mabrouk [1]). *If $x_0 \in \mathbb{C}^n$ is a nonzero vector, then a map φ on $\mathcal{M}_n(\mathbb{C})$ satisfies*

(8.79) $$\sigma_{[\varphi(T),\varphi(S)]_*}(x_0) = \sigma_{[T,S]_*}(x_0), \ (T, S \in \mathcal{M}_n(\mathbb{C})),$$

if and only if there exists a unitary matrix $U \in \mathcal{M}_n(\mathbb{C})$ and a nonzero scalar $\alpha \in \mathbb{C}$ such that $Ux_0 = \alpha x_0$ and either $\varphi(T) = UTU^*$ for all $T \in \mathcal{M}_n(\mathbb{C})$, or $\varphi(T) = -UTU^*$ for all $T \in \mathcal{M}_n(\mathbb{C})$.

One of the important steps in the proof of Theorem 8.5 is to show that if a map φ on $\mathcal{M}_n(\mathbb{C})$ satisfies (8.79) then when restricted on an appropriate open subset of $\mathcal{M}_n(\mathbb{C})$ the map φ coincides with an invertible linear map L. Roughly speaking, to achieve this, the identity (6.47) that asserts that $\sigma_T(x_0) = \sigma(T)$ for all $T \in \mathcal{C}_{n,x_0}$ was used to show that there are two nonempty open subsets $\mathcal{N}_1$ and $\mathcal{N}_2$ such that

(8.80) $$\operatorname{tr}(S(T-T^*)) = \operatorname{tr}([T,S]_*) = \operatorname{tr}([\varphi(T),\varphi(S)]_*) = \operatorname{tr}(\varphi(S)(\varphi(T)-\varphi(T)^*))$$

for all $S \in \mathcal{N}_1$ and $T \in \mathcal{N}_2$. This and the fact that the set $\{X - X^* : X \in \mathcal{O}\}$ spans $\mathcal{M}_n(\mathbb{C})$ whenever $\mathcal{O}$ is a nonempty open subset of $\mathcal{M}_n(\mathbb{C})$ together with an argument borrowed from [53] imply the existence of such an invertible linear map L on $\mathcal{M}_n(\mathbb{C})$ and an open dense subset Ω of $\mathcal{M}_n(\mathbb{C})$ containing ($i\mathbf{1}$) such that φ coincides with L on Ω and

(8.81) $$\sigma(TS - ST^*) = \sigma(\varphi(T)\varphi(S) - \varphi(S)\varphi(T)^*)$$

for all S and T in Ω. This together with the density of Ω in $\mathcal{M}_n(\mathbb{C})$ and the continuity of L and the spectrum, it was therefore deduced that

(8.82) $$\sigma(TS - ST^*) = \sigma(L(T)L(S) - L(S)L(T)^*)$$

for all S and T in $\mathcal{M}_n(\mathbb{C})$. A complete description of such a map L was therefore given in [1, Theorem 2.3] where it was proved that there is a unitary matrix $U \in$

$\mathcal{M}_n(\mathbb{C})$ such that
(8.83) $$L(T) = \pm UTU^{-1}, \ (T \in \mathcal{M}_n(\mathbb{C})).$$
This and (8.79) imply that
$$2i\sigma_T(x_0) = \sigma_{[(i\mathbf{1}),T]_*}(x_0) = \sigma_{[\varphi(i\mathbf{1}),\varphi(T)]_*}(x_0) = \sigma_{[L(i\mathbf{1}),L(T)]_*}(x_0) = 2i\sigma_{UTU^*}(x_0)$$
for all $T \in \Omega$, and thus there is a nonzero scalar $\alpha \in \mathbb{C}$ such that $Ux_0 = \alpha x_0$. To conclude, observe that, for every T in Ω and $S \in \mathcal{M}_n(\mathbb{C})$, one has
$$\begin{aligned}\sigma_{[T,\varphi(S)]_*}(x_0) &= \pm \sigma_{[\varphi(U^*TU),\varphi(S)]_*}(x_0) \\ &= \pm \sigma_{[U^*TU,S]_*}(x_0) \\ &= \pm \sigma_{U^*[T,USU^*]_*U}(x_0) \\ &= \sigma_{[T,L(S)]_*}(x_0).\end{aligned}$$
Then apply the following lemma to get that $\varphi(S) = L(S) = \pm USU^*$ for all $S \in \mathcal{M}_n(\mathbb{C})$.

Lemma 8.6 (Abdelali–Bourhim–Mabrouk [1]). *If Ω is a dense subset of $\mathcal{M}_n(\mathbb{C})$, then two matrices A and B in $\mathcal{M}_n(\mathbb{C})$ coincide if and only if*
(8.84) $$\sigma_{[T,A]_*}(x_0) = \sigma_{[T,B]_*}(x_0)$$
for all $T \in \Omega$.

This approach can't be used to treat Problem 1. Indeed, if φ satisfies (8.78) then
(8.85) $$\operatorname{tr}([S,T]) = \operatorname{tr}(S(T-T)) = 0 = \operatorname{tr}(\varphi(S)(\varphi(T)-\varphi(T))) = \operatorname{tr}([\varphi(S),\varphi(T)])$$
for all $S, T \in \mathcal{M}_n(\mathbb{C})$. This shows that the above described techniques won't help to show that, when restricted on an appropriate open subset of $\mathcal{M}_n(\mathbb{C})$, the map φ coincides with a map of a particular form that leads to the full description of φ. For the same reason, these methods won't be helpful as well to characterize maps on $\mathcal{M}_n(\mathbb{C})$ preserving the spectrum of Lie product of matrices, and therefore the following natural problem suggests itself.

PROBLEM 2. Which maps φ on $\mathcal{M}_n(\mathbb{C})$ satisfy
(8.86) $$\sigma([\varphi(T),\varphi(S)]) = \sigma([T,S]) \qquad (T, S \in \mathcal{M}_n(\mathbb{C}))?$$

It should be pointed out that even in the infinite-dimensional case and with the surjectivity condition the full description of maps from $\mathcal{B}(X)$ onto $\mathcal{B}(Y)$ preserving the spectrum of Lie product of operators is unknown.

PROBLEM 3. Characterize all maps φ from $\mathcal{B}(X)$ onto $\mathcal{B}(Y)$ satisfying
(8.87) $$\sigma([\varphi(T),\varphi(S)]) = \sigma([T,S]) \qquad (T, S \in \mathcal{B}(X)).$$

9. Concluding remarks and open problems

In the sequel, let X and Y be two complex Banach spaces, and let $x_0 \in X$ and $y_0 \in Y$ be two nonzero vectors. For any operators T and S in $\mathcal{B}(X)$, let $T \star S$ stands either for

(a) the usual product "TS", or
(b) the triple product "STS", or
(c) the Jordan product "$TS + ST$", or
(d) the Lie product "$[T,S]$", or

(e) the skew-Lie product "$[T,S]_*$" when $X = \mathcal{H}$ is a complex Hilbert space.

In this section, we make some remarks and comments on linear and nonlinear preservers of spectra and local spectra, and then discuss further challenging problems, which are suggested by the main results of this paper.

A number of authors described maps not assumed to be linear that preserve the spectrum of Banach space operators or matrices; see for [16, 32, 53, 61, 63, 64, 70, 112, 134, 135] and the references given there. However, the study of nonlinear maps preserving the spectral radius of Banach space operators or matrices has been considered only by a few authors; see for instance [27, 66, 123].

PROBLEM 4. Which maps φ from $\mathcal{B}(X)$ onto $\mathcal{B}(Y)$ satisfy

(9.88) $\qquad \mathrm{r}\left(\varphi(T) \star \varphi(S)\right) = \mathrm{r}(T \star S) \qquad (T,\, S \in \mathcal{B}(X)).$

When $X = Y = \mathbb{C}^n$, Bhatia, Šemrl and Sourour described in [27] the form of all surjective maps defined on the algebra $\mathcal{M}_n(\mathbb{C})$ of all complex $n \times n$-matrices and preserving the spectral radius of the difference of matrices. They showed that a surjective map φ on $M_n(\mathbb{C})$ satisfies

$$\mathrm{r}\left(\varphi(T) - \varphi(S)\right) = \mathrm{r}(T-S), \qquad T,\, S \in M_n(\mathbb{C}),$$

if and only if there exist a unimodular scalar $\alpha \in \mathbb{C}$, and matrices $A,\, B \in M_n(\mathbb{C})$ such that A is invertible and

$$\varphi(T) = \alpha A T^{\#} A^{-1} + B$$

for all $T \in M_n(\mathbb{C})$, where $T^{\#}$ stands for one of the mappings T, or T^{tr}, or T^*, or $\overline{T}$, the complex conjugation of T. A variant of the above problem is the following one that demands the full description of maps from $\mathcal{B}(X)$ onto $\mathcal{B}(Y)$ preserving the spectral radius of the difference or sum of operators.

PROBLEM 5. Which surjective maps φ from $\mathcal{B}(X)$ onto $\mathcal{B}(Y)$ satisfy

(9.89) $\qquad \mathrm{r}\left(\varphi(T) \pm \varphi(S)\right) = \mathrm{r}(T \pm S) \qquad (T,\, S \in \mathcal{B}(X)).$

We also would like to point out that one may ask similar questions when the spectral radius is replaced by the local spectral radius or by the inner local spectral radius. For a positive scalar r, let $\overline{\mathrm{D}}(0,r)$ (resp. $\mathrm{D}(0,r)$) denote the closed (resp. the open) disc centered at the origin with radius r, and for a closed subset F of $\mathbb{C}$ and an operator $T \in \mathcal{B}(X)$, the subspace

$$\mathcal{X}_T(F) := \{x \in X : (T - \lambda)f(\lambda) = x \text{ has an analytic solution } f \text{ on } \mathbb{C}\backslash F\}$$

is the so-called *glocal spectral subspace* of T. Recall that the local spectral radius of T at any vector $x \in X$ coincides with

$$\mathrm{r}_T(x) = \inf\left\{r \geq 0 : x \in \mathcal{X}_T(\overline{\mathrm{D}}(0,r))\right\};$$

see [121, Proposition 3.3.13]. The *inner local spectral radius* of an operator $T \in \mathcal{B}(X)$ at any vector $x \in X$ is defined, in a similar way, by

$$\iota_T(x) := \sup\left\{r \geq 0 : x \in \mathcal{X}_T(\mathbb{C}\backslash\mathrm{D}(0,r))\right\},$$

and note that $\iota_T(x) = 0$ if and only if $0 \in \sigma_T(x)$; see [108].

PROBLEM 6. Which maps φ from $\mathcal{B}(X)$ onto $\mathcal{B}(Y)$ satisfy

(9.90) $\qquad \mathrm{r}_{\varphi(T)\star\varphi(S)}(y_0) = \mathrm{r}_{T\star S}(x_0) \qquad (T,\, S \in \mathcal{B}(X))?$

or

(9.91) $$\mathrm{r}_{\varphi(T)\pm\varphi(S)}(y_0) = \mathrm{r}_{T\pm S}(x_0) \quad (T, S \in \mathcal{B}(X))?$$

What about those maps φ satisfying (9.90) or (9.91) but when the local spectral radius is replaced by the inner local spectral radius?

In [**59**], Costara described surjective maps φ on $M_n(\mathbb{C})$ preserving the local spectral radius distance. He proved that if $\overline{x_0}$ denotes the complex conjugation of x_0 then a surjective map φ on $M_n(\mathbb{C})$ satisfies $\varphi(0) = 0$ and (9.91) if and only if there exists an invertible matrix $A \in M_n(\mathbb{C})$ and unimodular scalar $\alpha \in \mathbb{C}$ such that either $Ax_0 = x_0$ and $\varphi(T) = ATA^{-1}$ for all $T \in M_n(\mathbb{C})$ or $A\overline{x_0} = x_0$ and $\varphi(T) = A\overline{T}A^{-1}$ for all $T \in M_n(\mathbb{C})$. But the above problem remains unsolved in the infinite-dimensional case. It is also worth mentioning that Jari showed in [**100**, Theorem 4.1 & Theorem 4.2] that a bicontinuous bijective surjective map φ from $\mathcal{B}(X)$ onto $\mathcal{B}(Y)$ satisfies $\varphi(0) = 0$ and

(9.92) $$\iota_{\varphi(T)\pm\varphi(S)}(y_0) = \iota_{T\pm S}(x_0) \quad (T, S \in \mathcal{B}(X)),$$

if and only if a linear or conjugate linear bijection $A : X \to Y$ and a unimodular scalar α such that $Ax_0 = y_0$ and $\varphi(T) = \alpha ATA^{-1}$ for all $T \in \mathcal{B}(X)$. He also described bicontinuous bijective maps φ from $\mathcal{B}(X)$ onto $\mathcal{B}(Y)$ preserving the local spectrum of the difference and sum operators at nonzero fixed vectors. But wether his results remain valid without the bicontinuity condition on φ remain unknown except when $X = Y = \mathbb{C}^n$; see [**19, 23, 24**]. In [**19**], Benbouziane and El Kettani obtained a complete description of all maps φ on $\mathcal{M}_n(\mathbb{C})$ (not linear or surjective) satisfying

(9.93) $$\sigma_{T\pm S}(x_0) \subseteq \sigma(\varphi(T) \pm \varphi(S)), \ (T, S \in \mathcal{M}_n(\mathbb{C})).$$

As a consequence, they recaptured the main result of [**61**] where Costara characterized maps φ on $\mathcal{M}_n(\mathbb{C})$ satisfying

(9.94) $$\sigma(T \pm S) \subseteq \sigma(\varphi(T) \pm \varphi(S)), \ (T, S \in \mathcal{M}_n(\mathbb{C})).$$

They also recaptured the main results of [**23, 24**] where maps φ on $\mathcal{M}_n(\mathbb{C})$ satisfying

(9.95) $$\sigma_{T\pm S}(x_0) \subseteq \sigma_{\varphi(T)\pm\varphi(S)}(x_0), \ (T, S \in \mathcal{M}_n(\mathbb{C})),$$

are characterized. They also obtained a similar description by supposing that φ is surjective and

(9.96) $$\sigma_{\varphi(T)\pm\varphi(S)}(x_0) \subseteq \sigma(T \pm S), \ (T, S \in \mathcal{M}_n(\mathbb{C})).$$

But their proofs use Costara's approach and ideas. Note that φ satisfies (9.93) provided that (9.94) or (9.95) holds.

PROBLEM 7. Which maps φ from $\mathcal{B}(X)$ onto $\mathcal{B}(Y)$ satisfy

(9.97) $$\sigma_{\varphi(T)\pm\varphi(S)}(y_0) = \sigma_{T\pm S}(x_0) \quad (T, S \in \mathcal{B}(X))?$$

In [**60**], Costara described surjective linear maps on $\mathcal{B}(X)$ which preserve operators of local spectral radius zero at points of X. He showed, in particular, that if φ is a surjective linear map on $\mathcal{B}(X)$ such that for every $x \in X$ and $T \in \mathcal{B}(X)$, we have

(9.98) $$\mathrm{r}_T(x) = 0 \quad \text{if and only if} \quad \mathrm{r}_{\varphi(T)}(x) = 0,$$

then there exists a nonzero scalar c such that $\varphi(T) = c\,T$ for all $T \in \mathcal{B}(X)$; see [**3, 37, 38, 78–80, 100**] for related results. In [**29**], Bourhim and Costara considered a more general problem of describing linear maps φ from $\mathcal{B}(X)$ onto $\mathcal{B}(Y)$ such that

(9.99) $\quad\quad \mathrm{r}_T(x_0) = 0 \quad$ if and only if $\quad \mathrm{r}_{\varphi(T)}(y_0) = 0 \quad (T \in \mathcal{B}(X))$.

When $X = Y = \mathbb{C}^n$, they showed that if $n \geq 3$ and $x_0 = y_0 \in \mathbb{C}^n$ is a nonzero vector then a linear map φ on $\mathcal{M}_n(\mathbb{C})$ satisfies (9.99) if and only if there exists a nonzero scalar α and an invertible matrix $U \in \mathcal{M}_n(\mathbb{C})$ such that $Ux_0 = x_0$ and

(9.100) $\quad\quad\quad\quad\quad\quad \varphi(T) = \alpha U T U^{-1}$

for all $T \in \mathcal{M}_n(\mathbb{C})$. If, however, $n = 2$, they proved that a linear map φ on $\mathcal{M}_2(\mathbb{C})$ satisfies (9.99) if and only if there exists a nonzero scalar α, an invertible matrix $U \in \mathcal{M}_2(\mathbb{C})$ for which $Ux_0 = x_0$ and a matrix $Q \in \mathcal{M}_2(\mathbb{C})$ satisfying $Qx_0 = 0$ and $\mathrm{tr}(Q) \neq -1$ such that

(9.101) $\quad\quad\quad\quad\quad \varphi(T) = \alpha \left(UTU^{-1} + \mathrm{tr}(T) \cdot Q \right)$

for all $T \in \mathcal{M}_2(\mathbb{C})$. Recently, Costara described in [57] the form of all surjective additive maps on $\mathcal{M}_n(\mathbb{C})$ which preserve matrices of inner local spectral radius zero at fixed nonzero vector. But it is still unknown wether the main results of [29, 57] remain true in the infinite-dimensional case and thus following problem suggests itself.

PROBLEM 8. Which linear maps φ from $\mathcal{B}(X)$ onto $\mathcal{B}(Y)$ satisfy (9.99)? What about those linear maps φ satisfying (9.99) but when the local spectral radius is replaced by the inner local spectral radius?

For a recent set of questions on maps (not assumed to be linear) preserving operators or matrices of local and inner local spectral radii zero at fixed nonzero vector and their motivations, we refer the interested reader to [29]. Here, we state one of them.

PROBLEM 9. Which maps φ from $\mathcal{B}(X)$ onto $\mathcal{B}(Y)$ satisfy

(9.102) $\quad\quad \mathrm{r}_{\varphi(T) \star \varphi(S)}(y_0) = 0 \iff \mathrm{r}_{T \star S}(x_0) = 0 \quad (T, S \in \mathcal{B}(X))$?

What about those maps φ satisfying (9.102) but when the local spectral radius is replaced by the inner local spectral radius?

One may naturally ask if the answer to the above problems is known when the local spectral radius is replaced by the spectral radius. It should be noted that even the question of which linear maps φ from $\mathcal{B}(X)$ onto $\mathcal{B}(Y)$ preserving the quasinilpotency in both directions is still unknown and remains an open problem. We therefore close this paper with the following problem that suggests itself. It is more general than the problem of describing quasinilpotency preserving linear maps.

PROBLEM 10. Describe all maps φ from $\mathcal{B}(X)$ onto $\mathcal{B}(Y)$ satisfying

(9.103) $\quad\quad \mathrm{r}\left(\varphi(T) \star \varphi(S)\right) = 0 \iff \mathrm{r}(T \star S) = 0 \quad (T, S \in \mathcal{B}(X))$,

or

(9.104) $\quad\quad \mathrm{r}\left(\varphi(T) \pm \varphi(S)\right) = 0 \iff \mathrm{r}(T \pm S) = 0 \quad (T, S \in \mathcal{B}(X))$.

References

[1] Z. Abdelali, A. Bourhim and M. Mabrouk, *Spectrum and local spectrum preservers of skew Lie products of matrices*, accepted for publication.

[2] Z. E. A. Abdelali, *Maps preserving the spectrum of polynomial products of matrices*, J. Math. Anal. Appl. **480** (2019), no. 2, 123392, 20, DOI 10.1016/j.jmaa.2019.123392. MR4000080

[3] Z. E. A. Abdelali, A. Achchi, and R. Marzouki, *Maps preserving the local spectral radius zero of generalized product of operators*, Linear Multilinear Algebra **67** (2019), no. 10, 2021–2029, DOI 10.1080/03081087.2018.1479371. MR3987577

[4] Z. E. A. Abdelali and A. Bourhim, *Maps preserving the local spectrum of quadratic products of matrices*, Acta Sci. Math. (Szeged) **84** (2018), no. 1-2, 49–64. MR3792765

[5] Z. Abdelali, A. Bourhim, and M. Mabrouk, *Lie product and local spectrum preservers*, Linear Algebra Appl. **553** (2018), 328–361, DOI 10.1016/j.laa.2018.05.013. MR3809383

[6] Z. E. A. Abdelali, A. Achchi, and R. Marzouki, *Maps preserving the local spectrum of some matrix products*, Oper. Matrices **12** (2018), no. 2, 549–562, DOI 10.7153/oam-2018-12-34. MR3812190

[7] Z. Abdelali, A. Achchi, and R. Marzouki, *Maps preserving the local spectrum of skew-product of operators*, Linear Algebra Appl. **485** (2015), 58–71, DOI 10.1016/j.laa.2015.07.019. MR3394138

[8] P. Aiena, *Fredholm and local spectral theory, with applications to multipliers*, Kluwer Academic Publishers, Dordrecht, 2004. MR2070395

[9] J. Alaminos, M. Brešar, J. Extremera, Š. Špenko, and A. R. Villena, *Determining elements in C^*-algebras through spectral properties*, J. Math. Anal. Appl. **405** (2013), no. 1, 214–219, DOI 10.1016/j.jmaa.2013.04.009. MR3053501

[10] G. An, *Characterizations of n-Jordan homomorphisms*, Linear Multilinear Algebra **66** (2018), no. 4, 671–680, DOI 10.1080/03081087.2017.1318818. MR3779141

[11] R. L. An and J. C. Hou, *Jordan ring isomorphism on the space of symmetric operators* (Chinese, with English and Chinese summaries), Acta Math. Sinica (Chin. Ser.) **55** (2012), no. 6, 991–1000. MR3058248

[12] R. An and J. Hou, *Additivity of Jordan multiplicative maps on Jordan operator algebras*, Taiwanese J. Math. **10** (2006), no. 1, 45–64, DOI 10.11650/twjm/1500403798. MR2186161

[13] B. Aupetit, *Spectrum-preserving linear mappings between Banach algebras or Jordan-Banach algebras*, J. London Math. Soc. (2) **62** (2000), no. 3, 917–924, DOI 10.1112/S0024610700001514. MR1794294

[14] B. Aupetit, *A primer on spectral theory*, Universitext, Springer-Verlag, New York, 1991. MR1083349

[15] Z. Bai and S. Du, *Maps preserving products $XY - YX^*$ on von Neumann algebras*, J. Math. Anal. Appl. **386** (2012), no. 1, 103–109, DOI 10.1016/j.jmaa.2011.07.052. MR2834869

[16] L. Baribeau and T. Ransford, *Non-linear spectrum-preserving maps*, Bull. London Math. Soc. **32** (2000), no. 1, 8–14, DOI 10.1112/S0024609399006426. MR1718765

[17] B. A. Barnes, *Common operator properties of the linear operators RS and SR*, Proc. Amer. Math. Soc. **126** (1998), no. 4, 1055–1061, DOI 10.1090/S0002-9939-98-04218-X. MR1443814

[18] W. E. Baxter and W. S. Martindale III, *Jordan homomorphisms of semiprime rings*, J. Algebra **56** (1979), no. 2, 457–471, DOI 10.1016/0021-8693(79)90349-1. MR528587

[19] H. Benbouziane and M. E.-C. El Kettani, *Maps on matrices compressing the local spectrum in the spectrum*, Linear Algebra Appl. **475** (2015), 176–185, DOI 10.1016/j.laa.2015.02.015. MR3325227

[20] M. Bendaoud, *Preservers of local spectrum of matrix skew products*, Linear Multilinear Algebra **66** (2018), no. 8, 1530–1537, DOI 10.1080/03081087.2017.1363152. MR3806237

[21] M. Bendaoud, *Preservers of local spectrum of matrix Jordan triple products*, Linear Algebra Appl. **471** (2015), 604–614, DOI 10.1016/j.laa.2015.01.022. MR3314355

[22] M. Bendaoud, M. Jabbar, and M. Sarih, *Preservers of local spectra of operator products*, Linear Multilinear Algebra **63** (2015), no. 4, 806–819, DOI 10.1080/03081087.2014.902944. MR3291565

[23] M. Bendaoud, *Preservers of local spectra of matrix sums*, Linear Algebra Appl. **438** (2013), no. 5, 2500–2507, DOI 10.1016/j.laa.2012.10.028. MR3005308

[24] M. Bendaoud, M. Douimi, and M. Sarih, *Maps on matrices preserving local spectra*, Linear Multilinear Algebra **61** (2013), no. 7, 871–880, DOI 10.1080/03081087.2012.716429. MR3175332

[25] C. Benhida and E. H. Zerouali, *Local spectral theory of linear operators RS and SR*, Integral Equations Operator Theory **54** (2006), no. 1, 1–8, DOI 10.1007/s00020-005-1375-3. MR2195227

[26] T. Bermudez and M. Gonzalez, *On the boundedness of the local resolvent function*, Integral Equations Operator Theory **34** (1999), no. 1, 1–8, DOI 10.1007/BF01332488. MR1690283

[27] R. Bhatia, P. Šemrl, and A. R. Sourour, *Maps on matrices that preserve the spectral radius distance*, Studia Math. **134** (1999), no. 2, 99–110. MR1688218

[28] N. Boudi and P. Šemrl, *Semigroup automorphisms of nest algebras*, Houston J. Math. **38** (2012), no. 4, 1197–1206. MR3019030

[29] A. Bourhim and C. Costara, *Linear maps preserving matrices of local spectral radius zero at a fixed vector*, Canad. J. Math. **71** (2019), no. 4, 749–771, DOI 10.4153/cjm-2018-017-0. MR3984020

[30] A. Bourhim and J. E. Lee, *Multiplicatively local spectrum-preserving maps*, Linear Algebra Appl. **549** (2018), 291–308, DOI 10.1016/j.laa.2018.03.042. MR3784351

[31] A. Bourhim and M. Mabrouk, *Jordan product and local spectrum preservers*, Studia Math. **234** (2016), no. 2, 97–120. MR3549185

[32] A. Bourhim, J. Mashreghi, and A. Stepanyan, *Maps between Banach algebras preserving the spectrum*, Arch. Math. (Basel) **107** (2016), no. 6, 609–621, DOI 10.1007/s00013-016-0960-9. MR3571153

[33] A. Bourhim and M. Mabrouk, *Maps preserving the local spectrum of Jordan product of matrices*, Linear Algebra Appl. **484** (2015), 379–395, DOI 10.1016/j.laa.2015.06.034. MR3385068

[34] A. Bourhim and J. Mashreghi, *A survey on preservers of spectra and local spectra*, Invariant subspaces of the shift operator, Contemp. Math., vol. 638, Amer. Math. Soc., Providence, RI, 2015, pp. 45–98, DOI 10.1090/conm/638/12810. MR3309349

[35] A. Bourhim and J. Mashreghi, *Maps preserving the local spectrum of product of operators*, Glasg. Math. J. **57** (2015), no. 3, 709–718, DOI 10.1017/S0017089514000585. MR3395343

[36] A. Bourhim and J. Mashreghi, *Maps preserving the local spectrum of triple product of operators*, Linear Multilinear Algebra **63** (2015), no. 4, 765–773, DOI 10.1080/03081087.2014.898299. MR3291562

[37] A. Bourhim and J. Mashreghi, *Local spectral radius preservers*, Integral Equations Operator Theory **76** (2013), no. 1, 95–104, DOI 10.1007/s00020-013-2041-9. MR3041723

[38] A. Bourhim, *Surjective linear maps preserving local spectra*, Linear Algebra Appl. **432** (2010), no. 1, 383–393, DOI 10.1016/j.laa.2009.08.020. MR2566487

[39] A. Bourhim and V. G. Miller, *Linear maps on $M_n(\mathbb{C})$ preserving the local spectral radius*, Studia Math. **188** (2008), no. 1, 67–75, DOI 10.4064/sm188-1-4. MR2430550

[40] A. Bourhim, *Spectral pictures of operator-valued weighted bi-shifts*, J. Operator Theory **59** (2008), no. 1, 193–210. MR2404470

[41] A. Bourhim and T. Ransford, *Additive maps preserving local spectrum*, Integral Equations Operator Theory **55** (2006), no. 3, 377–385, DOI 10.1007/s00020-005-1392-2. MR2244195

[42] A. Bourhim, *On the local spectral properties of weighted shift operators*, Studia Math. **163** (2004), no. 1, 41–69, DOI 10.4064/sm163-1-3. MR2047464

[43] G. Braatvedt and R. Brits, *Uniqueness and spectral variation in Banach algebras*, Quaest. Math. **36** (2013), no. 2, 155–165, DOI 10.2989/16073606.2013.779947. MR3060974

[44] J. Bračič and V. Müller, *Local spectrum and local spectral radius of an operator at a fixed vector*, Studia Math. **194** (2009), no. 2, 155–162, DOI 10.4064/sm194-2-3. MR2534182

[45] J. Bračič and V. Müller, *On bounded local resolvents*, Integral Equations Operator Theory **55** (2006), no. 4, 477–486, DOI 10.1007/s00020-005-1402-4. MR2250159

[46] M. Brešar and Š. Špenko, *Determining elements in Banach algebras through spectral properties*, J. Math. Anal. Appl. **393** (2012), no. 1, 144–150, DOI 10.1016/j.jmaa.2012.03.058. MR2921656

[47] M. Brešar, *Jordan homomorphisms revisited*, Math. Proc. Cambridge Philos. Soc. **144** (2008), no. 2, 317–328, DOI 10.1017/S0305004107000825. MR2405892

[48] M. Brešar and P. Šemrl, *An extension of the Gleason-Kahane-Żelazko theorem: a possible approach to Kaplansky's problem*, Expo. Math. **26** (2008), no. 3, 269–277, DOI 10.1016/j.exmath.2007.11.004. MR2437097

[49] M. Brešar, A. Fošner, and P. Šemrl, *A note on invertibility preservers on Banach algebras*, Proc. Amer. Math. Soc. **131** (2003), no. 12, 3833–3837, DOI 10.1090/S0002-9939-03-07192-2. MR1999931

[50] M. Brešar and M. Fošner, *On rings with involution equipped with some new product*, Publ. Math. Debrecen **57** (2000), no. 1-2, 121–134. MR1771679

[51] M. Brešar and P. Šemrl, *Linear maps preserving the spectral radius*, J. Funct. Anal. **142** (1996), no. 2, 360–368, DOI 10.1006/jfan.1996.0153. MR1423038

[52] B. A. Barnes, *Common operator properties of the linear operators RS and SR*, Proc. Amer. Math. Soc. **126** (1998), no. 4, 1055–1061, DOI 10.1090/S0002-9939-98-04218-X. MR1443814

[53] J.-T. Chan, C.-K. Li, and N.-S. Sze, *Mappings preserving spectra of products of matrices*, Proc. Amer. Math. Soc. **135** (2007), no. 4, 977–986, DOI 10.1090/S0002-9939-06-08568-6. MR2262897

[54] J.-T. Chan, C.-K. Li, and N.-S. Sze, *Mappings on matrices: invariance of functional values of matrix products*, J. Aust. Math. Soc. **81** (2006), no. 2, 165–184, DOI 10.1017/S1446788700015809. MR2267789

[55] M. A. Chebotar, Y. Fong, and P.-H. Lee, *On maps preserving zeros of the polynomial $xy - yx$*, Linear Algebra Appl. **408** (2005), 230–243, DOI 10.1016/j.laa.2005.06.015. MR2166866

[56] C. Chen and F. Lu, *Nonlinear maps preserving higher-dimensional numerical range of skew Lie product of operators*, Oper. Matrices **10** (2016), no. 2, 335–344, DOI 10.7153/oam-10-18. MR3517730

[57] C. Costara, *Additive maps preserving matrices of inner local spectral radius zero at some fixed vector*, accepted for publication.

[58] C. Costara, *Automatic continuity for linear surjective maps compressing the local spectrum at fixed vectors*, Proc. Amer. Math. Soc. **145** (2017), no. 5, 2081–2087, DOI 10.1090/proc/13364. MR3611322

[59] C. Costara, *Surjective maps on matrices preserving the local spectral radius distance*, Linear Multilinear Algebra **62** (2014), no. 7, 988–994, DOI 10.1080/03081087.2013.801967. MR3232673

[60] C. Costara, *Linear maps preserving operators of local spectral radius zero*, Integral Equations Operator Theory **73** (2012), no. 1, 7–16, DOI 10.1007/s00020-012-1953-0. MR2913657

[61] C. Costara, *Maps on matrices that preserve the spectrum*, Linear Algebra Appl. **435** (2011), no. 11, 2674–2680, DOI 10.1016/j.laa.2011.04.026. MR2825274

[62] C. Costara, *Automatic continuity for linear surjective mappings decreasing the local spectral radius at some fixed vector*, Arch. Math. (Basel) **95** (2010), no. 6, 567–573, DOI 10.1007/s00013-010-0191-4. MR2745466

[63] C. Costara and D. Repovš, *Nonlinear mappings preserving at least one eigenvalue*, Studia Math. **200** (2010), no. 1, 79–89, DOI 10.4064/sm200-1-5. MR2720208

[64] C. Costara and T. Ransford, *On local irreducibility of the spectrum*, Proc. Amer. Math. Soc. **135** (2007), no. 9, 2779–2784, DOI 10.1090/S0002-9939-07-08779-5. MR2317952

[65] G. Corach, B. Duggal, and R. Harte, *Extensions of Jacobson's lemma*, Comm. Algebra **41** (2013), no. 2, 520–531, DOI 10.1080/00927872.2011.602274. MR3011779

[66] S. Clark, C.-K. Li, and L. Rodman, *Spectral radius preservers of products of nonnegative matrices*, Banach J. Math. Anal. **2** (2008), no. 2, 107–120, DOI 10.15352/bjma/1240336297. MR2436871

[67] J. Cui, C.-K. Li, and Y.-T. Poon, *Preservers of unitary similarity functions on Lie products of matrices*, Linear Algebra Appl. **498** (2016), 160–180, DOI 10.1016/j.laa.2015.02.036. MR3478557

[68] J. Cui, Q. Li, J. Hou, and X. Qi, *Some unitary similarity invariant sets preservers of skew Lie products*, Linear Algebra Appl. **457** (2014), 76–92, DOI 10.1016/j.laa.2014.05.009. MR3230434

[69] J. Cui and C. Park, *Maps preserving strong skew Lie product on factor von Neumann algebras*, Acta Math. Sci. Ser. B (Engl. Ed.) **32** (2012), no. 2, 531–538, DOI 10.1016/S0252-9602(12)60035-6. MR2921895

[70] J. Cui and C.-K. Li, *Maps preserving peripheral spectrum of Jordan products of operators*, Oper. Matrices **6** (2012), no. 1, 129–146, DOI 10.7153/oam-06-09. MR2952440

[71] J. Cui and C.-K. Li, *Maps preserving product $XY - YX^*$ on factor von Neumann algebras*, Linear Algebra Appl. **431** (2009), no. 5-7, 833–842, DOI 10.1016/j.laa.2009.03.036. MR2535555

[72] J. Cui and J. Hou, *Linear maps preserving elements annihilated by the polynomial $XY-YX$*, Studia Math. **174** (2006), no. 2, 183–199, DOI 10.4064/sm174-2-5. MR2238461

[73] G. Dolinar, *Maps on upper triangular matrices preserving Lie products*, Linear Multilinear Algebra **55** (2007), no. 2, 191–198, DOI 10.1080/03081080600635484. MR2288901

[74] G. Dolinar, *Maps on M_n preserving Lie products*, Publ. Math. Debrecen **71** (2007), no. 3-4, 467–477. MR2361725

[75] M. B. Dollinger and K. K. Oberai, *Variation of local spectra*, J. Math. Anal. Appl. **39** (1972), 324–337, DOI 10.1016/0022-247X(72)90205-3. MR318950

[76] S. Du, J. Hou, and Z. Bai, *Nonlinear maps preserving similarity on $\mathcal{B}(H)$*, Linear Algebra Appl. **422** (2007), no. 2-3, 506–516, DOI 10.1016/j.laa.2006.11.008. MR2305136

[77] A. Ebadian, A. Jabbari, and N. Kanzi, *n-Jordan homomorphisms and pseudo n-Jordan homomorphisms on Banach algebras*, Mediterr. J. Math. **14** (2017), no. 6, Art. 241, 11, DOI 10.1007/s00009-017-1041-4. MR3735471

[78] M. Ech-Cherif El Kettani and H. Benbouziane, *Additive maps preserving operators of inner local spectral radius zero*, Rend. Circ. Mat. Palermo (2) **63** (2014), no. 2, 311–316, DOI 10.1007/s12215-014-0160-z. MR3274066

[79] M. Elhodaibi and A. Jaatit, *Inner local spectral radius preservers*, Rend. Circ. Mat. Palermo (2) **67** (2018), no. 2, 215–225, DOI 10.1007/s12215-017-0308-8. MR3833003

[80] M. Elhodaibi and A. Jaatit, *On maps preserving operators of local spectral radius zero*, Linear Algebra Appl. **512** (2017), 191–201, DOI 10.1016/j.laa.2016.10.001. MR3567521

[81] M. Eshaghi Gordji, *n-Jordan homomorphisms*, Bull. Aust. Math. Soc. **80** (2009), no. 1, 159–164, DOI 10.1017/S000497270900032X. MR2520532

[82] A. Fošner and B. Kuzma, *Preserving zeros of Lie product on alternate matrices*, Spec. Matrices **4** (2016), 80–100, DOI 10.1515/spma-2016-0009. MR3451272

[83] M. Fošner, *Prime rings with involution equipped with some new product*, Southeast Asian Bull. Math. **26**, no. 1, (2003) 27–31.

[84] H. Gao, *∗-Jordan-triple multiplicative surjective maps on $\mathcal{B}(H)$*, J. Math. Anal. Appl. **401** (2013), no. 1, 397–403, DOI 10.1016/j.jmaa.2012.12.019. MR3011280

[85] H.-L. Gau and C.-K. Li, *C^*-isomorphisms, Jordan isomorphisms, and numerical range preserving maps*, Proc. Amer. Math. Soc. **135** (2007), no. 9, 2907–2914, DOI 10.1090/S0002-9939-07-08807-7. MR2317968

[86] A. M. Gleason, *A characterization of maximal ideals*, J. Analyse Math. **19** (1967), 171–172, DOI 10.1007/BF02788714. MR213878

[87] M. González and M. Mbekhta, *Linear maps on $M_n(\mathbb{C})$ preserving the local spectrum*, Linear Algebra Appl. **427** (2007), no. 2-3, 176–182, DOI 10.1016/j.laa.2007.07.005. MR2351350

[88] M. González, *An example of a bounded local resolvent*, Operator theory, operator algebras and related topics (Timişoara, 1996), Theta Found., Bucharest, 1997, pp. 159–162. MR1728418

[89] P. R. Halmos, *Does mathematics have elements?*, Math. Intelligencer **3** (1980/81), no. 4, 147–153, DOI 10.1007/BF03022973. MR642132

[90] O. Hatori, T. Miura, and H. Takagi, *Unital and multiplicatively spectrum-preserving surjections between semi-simple commutative Banach algebras are linear and multiplicative*, J. Math. Anal. Appl. **326** (2007), no. 1, 281–296, DOI 10.1016/j.jmaa.2006.02.084. MR2277782

[91] O. Hatori, T. Miura, and H. Takagi, *Characterizations of isometric isomorphisms between uniform algebras via nonlinear range-preserving properties*, Proc. Amer. Math. Soc. **134** (2006), no. 10, 2923–2930, DOI 10.1090/S0002-9939-06-08500-5. MR2231616

[92] I. N. Herstein, *Jordan homomorphisms*, Trans. Amer. Math. Soc. **81** (1956), 331–341, DOI 10.2307/1992920. MR76751

[93] J. Hou and K. He, *Non-linear maps on self-adjoint operators preserving numerical radius and numerical range of Lie product*, Linear Multilinear Algebra **64** (2016), no. 1, 36–57, DOI 10.1080/03081087.2015.1007912. MR3433377

[94] J. Hou, C.-K. Li, and N.-C. Wong, *Maps preserving the spectrum of generalized Jordan product of operators*, Linear Algebra Appl. **432** (2010), no. 4, 1049–1069, DOI 10.1016/j.laa.2009.10.018. MR2577648

[95] J. Hou, C.-K. Li, and N.-C. Wong, *Jordan isomorphisms and maps preserving spectra of certain operator products*, Studia Math. **184** (2008), no. 1, 31–47, DOI 10.4064/sm184-1-2. MR2365474

[96] J. Hou and Q. Di, *Maps preserving numerical ranges of operator products*, Proc. Amer. Math. Soc. **134** (2006), no. 5, 1435–1446, DOI 10.1090/S0002-9939-05-08101-3. MR2199190

[97] M. Hosseini and F. Sady, *Multiplicatively range-preserving maps between Banach function algebras*, J. Math. Anal. Appl. **357** (2009), no. 1, 314–322, DOI 10.1016/j.jmaa.2009.04.008. MR2526831

[98] N. Jacobson and C. E. Rickart, *Jordan homomorphisms of rings*, Trans. Amer. Math. Soc. **69** (1950), 479–502, DOI 10.2307/1990495. MR38335

[99] A. A. Jafarian and A. R. Sourour, *Spectrum-preserving linear maps*, J. Funct. Anal. **66** (1986), no. 2, 255–261, DOI 10.1016/0022-1236(86)90073-X. MR832991

[100] T. Jari, *Nonlinear maps preserving the inner local spectral radius*, Rend. Circ. Mat. Palermo (2) **64** (2015), no. 1, 67–76, DOI 10.1007/s12215-014-0181-7. MR3324374

[101] A. Jiménez-Vargas, A. Luttman, and M. Villegas-Vallecillos, *Weakly peripherally multiplicative surjections of pointed Lipschitz algebras*, Rocky Mountain J. Math. **40** (2010), no. 6, 1903–1922, DOI 10.1216/RMJ-2010-40-6-1903. MR2764228

[102] R. V. Kadison, *Isometries of operator algebras*, Ann. Of Math. (2) **54** (1951), 325–338, DOI 10.2307/1969534. MR0043392

[103] J.-P. Kahane and W. Żelazko, *A characterization of maximal ideals in commutative Banach algebras*, Studia Math. **29** (1968), 339–343, DOI 10.4064/sm-29-3-339-343. MR226408

[104] D. Kokol Bukovšek and B. Mojškerc, *Jordan triple product homomorphisms on Hermitian matrices to and from dimension one*, Linear Multilinear Algebra **64** (2016), no. 8, 1669–1678, DOI 10.1080/03081087.2015.1112345. MR3503375

[105] S. Kowalski and Z. Słodkowski, *A characterization of multiplicative linear functionals in Banach algebras*, Studia Math. **67** (1980), no. 3, 215–223, DOI 10.4064/sm-67-3-215-223. MR592387

[106] M. Marcus and B. N. Moyls, *Linear transformations on algebras of matrices*, Canadian J. Math. **11** (1959), 61–66, DOI 10.4153/CJM-1959-008-0. MR99996

[107] W. S. Martindale III, *When are multiplicative mappings additive?*, Proc. Amer. Math. Soc. **21** (1969), 695–698, DOI 10.2307/2036449. MR240129

[108] T. L. Miller, V. G. Miller, and M. M. Neumann, *Local spectral properties of weighted shifts*, J. Operator Theory **51** (2004), no. 1, 71–88. MR2055805

[109] L. Molnár and P. Šemrl, *Transformations of the unitary group on a Hilbert space*, J. Math. Anal. Appl. **388** (2012), no. 2, 1205–1217, DOI 10.1016/j.jmaa.2011.11.007. MR2869819

[110] L. Molnár, *Multiplicative Jordan triple isomorphisms on the self-adjoint elements of von Neumann algebras*, Linear Algebra Appl. **419** (2006), no. 2-3, 586–600, DOI 10.1016/j.laa.2006.06.007. MR2277989

[111] L. Molnár, *Non-linear Jordan triple automorphisms of sets of self-adjoint matrices and operators*, Studia Math. **173** (2006), no. 1, 39–48, DOI 10.4064/sm173-1-3. MR2204461

[112] L. Molnár, *Some characterizations of the automorphisms of $B(H)$ and $C(X)$*, Proc. Amer. Math. Soc. **130** (2002), no. 1, 111–120, DOI 10.1090/S0002-9939-01-06172-X. MR1855627

[113] L. Molnár, *On isomorphisms of standard operator algebras*, Studia Math. **142** (2000), no. 3, 295–302, DOI 10.4064/sm-142-3-295-302. MR1792612

[114] L. Molnár, *Jordan ∗-derivation pairs on a complex ∗-algebra*, Aequationes Math. **54** (1997), no. 1-2, 44–55, DOI 10.1007/BF02755445. MR1466293

[115] L. Molnár, *A condition for a subspace of $\mathcal{B}(H)$ to be an ideal*, Linear Algebra Appl. **235** (1996), 229–234, DOI 10.1016/0024-3795(94)00143-X. MR1374262

[116] D. Mosić, *Extensions of Jacobson's lemma for Drazin inverses*, Aequationes Math. **91** (2017), no. 3, 419–428, DOI 10.1007/s00010-017-0476-9. MR3651555

[117] T. Miura and D. Honma, *A generalization of peripherally-multiplicative surjections between standard operator algebras*, Cent. Eur. J. Math. **7** (2009), no. 3, 479–486, DOI 10.2478/s11533-009-0033-4. MR2534467

[118] V. Müller and M. M. Neumann, *Localizable spectrum and bounded local resolvent functions*, Arch. Math. (Basel) **91** (2008), no. 2, 155–165, DOI 10.1007/s00013-008-2652-6. MR2430799

[119] V. Müller, *On smooth local resolvents*, Integral Equations Operator Theory **57** (2007), no. 2, 229–234, DOI 10.1007/s00020-006-1467-8. MR2296761

[120] M. M. Neumann, *On local spectral properties of operators on Banach spaces*, Rend. Circ. Mat. Palermo (2) Suppl. **56** (1998), 15–25. International Workshop on Operator Theory (Cefalù, 1997). MR1710819
[121] K. B. Laursen and M. M. Neumann, *An introduction to local spectral theory*, London Mathematical Society Monographs. New Series, vol. 20, The Clarendon Press, Oxford University Press, New York, 2000. MR1747914
[122] G. Lešnjak and P. Šemrl, *Continuous multiplicative mappings on $C(X)$*, Proc. Amer. Math. Soc. **126** (1998), no. 1, 127–133, DOI 10.1090/S0002-9939-98-03967-7. MR1402871
[123] C.-K. Li and L. Rodman, *Preservers of spectral radius, numerical radius, or spectral norm of the sum on nonnegative matrices*, Linear Algebra Appl. **430** (2009), no. 7, 1739–1761, DOI 10.1016/j.laa.2008.04.022. MR2494661
[124] C.-K. Li, P. Šemrl, and N.-S. Sze, *Maps preserving the nilpotency of products of operators*, Linear Algebra Appl. **424** (2007), no. 1, 222–239, DOI 10.1016/j.laa.2006.11.013. MR2324385
[125] W.-H. Lin, *Nonlinear $*$-Lie-type derivations on von Neumann algebras*, Acta Math. Hungar. **156** (2018), no. 1, 112–131, DOI 10.1007/s10474-018-0803-1. MR3856906
[126] C. Lin, Z. Yan, and Y. Ruan, *Common properties of operators RS and SR and p-hyponormal operators*, Integral Equations Operator Theory **43** (2002), no. 3, 313–325, DOI 10.1007/BF01255566. MR1902952
[127] F. Y. Lu and J. H. Xie, *Multiplicative mappings of rings*, Acta Math. Sin. (Engl. Ser.) **22** (2006), no. 4, 1017–1020, DOI 10.1007/s10114-005-0620-7. MR2245232
[128] F. Lu, *Jordan triple maps*, Linear Algebra Appl. **375** (2003), 311–317, DOI 10.1016/j.laa.2003.06.004. MR2013474
[129] F. Lu, *Multiplicative mappings of operator algebras*, Linear Algebra Appl. **347** (2002), 283–291, DOI 10.1016/S0024-3795(01)00560-2. MR1899895
[130] C.-K. Li, E. Poon, and N.-S. Sze, *Preservers for norms of Lie product*, Oper. Matrices **3** (2009), no. 2, 187–203, DOI 10.7153/oam-03-10. MR2522774
[131] J. Qian and P. Li, *Additivity of Lie maps on operator algebras*, Bull. Korean Math. Soc. **44** (2007), no. 2, 271–279, DOI 10.4134/BKMS.2007.44.2.271. MR2325029
[132] N. V. Rao and A. K. Roy, *Multiplicatively spectrum-preserving maps of function algebras. II*, Proc. Edinb. Math. Soc. (2) **48** (2005), no. 1, 219–229, DOI 10.1017/S0013091504000719. MR2117721
[133] N. V. Rao and A. K. Roy, *Multiplicatively spectrum-preserving maps of function algebras*, Proc. Amer. Math. Soc. **133** (2005), no. 4, 1135–1142, DOI 10.1090/S0002-9939-04-07615-4. MR2117215
[134] P. Šemrl, *Order and spectrum preserving maps on positive operators*, Canad. J. Math. **69** (2017), no. 6, 1422–1435, DOI 10.4153/CJM-2016-039-0. MR3715017
[135] P. Šemrl, *Characterizing Jordan automorphisms of matrix algebras through preserving properties*, Oper. Matrices **2** (2008), no. 1, 125–136, DOI 10.7153/oam-02-07. MR2392770
[136] P. Šemrl, *Maps on matrix spaces*, Linear Algebra Appl. **413** (2006), no. 2-3, 364–393, DOI 10.1016/j.laa.2005.03.011. MR2198941
[137] P. Šemrl, *Non-linear commutativity preserving maps*, Acta Sci. Math. (Szeged) **71** (2005), no. 3-4, 781–819. MR2206609
[138] P. Šemrl, *Isomorphisms of standard operator algebras*, Proc. Amer. Math. Soc. **123** (1995), no. 6, 1851–1855, DOI 10.2307/2161001. MR1242104
[139] P. Šemrl, *Jordan $*$-derivations of standard operator algebras*, Proc. Amer. Math. Soc. **120** (1994), no. 2, 515–518, DOI 10.2307/2159889. MR1186136
[140] P. Šemrl, *Quadratic functionals and Jordan $*$-derivations*, Studia Math. **97** (1991), no. 3, 157–165, DOI 10.4064/sm-97-3-157-165. MR1100685
[141] A. R. Sourour, *Invertibility preserving linear maps on $\mathcal{L}(X)$*, Trans. Amer. Math. Soc. **348** (1996), no. 1, 13–30, DOI 10.1090/S0002-9947-96-01428-6. MR1311919
[142] M. F. Smiley, *Jordan homomorphisms onto prime rings*, Trans. Amer. Math. Soc. **84** (1957), 426–429, DOI 10.2307/1992823. MR83484
[143] K. Yan and X. Fang, *Common properties of the operator products in spectral theory*, Ann. Funct. Anal. **6** (2015), no. 4, 60–69, DOI 10.15352/afa/06-4-60. MR3365981
[144] K. Yan and X. C. Fang, *Common properties of the operator products in local spectral theory*, Acta Math. Sin. (Engl. Ser.) **31** (2015), no. 11, 1715–1724, DOI 10.1007/s10114-015-5116-5. MR3406670

[145] X. Yu and F. Lu, *Maps preserving Lie product on $B(X)$*, Taiwanese J. Math. **12** (2008), no. 3, 793–806, DOI 10.11650/twjm/1500602436. MR2417148

[146] W. Zhang and J. Hou, *Maps preserving peripheral spectrum of generalized products of operators*, Linear Algebra Appl. **468** (2015), 87–106, DOI 10.1016/j.laa.2014.01.037. MR3293243

[147] W. Zhang, J. C. Hou, and X. F. Qi, *Maps preserving peripheral spectrum of generalized Jordan products of operators*, Acta Math. Sin. (Engl. Ser.) **31** (2015), no. 6, 953–972, DOI 10.1007/s10114-015-4367-5. MR3343962

[148] Q. Zeng and H. Zhong, *New results on common properties of the products AC and BA*, J. Math. Anal. Appl. **427** (2015), no. 2, 830–840, DOI 10.1016/j.jmaa.2015.02.037. MR3323010

[149] Q. Zeng and H. Zhong, *Common properties of bounded linear operators AC and BA: local spectral theory*, J. Math. Anal. Appl. **414** (2014), no. 2, 553–560, DOI 10.1016/j.jmaa.2014.01.021. MR3167980

[150] Q. P. Zeng and H. J. Zhong, *New results on common properties of bounded linear operators RS and SR*, Acta Math. Sin. (Engl. Ser.) **29** (2013), no. 10, 1871–1884, DOI 10.1007/s10114-013-1758-3. MR3096550

[151] A. Taghavi, F. Kolivand, and E. Tavakoli, *A note on strong (skew) η-Lie products preserving maps on some algebras*, Linear Multilinear Algebra **67** (2019), no. 5, 886–895, DOI 10.1080/03081087.2018.1435624. MR3923033

[152] A. Torgašev, *On operators with the same local spectra*, Czechoslovak Math. J. **48** (1998), no. 1, 77–83.

[153] M. Wang and G. Ji, *Maps preserving $*$-Lie product on factor von Neumann algebras*, Linear Multilinear Algebra **64** (2016), no. 11, 2159–2168, DOI 10.1080/03081087.2016.1142497. MR3539569

Syracuse University, Department of Mathematics, 215 Carnegie Building, Syracuse, NY 13244, USA
Email address: abourhim@syr.edu

Université Laval, Département de mathématiques et de statistique, Québec, QC, G1V 0A6, Canada
Email address: javad.mashreghi@mat.ulaval.ca

Sadovskii-type fixed point results for edge-preserving mappings

M. R. Alfuraidan and N. Machrafi

ABSTRACT. In this work, we give some graphical versions of Sadovskii's fixed point theorem using measure of noncompactness properties and the new concept of monotone distance graph. Our results generalize a fixed point theorem of Guo and Lakshmikantham in ordered Banach spaces.

1. Introduction

Measures of noncompactness are very useful tools in nonlinear analysis and were introduced in the fundamental papers of Kuratowski [8] and Darbo [4]. In the latter paper, Darbo proved that if f is a self-mapping on a nonempty, bounded, closed and convex subset Ω of a (complex) Banach space X such that f is continuous and set-contraction, i.e., there exists $k \in (0,1)$ with

$$\alpha(f(A)) \leq k\alpha(A) \text{ for all } A \subset \Omega,$$

then f has a fixed point in Ω, where α is a measure of noncompactness. Recall that a bounded mapping $f : X \to X$ is said to be condensing if $\alpha(f(A)) < \alpha(A)$ for any bounded subset $A \subset X$ with $\alpha(A) > 0$.

Sadovskii proved the following generalization of Darbo's fixed point theorem:

THEOREM 1.1. ([11]) *If f is a self-mapping on a nonempty, bounded, closed and convex subset Ω of a (complex) Banach space X such that f is continuous and condensing, then f has a fixed point in Ω.*

In the ordered Banach space setting, Guo and Lakshmikantham [5, Theorem 2.1.1] proved the following fixed point theorem for increasing mappings.

THEOREM 1.2. *Let $(X, \leq)$ be an ordered Banach space. Let $x_0, y_0 \in X$, $x_0 < y_0$ and $f : [x_0, y_0] \to X$ be an increasing mapping such that*

(1.1) $$x_0 \leq f(x_0) \text{ and } f(y_0) \leq y_0.$$

Suppose that one of the following two conditions is satisfied:

(H_1) *the cone X^+ is normal and f is continuous and condensing.*

2010 *Mathematics Subject Classification.* Primary 47H10, 46B20, 05C99.

Key words and phrases. Sadovskii's fixed point theorem, measure of noncompactness, condensing mapping, monotone distance graph.

©2020 American Mathematical Society

(H_2) the cone X^+ is reguler and f is semicontinuous, i.e., $x_n \to x$ strongly implies $f(x_n) \to f(x)$ weakly.

Then, f has a minimal fixed point x_* and a maximal fixed point x^* in $[x_0, y_0]$; moreover,
$$x_* = \lim_{n \to \infty} x_n, \quad x^* = \lim_{n \to \infty} y_n$$
where $x_n = f(x_{n-1})$ and $y_n = f(y_{n-1})$ $(n = 1, 2, 3, \ldots)$, and
$$x_0 \leq x_1 \leq \ldots \leq x_n \leq \ldots \leq y_n \leq \ldots \leq y_1 \leq y_0.$$

The existence part of Theorem 1.2 under condition (H_1) is followed directly from Sadovskii's fixed point theorem since the order interval $[x_0, y_0]$ is f-invariant. To extend Guo and Lakshmikantham result to metric spaces, we need the setting of nonconvexity in which the notion of condensing mapping was initially introduced there, see [11].

Here, we will consider the measure of noncompactness introduced by Kuratowski. However, our results work in the the general measure theory of noncompactness. The interested reader may consult [3]. Let (X, d) be a metric space and let $\mathcal{B}(X)$ denote the collection of nonempty, bounded subsets of X. The Kuratowski measure of noncompactness $\alpha : \mathcal{B}(X) \to [0, \infty)$ is defined by
$$\alpha(A) := \inf \left\{ \varepsilon > 0 \ : \ A \subset \bigcup_{i=1}^{n} A_i \text{ with } A_i \in \mathcal{B}(X) \text{ and } diam(A_i) \leq \varepsilon \right\}.$$

In this work, we present graphical versions of Sadovskii's fixed point theorem. Our results extend both the standard Sadovskii's fixed point theorem and Guo and Lakshmikantham fixed point theorem. The key assumption of normality of the cone of an ordered Banach space in the latter fixed point theorem is replaced by a new concept involving the setting of metric spaces endowed with a graph, namely, the so-called monotone distance graphs. This main idea is inspired by the work of Jachymski [7] who is credited as being the first one to use the idea of replacing a partial order by a graph, and thus to extend the main fixed point results of [9, 10] from a metric space endowed with a partial order to the case of a metric space endowed with a graph, in which he proved a generalization of the Banach contraction principle. In fact, as Jachymski asserted in his above-mentioned paper, the language of graph theory is more convenient to state several metric fixed point results involving certain conditions that can be described in a unified way using this language.

2. Notations and preliminaries

Throughout the paper, and unless otherwise stated, we let (X, d) to be a metric space endowed with a directed graph G for which the set $V(G)$ of its vertices coincides with X and the set of its edges $E(G)$ is a binary relation on X ($E(G) \subset X \times X$) so that G has no multiple arcs and is antisymmetric i.e., $(x, y) \in E(G)$ & $(y, x) \in E(G) \implies x = y$. The graph G is said to be reflexive, if the set of its edges contains all loops, that is, $(x, x) \in E(G)$ for all $x \in X$. It is said to be transitive, if for all $x, y, z \in X$, $(x, y) \in E(G)$ & $(y, z) \in E(G) \implies (x, z) \in E(G)$.

An example illustrating a directed and antisymmetric graph is an order graph, that is a graph G such that $V(G) = X$ and the relation $\leq$ on X defined by
$$x \leq y \text{ if } (x, y) \in E(G)$$

is a partial order relation. In this case, G will be denoted by $G_{\leq}$. For two elements $x, y \in X$, we define the G-intervals
$$[x)_G := \{z \in X : (x, z) \in E(G)\},$$
$$(y]_G := \{z \in X : (z, y) \in E(G)\},$$
and
$$[x, y]_G := \{z \in X : (x, z) \in E(G) \& (z, y) \in E(G)\}.$$
For the graph $G_{\leq}$, where X is a partially order set, we get the standard order intervals $[x), (y],$ and $[x, y]$. Let us recall that a graph G is said to be closed, if the G-intervals are closed subsets of X.

The conversion of the graph G, denoted by G^{-1}, is the graph obtained from G by reversing the direction of edges, i.e. $G^{-1} = (V(G), E(G^{-1}))$ and
$$E(G^{-1}) = \{(x, y) \in X \times X : (y, x) \in E(G)\}.$$

We will use $\widetilde{G}$ to denote the graph defined by $V(\widetilde{G}) = V(G)$ and $E(\widetilde{G}) = E(G) \cup E(G^{-1})$. For $x, y \in V(G)$, a (directed) path in G from x to y of length N ($N \in \mathbb{N} \cup \{0\}$) is a sequence $(x_i)_{i=0}^N$ of $N + 1$ vertices such that
$$x_0 = x, \ x_N = y \text{ and } (x_{i-1}, x_i) \in E(G) \text{ for } i = 1, ..., N.$$

The graph G is connected if there is a path between any two vertices. The graph G is said to be weakly connected if $\widetilde{G}$ is connected. We call (Ω, E) a subgraph of G if $\Omega \subset V(G)$, $E \subset E(G)$ and for any edge $(x, y) \in E$, $x, y \in \Omega$. For a subset $A \subset X$, the subgraph of G induced by A and denoted by $G[A]$, is the directed graph defined by $V(G[A]) = A$ and $E(G[A]) = \{(x, y) \in E(G) : x, y \in A\}$. A subset $A \subset X$ is said to be (weakly) connected if the subgraph $G[A]$ is (weakly) connected. If G is such that $E(G)$ is symmetric and $x \in V(G)$, then the (connected) subgraph G_x consisting of all edges and vertices which are contained in some path beginning at x is called the connected component of G containing x. In this case, $V(G_x) = [x]_G$, where $[x]_G$ is the equivalence class of the following relation defined on $V(G)$ by

$$u \ \mathcal{R} \ v \text{ if and only if there is a path in } G \text{ from } u \text{ to } v.$$

A sequence $(x_n) \subset X$ is said to be G-increasing (resp. G-decreasing) if $(x_n, x_{n+1}) \in E(G)$, $n \in \mathbb{N}$ (resp. $(x_{n+1}, x_n) \in E(G)$, $n \in \mathbb{N}$). (x_n) is said to be G-monotone if it is either G-increasing or G-decreasing.

A nonempty subset $A \subset X$ is said to be weakly G-bounded if any G-monotone sequence $(x_n) \subset A$ is bounded, i.e.,
$$\text{diam}\,((x_n)) := \sup\,\{d(x_n, x_m) : n, m \in \mathbb{N}\} < +\infty.$$

A mapping $f : X \to X$ is said to be G-continuous if given $x \in X$ and a G-monotone sequence $(x_n) \subset X$, $x_n \to x \implies f(x_n) \to f(x)$.

A mapping $f : X \to X$ is said to be edge-preserving if for each $x, y \in X$
$$(x, y) \in E(G) \implies (f(x), f(y)) \in E(G).$$

Let us recall finally, that a subset A of an ordered vector space X is said to be order convex (or full) if for every $x, y \in A$, we have $[x, y] \subset A$. The cone X^+ of an ordered topological vector space (X, τ) is said to be normal if τ has a base at zero consisting of order convex sets. When $(X, \|\cdot\|)$ is an ordered normed space, the cone X^+ is normal if and only if X admits an equivalent monotone norm, i.e.,

there exists a norm $|||.|||$ on X such that $|||.|||$ is equivalent to $\|.\|$ and for every $x, y \in X$
$$0 \leq x \leq y \Rightarrow |||x||| \leq |||y|||.$$

For more detail on such concepts, see [2]. The following Lemma will be crucial to prove our last result in this paper.

LEMMA 2.1 ([2, Lemma 2.28]). *Assume that a Hausdorff ordered locally convex space (X, τ) has a τ-normal cone and that a net $(x_\alpha) \subseteq X^+$ is decreasing. Then $x_\alpha \xrightarrow{\tau} 0$ if and only if $x_\alpha \xrightarrow{w} 0$.*

For more details on graph theory and order-theoretic aspects in relation with metric fixed point theory, the reader can consult [1, 6] and the references therein.

3. Fixed points of condensing mappings on a metric space with a monotone distance graph

In this section, by using the concept of noncompactness measure, we prove some graphical versions of Sadovskii's fixed point theorem. Our results generalize a fixed point theorem of Guo and Lakshmikantham in ordered Banach spaces. To do this, we introduce the concepts of a set-contraction and a condensing mapping in the setting of a metric space with a graph.

DEFINITION 3.1. *Consider the triple (X, d, G) as described above. A bounded mapping $f : X \to X$ is said to be*

(i) *a G-set-contraction, if there exists $k \in (0, 1)$ such that*
$$\alpha(f(A)) \leq k\, \alpha(A)$$
for any weakly connected subset $A \in \mathcal{B}(X)$.

(ii) *G-condensing, if*
$$\alpha(f(A)) < \alpha(A)$$
for any weakly connected subset $A \in \mathcal{B}(X)$ with $\alpha(A) > 0$.

Note that, if G is defined by $V(G) = X$ and $E(G) = X \times X$, then we get the standard versions of a set-contraction and a condensing mapping. Furthermore, as for a set-contraction and a condensing mapping, it is clear that any G-set-contraction $f : X \to X$ is a G-condensing mapping.

Now, we give an example of a G-condensing mapping which is not condensing.

EXAMPLE 3.2. *Consider $X = \mathbb{N}$ endowed with the discrete distance. Then we have $\alpha(X) = 1$. Consider the graph G with $V(G) = X$ and*
$$E(G) = \{(0, 1)\} \cup \{(n, n); n \in \mathbb{N}\}.$$
Let $f : X \to X$ *be defined by*
$$f(0) = f(1) = 0, \text{ and } f(n) = n - 1 \text{ for } n \geq 2.$$
It is clear that f is onto, i.e., $f(X) = X$. Hence $\alpha(f(X)) = \alpha(X) = 1$. Therefore f is not condensing. However, it is easy to check that f is G-condensing.

Now, let us recall that a sequence $(x_n) \subset X$ is said to be weakly G-increasing (resp. weakly G-decreasing) if for every n there exists a direct path from x_n to x_{n+1} (resp. from x_{n+1} to x_n). Clearly, every G-increasing (resp. G-decreasing) sequence in X is weakly G-increasing (resp. weakly G-decreasing), and the converse holds if G is transitive. Following the way introduced by Nieto and Rodriguez-Lopez

[**9**] to improve the result of Ran and Reurings [**10**], we will use the following two properties for the triple (X, d, G):

(P_1) for any weakly G-increasing sequence $(x_n) \subset X$, if $x_n \to x$ for some $x \in X$, then $(x_n, x) \in E(G)$, $n = 0, 1, \dots$.

(P_2) for any weakly G-decreasing sequence $(x_n) \subset X$, if $x_n \to x$ for some $x \in X$, then $(x, x_n) \in E(G)$, $n = 0, 1, \dots$.

Following the characterization of normality of a cone in an ordered normed space, as mentioned in the preceding section, we introduce a similar concept for a metric space endowed with a graph.

DEFINITION 3.3. *For the triple (X, d, G), the distance d is said to be G-monotone if for any $x, y, z \in X$, we have*

$$y \in [x, z]_G \text{ implies } d(x, y) \leq d(x, z) \text{ and } d(y, z) \leq d(x, z).$$

In this case, the graph G is called a monotone distance graph.

If $(X, \leq)$ is a partially ordered metric space and the graph $G = G_\leq$, then we say simply that d is a monotone distance to mean a G-monotone distance. Note that a necessary condition for a graph G to be a monotone distance graph is that G is antisymmetric. Furthermore, if G is a weakly connected graph and d is the path metric in $\widetilde{G}$, i.e., for any $x, y \in X$, $d(x, y)$ is the length of a shortest path in $\widetilde{G}$ from x to y, then it is easy to show that (X, d, G) is a monotone distance graph.

We give in the following two other examples to illustrate the concept of a monotone distance graph.

EXAMPLE 3.4. *Let $(X, \leq)$ be an ordered normed space with a normal cone X^+, and let $G_\leq$ be the order graph with $V(G_\leq) = X$. Then, $(X, d, G_\leq)$ is a monotone distance graph, where d is the distance induced by a monotone norm $\|.\|$ on X. Indeed, for any $x, y, z \in X$, $y \in [x, z]_{G_\leq}$ means that $x \leq y \leq z$, and hence, since $0 \leq y - x \leq z - x$ and $0 \leq z - y \leq z - x$, we have $\|y - x\| \leq \|z - x\|$ and $\|z - y\| \leq \|z - x\|$.*

EXAMPLE 3.5. *Let $X = C^1[0, 1]$ be the real vector space of a continuously differentiable functions on $[0, 1]$ equipped with the distance d induced by the norm*

$$\|f\| = \|f\|_\infty + \|f'\|_\infty,$$

where $\|f\|_\infty = \sup_{t \in [0,1]} |f(t)|$ and f' is the derivative of f. It is well known that the cone X^+ is not normal. Hence, if the order graph $G_\leq$ is defined by $V(G_\leq) = X$ and

$$f \leq g \text{ if } f(t) \leq g(t) \text{ for each } t \in [0, 1],$$

then clearly $(X, d, G_\leq)$ is not a monotone distance graph.

The following proposition shows that Definition 3.3 extends naturally the normality of the cone of an ordered topological vector space in case X is a partially ordered metric space.

PROPOSITION 3.6. *Let $(X, d, \leq)$ be a partially ordered metric space with order convex balls. Then d is a monotone distance.*

PROOF. Let $x, y, z \in X$ such that $x \leq y \leq z$. Since the balls $B(x, d(x, z))$ and $B(z, d(x, z))$ are order convex and $x, z \in B(x, d(x, z)) \cap B(z, d(x, z))$, we have

$y \in B(x, d(x, z)) \cap B(z, d(x, z))$, that is, $d(x, y) \leq d(x, z)$ and $d(y, z) \leq d(x, z)$ as desired. □

The following example shows that a monotone distance graph G that satisfies properties (P_1) and (P_2) is not necessarily an order graph.

EXAMPLE 3.7. Let (X, d) be a metric space containing at least two elements. Let $x_0 \in X$ be fixed. Define the graph G_{x_0} by $V(G_{x_0}) = X$ and $E(G_{x_0}) = \{x_0\} \times X$. Then, it is easy to show that G_{x_0} is a monotone distance graph that is transitive and closed, and hence, (X, d, G_{x_0}) satisfies the properties (P_1) and (P_2). However, G_{x_0} is not an order graph since it is not reflexive.

Next, we prove some topological properties of monotone distance graphs that will be used in the sequel. We show also that the properties of being a monotone distance graph, (P_1), and (P_2) are preserved under Cartesian product of two metric spaces with a graph. The details are in the following two propositions. The proof of the second one is left for the reader.

PROPOSITION 3.8. Let the triple (X, d, G) be as described above such that G is a monotone distance graph. Then, the following hold:
(1) every set containing in some G-interval $[x, y]_G$, is bounded;
(2) for every sequences (x_n), (y_n), (z_n) in X, if $y_n \in [x_n, z_n]_G$ for each n and (x_n) and (z_n) converge to the same limit x then (y_n) converges to x too.

PROOF. (1) It suffices to show that every G-interval $[x, y]_G$ is bounded. Let $u, v \in [x, y]_G$. Then, since G is a monotone distance graph, we have
$$d(u, v) \leq d(u, x) + d(x, v) \leq 2d(x, y).$$
Hence, $\text{diam}[x, y]_G \leq 2d(x, y)$ and the result is proved.
(2) Since G is a monotone distance graph, for each n we have
$$d(y_n, x) \leq d(y_n, x_n) + d(x_n, x) \leq d(x_n, z_n) + d(x_n, x) \to 0.$$
This shows that (y_n) converges to x as desired. □

PROPOSITION 3.9. Let (X, d_1, G_1) (Y, d_2, G_2) be two triples as described above. Define on $X \times Y$ the distance d and the graph $G_1 \times G_2$ by
$$d((x, y), (x', y')) = \max(d_1(x, x'), d_2(y, y')),$$
$$V(G_1 \times G_2) = X \times Y \text{ and}$$
$$E(G_1 \times G_2) = \{((x, y), (x', y')) \in X \times Y : (x, x') \in E(G_1), (y, y') \in E(G_2)\}.$$

Then, if both (X, d_1, G_1) (Y, d_2, G_2) are a monotone distance graphs and satisfy the properties (P_1) and (P_2), then so is for $(X \times Y, d, G_1 \times G_2)$.

We are now in position to state our first main result.

THEOREM 3.10. *Let (X, d) be a complete metric space endowed with a monotone distance graph G, such that $V(G) = X$ is weakly G-bounded and the triple (X, d, G) satisfies the properties (P_1) and (P_2). Let $f : X \to X$ be a bounded, edge-preserving, G-continuous and G-condensing mapping such that there exists $x_0 \in X$ with $(x_0, f(x_0)) \in E(\widetilde{G})$. Then, f has a fixed point x_* in the component $[x_0]_{\widetilde{G}}$ and the G-monotone sequence $(f^n x_0)$ converges to x_*.*

PROOF. Without loss of generality, we can assume that $(x_0, f(x_0)) \in E(G)$. The proof breaks down into three steps:

STEP 1. *There exists a Cauchy subsequence $(f^{\varphi(n)}(x_0))$ of $(f^n(x_0))$ that converges to some $x_* \in X$.*

For $n \in \mathbb{N}$, set $x_n = f^n(x_0)$. Since f is edge-preserving, then $(x_n, x_{n+1}) \in E(G)$ for each $n \in \{0\} \cup \mathbb{N}$. Let $\Theta(x_0) = \{x_n : n = 0, 1, ...\}$. Clearly, $\Theta(x_0) = f(\Theta(x_0)) \cup \{x_0\}$ is a weakly connected subset of X. Since f is G-condensing then it follows from

$$\begin{aligned} \alpha(\Theta(x_0)) &= \alpha(f(\Theta(x_0)) \cup \{x_0\}) \\ &\leq \alpha(\{x_0\}) + \alpha(f(\Theta(x_0))) \\ &= \alpha(f(\Theta(x_0))) \end{aligned}$$

that $\alpha(\Theta(x_0)) = 0$. Hence, there exists a Cauchy subsequence $(x_{\varphi(n)}) \subset (x_n)$ with $x_{\varphi(n)} \to x_*$ for some $x_* \in X$.

STEP 2. *The limit x_* satisfies $(x_n, x_*) \in E(G)$ for each $n = 0, 1, ...$ and hence $x_* \in [x_0]_{\widetilde{G}}$.*

In fact, let $n \in \{0\} \cup \mathbb{N}$ be fixed. Define the sequence $(u_k) \subset X$ by

$$\begin{aligned} u_k &= x_k \text{ if } k = 0, 1, ..., n; \\ u_k &= x_{\varphi(k)} \text{ if } k \geq n+1. \end{aligned}$$

Then, clearly (u_k) is a weakly G-increasing sequence that converges to x_*. Property (P_1) implies $(u_k, x_*) \in E(G)$ for each $k \in \{0\} \cup \mathbb{N}$, and hence $(x_n, x_*) \in E(G)$. Since $n \in \{0\} \cup \mathbb{N}$ was arbitrary, we have the desired conclusion.

STEP 3. *The sequence (x_n) converges to x_* and $f(x_*) = x_*$.*

In fact, if $n \in \mathbb{N}$ and $m > \varphi(n)$, then since G is a monotone distance graph and $(x_n, x_*) \in E(G)$ for every $n \in \{0\} \cup \mathbb{N}$, we see that

$$d(x_m, x_*) \leq d(x_{m-1}, x_*) \leq ... \leq d(x_{\varphi(n)}, x_*).$$

Therefore, $x_m \to x_*$ as $m \to \infty$. Taking the limit $n \to \infty$ in the equality $x_n = f x_{n-1}$, we get $f x_* = x_*$ since f is G-continuous. $\square$

We illustrate our previous main result by the following example.

EXAMPLE 3.11. *Let $X = [0,1]$ be endowed with the Euclidean metric d. Consider the graph G_0 defined in Example 3.7 for $x_0 = 0$, and note that $[x]_{\widetilde{G_0}} = [0,1]$ for every $x \in [0,1]$, since G_0 is weakly connected (in fact, for $x, y \in [0,1]$, $(x, 0, y)$ is a path in $\widetilde{G_0}$ from x to y). Let $f : X \to X$ be defined by $f(0) = 0$ and $f(x) = \frac{1}{2}$ if $x \neq 0$. It is easy to check that all conditions of Theorem 3.10 are fulfilled and clearly*

$Fix f = \{0, \frac{1}{2}\}$. Notice that neither the standard Sadovskii fixed point theorem nor the G_0-contraction principle of Jachymski [**7**, Theorem 3.2] may be applied here. Indeed, f is a discontinuous map and it is not a G_0-contraction (see [**7**, Definition 2.1]) since

$$\left(0, \frac{1}{2}\right) \in E(G_0) \text{ and } d\left(f(0), f\left(\frac{1}{2}\right)\right) = d\left(0, \frac{1}{2}\right) > k\, d\left(0, \frac{1}{2}\right)$$

for every $k \in (0, 1)$.

COROLLARY 3.12. *Let the triple (X, d, G) be such that X is a complete metric space and let $f : X \to X$ be bounded, edge-preserving, G-continuous and G-condensing. Assume that*
  (i) *G is a transitive and monotone distance graph.*
  (ii) *there exist $x_0, y_0 \in X$, $x_0 \neq y_0$ with $(x_0, y_0) \in E(G)$, $(x_0, f(x_0)) \in E(G)$ and $(f(y_0), y_0) \in E(G)$.*
  (iii) *the G-intervals $[x, y]_G$, $x, y \in X$, are closed.*

Then, there exist a minimal fixed point x_* and a maximal fixed point x^* of f in $[x_0, y_0]_G$, that is for every fixed point x of f in $[x_0, y_0]_G$ we have $(x_*, x) \in E(G)$ and $(x, x^*) \in E(G)$. Moreover, the G-increasing sequence $(f^n x_0)$ and the G-decreasing sequence $(f^n y_0)$ converge respectively to x_* and x^*.

PROOF. Since f is edge-preserving, it is easily seen that the set $\Omega = [x_0, y_0]_G$ is nonempty and $f(\Omega) \subset \Omega$. Since G is a monotone distance graph, by Proposition 3.8 Ω is a bounded subset of X. Let Ω be endowed with the subgraph $G[\Omega]$ induced by G. Now, to show that $G[\Omega]$ satisfies the properties (P_1) and (P_2) it suffices to show that $G[\Omega]$ is closed, since it is transitif. To this end, let $x \in \Omega$ and $(u_n) \subset \Omega$ such that $(x, u_n) \in E(G[\Omega])$, $n = 0, 1, ...$ and $u_n \to u$ for some $u \in \Omega$. Since $(u_n) \subset [x, y_0]_G$ and $[x, y_0]_G$ is closed, $u \in [x, y_0]_G$ and thus $(x, u) \in E(G[\Omega])$ as required. Similarly, we can see that the G-intervals $[x]_{G[\Omega]}$, $x \in \Omega$, are closed. Now, by Theorem 3.10, f admits a fixed point $x_* = \lim f^n(x_1)$ in $[x_1]_{\widetilde{G[\Omega]}} = \Omega$, where $x_1 = f(x_0)$ satisfies $x_1 \in \Omega$ and $(x_1, f(x_1)) \in E(G[\Omega])$. If x is another fixed point in Ω, then since $(x_0, x) \in E(G)$ we see that $(f^n(x_0), x) \in E(G)$ for each n and hence $(x_*, x) \in E(G)$ since the G-interval $[x_0, x]_G$ is closed. It follows that x_* is a minimal fixed point of f in Ω. For the existence of a maximal fixed point $x^* \in \Omega$ of f, we consider the graph $G[\Omega]^{-1}$ and by the same arguments such fixed point of f exists in $[y_0]_{\widetilde{G[\Omega]^{-1}}} = \Omega$. □

The following corollary is a direct consequence of the above corollary and presents a generalisation of Theorem 1.2 in the setting of partially ordered metric spaces.

COROLLARY 3.13. *Let the triple $(X, d, \leq)$ be a complete partially ordered metric space. Let $f : X \to X$ be an increasing mapping such that there exist $x_0, y_0 \in X$ with $x_0 < y_0$ and (1.1) holds. Assume that the following conditions are satisfied:*
  (1) *the distance d of X is a monotone distance;*
  (2) *the order intervals $[x, y]$ ($x \leq y$) of X are closed;*
  (3) *f is a condensing and monotone continuous mapping, i.e., for every monotone sequence $(x_n) \subset X$, $x_n \to x$ implies $f(x_n) \to f(x)$.*

Then, the conclusion of Theorem 1.2 holds.

Our next last result is an improvement version of Theorem 1.2. In fact, we provide a third condition under which we have neither the continuity of the mapping nor the regularity of the cone.

COROLLARY 3.14. *Let $(X, \|.\|, \leq)$ be an ordered Banach space with a monotone norm $\|.\|$. Let $f : X \to X$ be a bounded increasing mapping such that there exist $x_0, y_0 \in X$ with $x_0 < y_0$ and (1.1) holds. Assume that the following condition is satisfied:*

(H_3) *f is condensing and semicontinuous i.e., $x_n \to x$ strongly implies $f(x_n) \to f(x)$ weakly.*

Then, the conclusion of Theorem 1.2 holds.

PROOF. We only need to show that f is monotone continuous. To this end, let $(x_n) \subset X$ be such that (x_n) is a increasing sequence and $x_n \to x$ for some $x \in X$. Hence $x = \sup x_n$. Therefore, since f is increasing, $(f(x_n))$ is an increasing sequence and $f(x_n) \leq f(x)$ for each $n = 0, 1, \ldots$. Now, since f is semicontinuous, it follows from Lemma 2.1 that $f(x_n) \to f(x)$. Note that the proof is similar if (x_n) is a decreasing sequence of X. Hence, f is monotone continuous. The desired conclusion follows by applying Corollary 3.13 for the distance d induced by the norm $\|.\|$. □

Acknowledgements

The first author would like to acknowledge the support provided by King Fahd University of Petroleum & Minerals. The second author would like to thank King Fahd University of Petroleum & Minerals for hosting him during the preparation of this work.

References

[1] M. R. Alfuraidan and Q. H. Ansari (eds.), *Fixed point theory and graph theory*, Elsevier/Academic Press, Amsterdam, 2016. Foundations and integrative approaches. MR3470114

[2] C. D. Aliprantis and R. Tourky, *Cones and duality*, Graduate Studies in Mathematics, vol. 84, American Mathematical Society, Providence, RI, 2007. MR2317344

[3] J. Appell, *Measures of noncompactness, condensing operators and fixed points: an application-oriented survey*, Fixed Point Theory **6** (2005), no. 2, 157–229. MR2196709

[4] G. Darbo, *Punti uniti in trasformazioni a codominio non compatto* (Italian), Rend. Sem. Mat. Univ. Padova **24** (1955), 84–92. MR70164

[5] D. J. Guo and V. Lakshmikantham, *Nonlinear problems in abstract cones*, Notes and Reports in Mathematics in Science and Engineering, vol. 5, Academic Press, Inc., Boston, MA, 1988. MR959889

[6] J. Jachymski, *Order-theoretic aspects of metric fixed point theory*, Handbook of metric fixed point theory, Kluwer Acad. Publ., Dordrecht, 2001, pp. 613–641, DOI 10.1007/978-94-017-1748-9_18. MR1904289

[7] J. Jachymski, *The contraction principle for mappings on a metric space with a graph*, Proc. Amer. Math. Soc. **136** (2008), no. 4, 1359–1373, DOI 10.1090/S0002-9939-07-09110-1. MR2367109

[8] K. Kuratowski, *Sur les espaces complets*, Fund. Math., **15,** (1930), 301-309.

[9] J. J. Nieto and R. Rodríguez-López, *Contractive mapping theorems in partially ordered sets and applications to ordinary differential equations*, Order **22** (2005), no. 3, 223–239 (2006), DOI 10.1007/s11083-005-9018-5. MR2212687

[10] A. C. M. Ran and M. C. B. Reurings, *A fixed point theorem in partially ordered sets and some applications to matrix equations*, Proc. Amer. Math. Soc. **132** (2004), no. 5, 1435–1443, DOI 10.1090/S0002-9939-03-07220-4. MR2053350

[11] B. N. Sadovskiĭ, *On a fixed point principle* (Russian), Funkcional. Anal. i Priložen. **1** (1967), no. 2, 74–76. MR0211302

Monther Rashed Alfuraidan, Department of Mathematics & Statistics, King Fahd University of Petroleum and Minerals, Dhahran 31261, Saudi Arabia.
Email address: monther@kfupm.edu.sa

Nabil Machrafi, Mohammed V University in Rabat, Faculty of Sciences, Department of Mathematics, Team GrAAF, Laboratory LMSA, Center CeReMAR, B.P. 1014 RP, Rabat, Morocco.
Email address: nmachrafi@gmail.com

The joint numerical radius on C^*-algebras

Mohamed Mabrouk

ABSTRACT. Let $\mathfrak{A}$ be unital C^*-algebra with unit e and positive cone $\mathfrak{A}^+$ such that every irreducible representation is infinite dimensional. For every $\mathbf{a} = (a_1,\ldots,a_n) \in \mathfrak{A}^n$, the joint numerical radius of $\mathbf{a}$ is denoted by $\mathbf{v}(\mathbf{a})$. It is shown that an element $\mathbf{a} \in \mathfrak{A}^n$ satisfies $\sum_{j=1}^{n}|f(a_j)|^2 = 1$ for every pure state f of $\mathfrak{A}$ if and only if each a_j is in the center of $\mathfrak{A}$ and $\sum_{j=1}^{n} a_j a_j^* = e$. Furthermore, we characterize elements $\mathbf{a}_1,\ldots,\mathbf{a}_n \in \mathfrak{A}^n$ such that for any $\mathbf{x} \in (\mathfrak{A}^+)^n$ there exists $\alpha = (\alpha_1,\ldots,\alpha_n) \in \mathbb{R}^n$ such that $\sum_{j=1}^{j=n} \alpha_j^2 = 1$ and $\mathbf{v}\left(\sum_{j=1}^{j=n} \alpha_j \mathbf{a}_j + \mathbf{x}\right) = 1 + \mathbf{v}(\mathbf{x})$.

1. Introduction and preliminaries

Let $\mathfrak{A}$ be a unital C^*-algebra with unit e, and denote by $\mathfrak{A}^+$, $\mathcal{U}(\mathfrak{A})$ and $\mathcal{Z}(\mathfrak{A})$ the cone of positive elements in $\mathfrak{A}$, the group of all unitary elements in $\mathfrak{A}$ and the centre of $\mathfrak{A}$, respectively. For any element x of $\mathfrak{A}$, we denote by $\Re(x) = \frac{1}{2}(x + x^*)$ and $\Im(x) = \frac{1}{2i}(x - x^*)$ the real and the imaginary parts of x. Let $\mathfrak{A}'$ denote the topological dual space of $\mathfrak{A}$, and define the set of normalized states of $\mathfrak{A}$ by

$$\mathcal{S}(\mathfrak{A}) = \{f \in \mathfrak{A}' : f(e) = \|f\| = 1\}.$$

A linear functional $f \in \mathfrak{A}'$ is said to be positive, and write $f \geq 0$, if $f(xx^*) \geq 0$ for all $x \in \mathfrak{A}$. Recall that $f \geq 0$ if and only if f is bounded and $\|f\| = f(e)$ ([21, Corollary 3.3.4]). Recall also that a positive linear functional f on $\mathfrak{A}$ is said to be pure if for every positive functional g on $\mathfrak{A}$ such that $g \leq f$, there is a scalar $0 \leq \lambda \leq 1$ such that $g = \lambda f$. The set of pure states on $\mathfrak{A}$ is denoted by $\mathcal{P}(\mathfrak{A})$. It is well known that $\mathcal{P}(\mathfrak{A})$ coincides with the set of all extremal points of $\mathcal{S}(\mathfrak{A})$. Note that if $f \in \mathcal{P}(\mathfrak{A})$ then $f(ax) = f(a)f(x)$ for all $a \in \mathcal{Z}(\mathfrak{A})$ and $x \in \mathfrak{A}$. Moreover, the restriction f on $\mathcal{Z}(\mathfrak{A})$ is a pure state of $\mathcal{Z}(\mathfrak{A})$, see for instance [16, Proposition 4.3.14].

For any positive integer n, let $\mathcal{C}_n = \{\lambda = (\lambda_1,\ldots,\lambda_n) \in \mathbb{C}^n : \sum_{k=1}^{n} |\lambda_k|^2 = 1\}$ and note that $\mathcal{C}_1$ is nothing but the unit circle of $\mathbb{C}$. Let $\mathfrak{A}^n$ denote the product of

---

2010 *Mathematics Subject Classification.* Primary 15A86; 46L05; Secondary 47A12; 47B49.
Key words and phrases. Linear preservers, Joint Numerical range, Joint Numerical radius.

n-copies of $\mathfrak{A}$, i.e.
$$\mathfrak{A}^n = \{\mathbf{a} = (a_1, \ldots, a_n) : a_1, \ldots, a_n \in \mathfrak{A}\}.$$
When $\mathfrak{A} = \mathcal{B}(\mathcal{H})$ is the C^*-algebra of all bounded linear operators on a complex Hilbert space $(\mathcal{H}, \langle .,. \rangle)$ and $\mathbf{A} = (A_1, \ldots, A_n) \in \mathcal{B}(\mathcal{H})^n$, define the spacial joint numerical range of $\mathbf{A}$ by
$$\mathbf{W}(\mathbf{A}) = \{(\langle A_1 x, x\rangle, \ldots, \langle A_n x, x\rangle) \in \mathbb{C}^n : \|x\| = 1\}.$$
Using the Euclidean norm of $\mathbb{C}^n$ the joint numerical radius of $\mathbf{A}$ is given by
$$\mathbf{w}(\mathbf{A}) = \sup\{\|(\alpha_1, \ldots, \alpha_n)\|_2 : (\alpha_1, \ldots, \alpha_n) \in \mathbf{W}(\mathbf{A})\},$$
where $\|(\alpha_1, \ldots, \alpha_n)\|_2 = \sqrt{|\alpha_1|^2 + \ldots + |\alpha_n|^2}$. Similarly, for every $\mathbf{a} \in \mathfrak{A}^n$, the algebraic joint numerical range is defined by
$$\mathbf{V}(\mathbf{a}) = \{(f(a_1), \ldots, f(a_n)) \in \mathbb{C}^n : f \in \mathcal{S}(\mathfrak{A})\}$$
and the joint numerical radius of $\mathbf{a}$ is defined by
$$\mathbf{v}(\mathbf{a}) = \sup\{\|(\alpha_1, \ldots, \alpha_n)\|_2 : (\alpha_1, \ldots, \alpha_n) \in \mathbf{V}(\mathbf{a})\}.$$
For any $\mathbf{a} = (a_1, \ldots, a_n) \in \mathfrak{A}^n$ and $\lambda \in \mathcal{C}_n$, define the product $\lambda.\mathbf{a} = (\lambda_1 a_1, \ldots, \lambda_n a_n)$. Also, for any $x \in \mathfrak{A}$, set $\mathbf{S}_x = (x, \ldots, x) \in \mathfrak{A}^n$. Finally, for any $\mathbf{a} = (a_1, \ldots, a_n) \in \mathfrak{A}^n$ and any state $f \in \mathcal{S}(\mathfrak{A})$, let
$$f(\mathbf{a}) = (f(a_1), \ldots, f(a_n)).$$

The spatial joint (resp. algebraic) numerical range is a generalization of the classical spatial (resp. algebraic) numerical range $W(A) = \{\langle Ax, x\rangle : \|x\| = 1\}$ (resp. $V(a) = \{f(a) : f \in \mathcal{S}(\mathfrak{A})\}$). These concepts has been studied by many researchers in order to understand the joint behaviour of several operators $A_1, \ldots, A_n$. Since then, many properties have been discovered. For instance, it is well known that if $a \in \mathfrak{A} \subset \mathcal{B}(\mathcal{H})$, then $V(a)$ and $W(a)$ are convex and $\text{cl}(W(a)) = V(a)$, see [1] and [15]. Here $\text{cl}(S)$ denotes the closure of the set S. For the joint numerical range, we note that $\mathbf{V}(\mathbf{a})$ is a compact and convex subset of $\mathbb{C}^n$, [4]. On the other hand, unlike the single operator case (i.e when $n = 1$), $\mathbf{W}(\mathbf{A})$ is, in general, neither convex nor closed for $n \geq 2$. One may consult [14, 22] for a discussion on the convexity of this set. However, an important fact when $\mathcal{H}$ is infinite-dimensional, is that the closure of $\mathbf{W}(\mathbf{A})$ is star-shaped. See [17, Theorem 3.1]. In fact, in a subsequent paper [18] the same authors showed that $\mathbf{W}(\mathbf{A})$ is also star-shaped. Another important result connecting the spatial and algebraic numerical ranges was given in [20, Theorem 1]. It was established that for any $\mathbf{A} \in \mathcal{B}(\mathcal{H})^n$ we have $\mathbf{V}(\mathbf{A}) = \overline{\text{conv}}\mathbf{W}(\mathbf{A})$, where $\overline{\text{conv}}M$ stands for the closed convex hull of any subset M of $\mathbb{C}^n$. For more properties and applications of the joint numerical range, one may consult [2, 4, 12, 14, 15, 17–20, 22] and the references therein.

In recent years, characterizations of some elements in a C^*-algebra through spectral properties have been investigated by several authors [6–8, 10]. Some of them are concerned with the numerical range. For instance, in [10] and [6] it is shown among other things that the following assertions are equivalent for an element $a \in \mathfrak{A}$:

(1) $a \in \mathcal{Z}(\mathfrak{A}) \cap \mathcal{U}(\mathfrak{A})$.
(2) $f(a) \in \mathcal{C}_1$, for any $f \in \mathcal{P}(\mathfrak{A})$.
(3) For any $x \in \mathfrak{A}$ there exists $\lambda \in \mathcal{C}_1$ such that $v(a + \lambda x) = 1 + v(x)$.

Motivated by this, the following natural questions suggest themselves: For $n \geq 2$, characterize elements $\mathbf{a} \in \mathfrak{A}^n$, satisfying $f(\mathbf{a}) \in \mathcal{C}_n$, for any $f \in \mathcal{P}(\mathfrak{A})$. In particular if $\mathbf{A} = (A_1, \ldots, A_n) \in \mathcal{B}(\mathcal{H})^n$ is such that $\mathbf{W}(\mathbf{A}) \subset \mathcal{C}_n$. Does $\mathbf{A} = (c_1 I, \ldots, c_n I) \in \mathcal{B}(\mathcal{H})^n$ for some $(c_1, \ldots, c_n) \in \mathcal{C}_n$.

The main concern of this paper is to study these questions in the general setting of infinite dimensional C^*-algebras. Firstly, we characterize elements $\mathbf{a} \in \mathfrak{A}^n$ satisfying $\sum_{j=1}^{n} |f(a_j)|^2 = 1$ for every pure state f of $\mathfrak{A}$. We then apply this to characterize elements $\mathbf{a}_1, \ldots, \mathbf{a}_n \in \mathfrak{A}^n$ for which for any $\mathbf{x} \in (\mathfrak{A}^+)^n$ there exist $\alpha = (\alpha_1, \ldots, \alpha_n) \in \mathbb{R}^n$ such that $\sum_{j=1}^{n} \alpha_j^2 = 1$ and $\mathbf{v}\left(\sum_{j=1}^{j=n} \alpha_j \mathbf{a}_j + \mathbf{x}\right) = 1 + \mathbf{v}(\mathbf{x})$. In particular, we recapture some results of [5, 6, 9–11].

2. Statement of the main results.

In the sequel $\mathcal{H}$ (resp. $\mathfrak{A}$) denotes an infinite dimensional Hilbert space (resp. C^*-algebra such that every irreducible representation is infinite dimensional) and I stands for the identity operator on $\mathcal{H}$.

Since for any $x \in \mathcal{H}$ the functional $A \in \mathcal{B}(\mathcal{H}) \longmapsto \langle Ax, x \rangle$ is a pure state of $\mathcal{B}(\mathcal{H})$, we start our work with the following particular case which will be useful in other contexts as well.

Theorem 2.1. *For $\mathbf{A} = (A_1, \ldots, A_n) \in \mathcal{B}(\mathcal{H})^n$, the following assertions are equivalent.*

(1) *There exists $c = (c_1, \ldots, c_n) \in \mathcal{C}_n$ such that $\mathbf{A} = (c_1 I, \ldots, c_n I)$.*

(2) *For any unit vector $x \in \mathcal{H}$, we have*

(2.1) $$\sum_{k=1}^{n} |\langle A_k x, x \rangle|^2 = |\langle A_1 x, x \rangle|^2 + \ldots + |\langle A_n x, x \rangle|^2 = 1.$$

(3) *For any positive operator $X \in \mathcal{B}(\mathcal{H})^+$ there exists $\lambda \in \mathcal{C}_n$ such that*

(2.2) $$\mathbf{w}(\mathbf{A} + \lambda \cdot \mathbf{S}_X) = \mathbf{w}(\mathbf{A}) + w(X) = 1 + w(X).$$

REMARK 2.2. Note that Condition (2.1) of Theorem 2.1 does not imply in general that for any n-tuple $X = (X_1, \ldots, X_n) \in (\mathcal{B}(\mathcal{H})^+)^n$ there exists $\lambda \in \mathcal{C}_n$ such that $\mathbf{w}(\mathbf{A} + \lambda \cdot \mathbf{X}) = 1 + \mathbf{w}(\mathbf{X})$. Indeed: take $\mathbf{A} = (\frac{1}{\sqrt{2}}I, \frac{1}{\sqrt{2}}I)$ and $\mathbf{X} = (I, I)$. Then $1 + \mathbf{w}(\mathbf{X}) = 1 + \sqrt{2}$ but for any $\lambda \in \mathcal{C}_2$, we have

$$\mathbf{w}(\mathbf{A} + \lambda \cdot \mathbf{X}) = \sup_{\|x\|=1} \sqrt{\left|\frac{1}{\sqrt{2}} + \lambda_1\right|^2 + \left|\frac{1}{\sqrt{2}} + \lambda_2\right|^2} \leq 2 < 1 + \sqrt{2} = 1 + \mathbf{w}(\mathbf{X}).$$

Now, we are in a position to state a useful characterization of the joint numerical radius for infinite dimensional C^*-algebras.

Theorem 2.3. *Let $\mathbf{a} = (a_1, \ldots, a_n)$ in $\mathfrak{A}^n$. The following assertions are equivalent.*

(1) $\mathbf{a} \in \mathcal{Z}(\mathfrak{A})^n$ *with* $\sum_{i=1}^{n} a_k a_k^* = e.$

(2) *For any pure state $f \in \mathcal{P}(\mathfrak{A})$, we have*

$$(2.3) \qquad \sum_{i=1}^{n} |f(a_i)|^2 = 1.$$

(3) *For any positive element $x \in \mathfrak{A}^+$, there exists $\lambda \in \mathcal{C}_n$ such that*

$$(2.4) \qquad \mathbf{v}(\mathbf{a} + \lambda \cdot \mathbf{S}_x) = 1 + v(x).$$

As a consequence of Theorem 2.3, we have the following result.

Corollary 2.4. *([6, Corollary 2.4], [9, Proposition 1].)*
For an element u in $\mathfrak{A}$, the following assertions are equivalent.

(1) *u is unitary in the centre of $\mathfrak{A}$*
(2) *For any $x \in \mathfrak{A}^+$ there exists $\lambda \in \mathcal{C}_1$ such that $v(u + \lambda x) = 1 + v(x)$.*
(3) *$|f(u)| = 1$ for every $f \in \mathcal{P}(\mathfrak{A})$.*

Our next result is a variant of the main result of [6] and is an extension of Corollary 2.4.

Theorem 2.5. *Let $\mathbf{a}_1, \ldots, \mathbf{a}_n \in \mathfrak{A}^n$ and suppose that $\mathbf{v}\left(\sum_{s=1}^{s=n} \alpha_s \mathbf{a}_s\right) \leq 1$ whenever $\alpha = (\alpha_1, \ldots, \alpha_n) \in \mathcal{C}_n \cap \mathbb{R}^n$. Then the following conditions are equivalent.*

(1) $\mathbf{a}_j \in \mathcal{Z}(\mathfrak{A})^n \cap \text{Her}(\mathfrak{A})^n$ *and* $\sum_{l=1}^{n} a_{jl} a_{kl} = \delta_{jk} e$ *for any* $1 \leq j, k \leq n$.
(2) *For any $\mathbf{x} \in (\mathfrak{A}^+)^n$, there exist $\alpha = (\alpha_1, \ldots, \alpha_n) \in \mathcal{C}_n \cap \mathbb{R}^n$ such that*

$$\mathbf{v}\left(\sum_{j=1}^{j=n} \alpha_j \mathbf{a}_j + \mathbf{x}\right) = 1 + \mathbf{v}(\mathbf{x}).$$

3. Proof of theorem 2.1

Let i be the complex number so that $i^2 = -1$. Given a unit vector x in $\mathcal{H}$, let $x \otimes x$ denote the rank one projection defined by

$$(x \otimes x)h := \langle h, x \rangle x.$$

Since any operator $T \in \mathcal{B}(\mathcal{H})$ can be written as $T = \Re(T) + i\Im(T)$, an element $\mathbf{A} = (A_1, \ldots, A_n) \in \mathcal{B}(\mathcal{H})^n$ can be then identified with the $(2n)$-tuple $(\Re(A_1), \Im(A_1), \ldots, \Re(A_n), \Im(A_n))$ of self-adjoint operators and the joint numerical range $\mathbf{W}(\mathbf{A}) \subset \mathbb{C}^n$ can be identified with

$$\mathbf{W}(\Re(A_1), \Im(A_1), \ldots, \Re(A_n), \Im(A_n)) \subset \mathbb{R}^{2n}.$$

So, in the sequel we may and shall assume that the operators $A_1, \ldots, A_n$ are self-adjoint. Clearly, (1) implies (2) and (3). Now, assume that (2) holds. By a result of [17] (see also [18]), the closure of the joint numerical range is star-shaped, then $\mathbf{W}(\mathbf{A})$ is a singleton. Hence $\mathbf{A} = (c_1 I, \ldots, c_n I)$ for $c = (c_1, \ldots, c_n) \in \mathcal{C}_n$. The implication (2) $\Longrightarrow$ (1) is proved.

Let us finally show that (3) $\Longrightarrow$ (2). It suffices to show that condition (2.1) holds. To that end, we borrow some ideas from the proof of [9, Lemma 1]. For the sake of contradiction, suppose the contrary that $\sum_{k=1}^{n} |\langle A_k x, x \rangle|^2 \neq 1$ for some

unit vector $x \in \mathcal{H}$ and note that, since $\mathbf{w}(\mathbf{A}) = 1$, one has $\sum_{k=1}^{n} |\langle A_k x, x \rangle|^2 < 1$.
To generate a contradiction, we shall construct an element $X \in \mathcal{B}(\mathcal{H})^+$ such that $\mathbf{w}(\mathbf{A} + \lambda \cdot \mathbf{S}_X) < 1 + w(X)$ for all $\lambda \in \mathcal{C}_n$. Indeed let $X = x \otimes x$, then it is clear that $X \in \mathcal{B}(\mathcal{H})^+$ and $w(X) = 1$. Fix any r such that $\sum_{k=1}^{n} |\langle A_k x, x \rangle|^2 < r^2 < 1$.
Then we can find an $\epsilon > 0$ such that $\sum_{k=1}^{n} |\langle A_k y, y \rangle|^2 < r^2$ whenever $\|y - x\| < \epsilon$.
Let y be an arbitrary unit vector in $\mathcal{H}$. We have two cases. If there exists $\gamma \in \mathcal{C}_1$ such that $\|\gamma y - x\| < \epsilon$, then we must have $\sum_{k=1}^{n} |\langle A_k y, y \rangle|^2 < r^2$ and hence

$$\|\langle(\mathbf{A} + \lambda \cdot \mathbf{S}_X) y, y \rangle\|_2 \leq \|\langle \mathbf{A} y, y \rangle\|_2 + \|\langle (\lambda \cdot \mathbf{S}_X) y, y \rangle\|_2$$
$$= \|\langle \mathbf{A} y, y \rangle\|_2 + |\langle x, y \rangle| < r + 1,$$

whenever $\lambda \in \mathcal{C}_n$. Now, suppose that $\|y - \gamma x\| \geq \epsilon$ for any $\gamma \in \mathcal{C}_1$, . Then

$$\epsilon^2 \leq \langle y - \gamma x, y - \gamma x \rangle = 2 - 2\Re(\langle y, \gamma x \rangle) \text{ for every } \gamma \in \mathcal{C}_1.$$

It yields that $|\langle y, x \rangle| \leq 1 - \frac{1}{2}\epsilon$.

Let $k = \min(r + 1, 2 - \frac{1}{2}\epsilon^2)$, then for every $\lambda \in \mathcal{C}_n$ and $y \in \mathcal{H}$ with $\|y\| = 1$, we have

$$\|\langle(\mathbf{A} + \lambda \cdot \mathbf{S}_X) y, y \rangle\|_2 \leq \|\langle \mathbf{A} + y, y \rangle\|_2 + \|\langle \lambda \cdot \mathbf{S}_X y, y \rangle\|_2$$
$$= \|\langle \mathbf{A} y, y \rangle\|_2 + |\langle x, y \rangle|$$
$$\leq k.$$

Hence, $\mathbf{w}(\mathbf{A} + \lambda \cdot \mathbf{S}_X) < 1 + w(X)$, a contradiction.

4. Proof of theorem 2.3

We will need the following lemma which can be found in [13] and gives a characterization of pure states of $\mathfrak{A}$. We endow the set $\mathcal{P}(\mathfrak{A})$ with the relative w^*-topology.

Lemma 4.1. [13, Theorem 9.]
For any state f of $\mathfrak{A}$, the followings are equivalent:

(1) *f is a pure state.*
(2) *For each open neighborhood U of φ in $\mathcal{S}(\mathfrak{A})$ there is a positive δ and an $x \in \mathfrak{A}$ such that $0 \leq x \leq e$, $f(x) = 1$ and $g(x) < 1 - \delta$ for all $g \in \mathcal{S}(\mathfrak{A}) \backslash U$.*
(3) *For each closed G_δ set Δ containing f, there is a positive element $b \in \mathfrak{A}$ such that $\|b\| = f(b)$ and $\{g : g \in \mathcal{S}(\mathfrak{A}), g(b) = \|b\|\} \subset \Delta$.*

Now, we are in a position to prove Theorem 2.3.

The implications (1)$\iff$(2): First, assume that $\mathbf{a} \in \mathcal{Z}(\mathfrak{A})^n$ with $\sum_{j=n}^{n} a_j a_j^* = e$. Let f is a pure state of $\mathfrak{A}$. By [21, Theorem 5.1.6] the representation $\pi_f : \mathfrak{A} \longrightarrow \mathcal{B}(\mathcal{H}_f)$ produced by the usual GNS construction is irreducible. Hence $\pi_f(a_j)$ is in the centre of $\mathcal{B}(\mathcal{H}_f)$, and must be a scalar multiple of the identity. Say $\pi_f(a_j) = c_j(f) I_f$, where $c_j \in \mathbb{C}$ and I_f denotes the identity operator of $\mathcal{H}_f$. Since

$f(a_j) = \langle \pi(a_j)\xi, \xi \rangle$ for some unit vector $\xi \in \mathcal{H}_f$ and $a_j \in \mathcal{Z}(\mathfrak{A})$, it yields that $f(a_j a_j^*) = |c_j(f)|^2 = |f(a_j)|^2$. Hence

$$\sum_{j=1}^n |f(a_j)|^2 = \sum_{j=1}^n f(a_j a_j^*) = f(e) = 1.$$

Conversely, assume that $\sum_{j=1}^n |f(a_j)|^2 = 1$ for any pure state $f \in \mathcal{P}(\mathfrak{A})$. The case where $\mathfrak{A} = \mathcal{B}(\mathcal{H})$ follows from Theorem 2.1. For any $f \in \mathcal{P}(\mathfrak{A})$, by [21, Theorem 5.1.7] the GNS representation $\pi_f : \mathfrak{A} \longrightarrow \mathcal{B}(\mathcal{H}_f)$ attached to f is irreducible and for each unit vector x in $\mathcal{H}_f$, the state $w_x \circ \pi_f$ is pure. The hypothesis yields

$$\sum_{j=1}^n |w_x \circ \pi_f(a_j)|^2 = \sum_{j=1}^n |\langle \pi_f(a_j)x, x \rangle|^2 = 1$$

for any $x \in \mathcal{H}_f$. Hence by Theorem 2.1 there exist complex scalars $(c_1(f), \ldots, c_n(f)) \in \mathcal{C}_n$ such that

$$(\pi_f(a_1), \ldots, \pi_f(a_n)) = (c_1(f)I_{\mathcal{H}_f}, \ldots, c_n(f)I_{\mathcal{H}_f}).$$

This shows by [3, Proposition II 6.4.13] that $a_j \in \mathcal{Z}(\mathfrak{A})$ for any $1 \leq j \leq n$. Moreover, the fact that the representation $\Phi = \bigoplus_{f \in \mathcal{P}(\mathfrak{A})} \pi_f$ is faithful yields that $\sum_{j=1}^n a_j a_j^* = e$. The proof of the implication (2) $\Longrightarrow$ (1) is then complete.

The implications (1)$\Longleftrightarrow$(3): Let $x \in \mathfrak{A}^+$ and $f \in \mathcal{P}(\mathfrak{A})$ such that $f(x) = v(x)$. Set $\lambda = (f(a_1), \ldots, f(a_n))$. Clearly $\lambda \in \mathcal{C}_n$ and

$$\|f(\mathbf{a} + \lambda \cdot \mathbf{S}_x)\|_2 = 1 + v(x) \leq \mathbf{v}(\mathbf{a} + \lambda \cdot \mathbf{S}_x) \leq 1 + v(x).$$

Hence $\mathbf{v}(\mathbf{a} + \lambda \cdot \mathbf{S}_x) = 1 + v(x)$. To prove the converse, suppose that (3) is satisfied. Based on the aforesaid it suffices to show that $\sum_{i=1}^n |f(a_i)|^2 = 1$ for any $f \in \mathcal{P}(\mathfrak{A})$. Note that $\sum_{i=1}^n |f(a_i)|^2 \leq 1$ since $\mathbf{v}(\mathbf{a}) = 1$. Assume that $\sum_{i=1}^n |f(a_i)|^2 < 1$ for some $f \in \mathcal{P}(\mathfrak{A})$. Let r and ε be two positive real numbers such that $\sum_{i=1}^n |f(a_i)|^2 < r^2 < 1$ and $\varepsilon < 1 - r$ and consider the open neighborhood Ω of f in $S(\mathfrak{A})$ defined by

$$\Omega := \left\{ g \in S(\mathfrak{A}) : |f(a_i) - g(a_i)| < \frac{\epsilon^2}{n}, \forall 1 \leq i \leq n \right\}.$$

Lemma 4.1 provides a positive scalar δ and an element $x \in \mathfrak{A}$ such that $0 \leq x \leq \mathbf{1}$, $f(x) = 1$ and $g(x) < 1 - \delta$ for all $g \in S(\mathfrak{A}) \setminus \Omega$. Observe that, since $0 \leq x \leq e$ and $f(x) = 1 \leq v(x)$, we have $v(x) = \|x\| = 1$. The argument now splits into two cases.

(1) If $g \in \Omega$, then $|f(a_i) - g(a_i)| < \frac{\varepsilon}{n}$. Accordingly $\|g(\mathbf{a})\|_2 < r + \varepsilon$. Hence for any $\lambda \in \mathcal{C}_n$, we have

(4.5)
$$\begin{aligned}
\|g(\mathbf{a} + \lambda \cdot \mathbf{S}_x)\|_2 &= \|g(\mathbf{a}) + \lambda g(\lambda \cdot \mathbf{S}_x)\|_2 \\
&\leq \|g(\mathbf{a})\|_2 + \|g(\lambda \cdot \mathbf{S}_x)\|_2 \\
&\leq \|g(\mathbf{a})\|_2 + |g(x)| < \varepsilon + r + v(x) < 1 + v(x).
\end{aligned}$$

(2) If $g \notin \Omega$, then $g(x) \leq 1-\delta$. Therefore $\|g(\lambda \cdot \mathbf{S}_x)\|_2 = g(x) \leq 1-\delta$ and

(4.6) $$\|g(\mathbf{a}+\lambda \cdot \mathbf{S}_x)\|_2 < 1 + v(x).$$

Putting Equations (4.5) and (4.6) all together, we conclude that

(4.7) $$\|g(\mathbf{a}+\lambda \cdot \mathbf{S}_x)\|_2 < 1 + v(x)$$

for every $\lambda \in \mathcal{C}_n$ and $g \in S(\mathfrak{A})$. Accordingly

(4.8) $$\mathbf{v}(\mathbf{a}+\lambda \cdot \mathbf{S}_x) < 1 + v(x)$$

for every $\lambda \in \mathcal{C}_n$. This contradicts our assumption and thus $\sum_{i=1}^{n} |f(a_i)|^2 = 1$, for any $f \in \mathcal{P}(\mathfrak{A})$. The proof is thus complete.

5. Proof of theorem 2.5

Before embarking the proof, we introduce some notation. Let $\mathbf{a}_j = (a_{j1}, \ldots, a_{jn}) \in \mathfrak{A}^n$ for any $1 \leq j \leq n$. For a be pure state $f \in \mathcal{P}(\mathfrak{A})$, set

$$\Lambda(f) = (f(a_{jk}))_{1 \leq k,j \leq n} = \begin{pmatrix} f(a_{11}) & \cdots & f(a_{1n}) \\ f(a_{21}) & \cdots & f(a_{2n}) \\ \vdots & \ddots & \vdots \\ f(a_{n1}) & \cdots & f(a_{nn}) \end{pmatrix}.$$

By $\mathrm{Her}(\mathfrak{A})$ we denote the set of all hermitian elements of $\mathfrak{A}$. Finally, recall that the Kronecker delta is defined as:

$$\delta_{jk} = \begin{cases} 1 & \text{if } j=k \\ 0 & \text{if } j \neq k \end{cases}$$

We return now to the proof of Theorem 2.5. Assume that $\mathbf{a}_j = (a_{j1}, \ldots, a_{jn}) \in \mathcal{Z}(\mathfrak{A})^n \cap \mathrm{Her}(\mathfrak{A})^n$ and $\sum_{i=1}^{n} a_{ji} a_{ki} = \delta_{jk} e$ for any $1 \leq j, k \leq n$. Note that $f(a_{ji}a_{ki}) = f(a_{ji})f(a_{ki})$ for any $f \in \mathcal{P}(\mathfrak{A})$. This tells us that $\mathbf{v}(\mathbf{a}_j) = 1$ and the matrix $\Lambda(f)$ is an orthogonal matrix. Pick up an element $\mathbf{x} \in (\mathfrak{A}^+)^n$. If $\mathbf{x} = 0$, then (2) trivially holds with $\alpha = (\alpha_1, \ldots, \alpha_n) = (1, 0, \ldots, 0)$. So, assume that $\mathbf{x} \neq 0$. Let $f \in \mathcal{P}(\mathfrak{A})$ such that $\|f(\mathbf{x})\|_2 = \sqrt{|f(x_1)|^2 + \ldots + |f(x_n)|^2} = \mathbf{v}(\mathbf{x})$. Put $\beta_j = \frac{f(x_j)}{\mathbf{v}(\mathbf{x})}$. As the matrix $\Lambda(f)$ is orthogonal, there exists $\alpha = (\alpha_1, \ldots, \alpha_n) \in \mathcal{C}_n \cap \mathbb{R}^n$ such that $(\alpha_1 \ \cdots \ \alpha_n) \Lambda(f) = (\beta_1 \ \cdots \ \beta_n) = \beta$. Thus

$$\left\| f\left(\sum_{j=1}^{j=n} \alpha_j \mathbf{a}_j + \mathbf{x}\right) \right\|_2 = \|\beta + \mathbf{v}(\mathbf{x})\beta\|_2 = \|\beta\|_2(1 + \mathbf{v}(\mathbf{x})) = 1 + \mathbf{v}(\mathbf{x}).$$

Hence the implication (1)$\Longrightarrow$(2) is established. For the implication (2) $\Longrightarrow$(1), assume that for any $\mathbf{x} \in (\mathfrak{A}^+)^n$ there exist $\alpha_1, \ldots, \alpha_n \in \mathbb{R}$ such that $\alpha_1^2 + \ldots + \alpha_n^2 = 1$ and

$$\mathbf{v}\left(\sum_{j=1}^{n} \alpha_j \mathbf{a}_j + \mathbf{x}\right) = 1 + \mathbf{v}(\mathbf{x}).$$

The proof will be completed by proving two claims.

CLAIM 1. For any $(f, b) \in \mathcal{P}(\mathfrak{A}) \times \mathcal{C}_n \cap \mathbb{R}^n$ there exists $\alpha = (\alpha_1, \ldots, \alpha_n) \in \mathcal{C}_n \cap \mathbb{R}^n$ such that $\left\| f(\sum_{j=1}^n \alpha_j \mathbf{a}_j) + b \right\|_2 = 2.$

Assume for the sake of contradiction that there exists a pair $(f, b) \in \mathcal{P}(\mathfrak{A}) \times \mathcal{C}_n \cap \mathbb{R}^n$ such that $\|f(\sum_{j=1}^n \alpha_j \mathbf{a}_j) + b\|_2 < 2$ for all $\alpha \in \mathcal{C}_n \cap \mathbb{R}^n$. Choose two positive real numbers r and ε such that $\varepsilon < 2 - r$ and

$$\max \left\{ \|f(\sum_{j=1}^n \alpha_j \mathbf{a}_j) + b\|_2 : \alpha \in \mathcal{C}_n \cap \mathbb{R}^n \right\} < r < 2$$

Consider the set $\Delta = \bigcap_{s \in \mathcal{C} \cap \mathbb{Q}^n} \Delta_s$ in $S(\mathfrak{A})$, where Δ_s is the set defined by

$$\Delta_s := \left\{ g \in S(\mathfrak{A}) : \|f(\sum_{j=1}^n \alpha_j \mathbf{a}_j) - g(\sum_{j=1}^n \alpha_j \mathbf{a}_j)\|_2 \leq \epsilon, (s, t) \in \mathbb{T} \right\}.$$

It is clear that Δ is a closed subset relative to the w^*-topology containing f. Also, straightforward computation entails that Δ_s is a G_δ set. Since a countable intersection of G_δ sets is also a G_δ set, we infer that Δ is also a closed G_δ set. So, by Lemma 4.1 there is a positive element $d \in \mathfrak{A}$ such that $\|d\| = f(d)$ and $\{g \in S(\mathfrak{A}) : |g(d)| = \|d\|\} \subset \Delta$. Let $\mathbf{d} = (b_1 d, \ldots, b_n d)$. We have

$$\|f(\mathbf{d} + b \cdot \mathbf{e})\|_2 = 1 + \|b\| \leq \mathbf{v}(\mathbf{d} + b \cdot \mathbf{e}) \leq \|d\| + 1.$$

Accordingly $\mathbf{v}(\mathbf{d} + b \cdot \mathbf{e}) = 1 + \|d\| = f(d) + 1$. Now, to get a contradiction and complete the proof of this claim, we shall prove that

(5.9) $$\left\| g(\sum_{j=1}^n \alpha_j \mathbf{a}_j + \mathbf{d} + b \cdot \mathbf{e}) \right\|_2 < 1 + \mathbf{v}(\mathbf{d} + b \cdot \mathbf{e}) = 2 + \|b\|.$$

for all $g \in S(\mathfrak{A})$ and α. Due to the compactness of $\mathbf{W}\left(\sum_{j=1}^n \alpha_j \mathbf{a}_j + \mathbf{d} + b \cdot \mathbf{e}\right)$, certainly, this would contradicts condition (2) in Theorem 2.5, and completes the proof of this claim. To that end, pick an element $g \in S(\mathfrak{A})$. We shall distinguish two cases. If $g \in \Delta$, for all $s = (s_1, \ldots, s_n) \in \mathcal{C} \cap \mathbb{Q}^n$ it yields

$$\begin{aligned}
\left\| g(\sum_{j=1}^n s_j \mathbf{a}_j + \mathbf{d} + b \cdot \mathbf{e}) \right\|_2 &\leq \left\| g(\sum_{j=1}^n s_j \mathbf{a}_j + b \cdot \mathbf{e})\|_2 + \|g(d) \right\|_2 \\
&\leq \left\| g(\sum_{j=1}^n s_j \mathbf{a}_j) - f(\sum_{j=1}^n s_j \mathbf{a}_j) \right\|_2 + \left\| f(\sum_{j=1}^n s_j \mathbf{a}_j) + b\|_2 + \|d \right\|_2 \\
&< \epsilon + r + \|d\| \\
&< 2 + \|d\|
\end{aligned}$$

for all $s \in \mathcal{C} \cap \mathbb{Q}^n$. Therefore

$$\left\| g(\sum_{j=1}^{n} \alpha_j \mathbf{a}_j + \mathbf{d} + b \cdot \mathbf{e}) \right\|_2 < 2 + \|d\|$$

for all $\alpha \in \mathcal{C} \cap \mathbb{R}^n$, and (5.9) holds.
If $g \in S(\mathfrak{A}) \setminus \Delta$, then $g(b) < \|b\|$ and thus

$$\left\| g(\sum_{j=1}^{n} s_j \mathbf{a}_j + \mathbf{d} + b \cdot \mathbf{e}) \right\|_2 \leq \left\| g(\sum_{j=1}^{n} s_j \mathbf{a}_j) \right\|_2 + \|g(\mathbf{d} + b \cdot \mathbf{e})\|_2$$
$$\leq 2 + g(d) < 2 + \|d\|$$

for all $\alpha \in \mathcal{C} \cap \mathbb{R}^n$. Clearly, (5.9) holds in this case too, and the proof is thus complete.

CLAIM 2. $(\mathbf{a}_1, \ldots, \mathbf{a}_n)$ has the desired forms.

Firstly, we claim that $\Lambda(f)$ is an orthogonal matrix for any pure state $f \in \mathcal{P}(\mathfrak{A})$. To that end pick up pure state $f \in \mathcal{P}(\mathfrak{A})$ and let $b \in \mathcal{C} \cap \mathbb{R}^n$. By Claim 1, there exists $\alpha = (\alpha_1, \ldots, \alpha_n) \in \mathcal{C} \cap \mathbb{R}^n$ such that

$$\left\| f(\sum_{j=1}^{n} \alpha_j \mathbf{a}_j) + b \right\|_2 = 2.$$

Since $\left\| f(\sum_{j=1}^{n} \alpha_j \mathbf{a}_j) \right\|_2 \leq 1$, we infer that $\left\| f(\sum_{j=1}^{n} \alpha_j \mathbf{a}_j) \right\|_2 = 1$. As the norm $\|.\|_2$ is strictly convex we infer that

$$f\left(\sum_{j=1}^{n} \alpha_j \mathbf{a}_j\right) = \sum_{j=1}^{n} \alpha_j f(\mathbf{a}_j) = \sum_{j=1}^{n} \alpha_j (f(a_{j1}), \ldots, f(a_{jn}))$$
$$= (\alpha_1 \ldots \alpha_n) \Lambda(f)$$
$$= b = (b_1, \ldots, b_n).$$

This tells us that $\Lambda(f)$ is a linear maps on $\mathbb{R}^n$, maps the unit ball of $\mathbb{R}^n$ onto itself. Hence $\Lambda(f)$ is orthogonal as claimed. In particular $\Lambda(f)$ is a real matrix. Then by [16, Theorem 4.3.8], $\mathbf{a}_j \in \text{Her}(\mathfrak{A})^n$. Moreover Theorem 2.3 entails that $\mathbf{a}_j \in \mathcal{Z}(\mathfrak{A})^n$ and $\sum_{l=1}^{n} a_{jl} a_{kl}^* = \delta_{jk} e$ for any $1 \leq j, k \leq n$. The proof of this theorem is thus complete.

6. Concluding remarks.

Similar proof to the one of Theorem 2.5 yields the following result. The details are omitted. Note that unlike of Lemma 1 in [11], we do not suppose that Note that we do not assume that A_{ij} is self adjoint for any j. But such a condition is a part of the conclusion of the theorem bellow rather than being a part of its hypothesis.

Theorem 6.1. Let $\mathbf{A}_1, \ldots, \mathbf{A}_n \in \mathcal{B}(\mathcal{H})^n$ and suppose that $\mathbf{w}\left(\sum_{s=1}^{s=n} \alpha_s \mathbf{A}_s\right) \leq 1$ whenever $\alpha = (\alpha_1, \ldots, \alpha_n) \in \mathcal{C}_n \cap \mathbb{R}^n$. Then the following conditions are equivalent.

(1) There is an orthogonal matrix (c_{jk}) such that $\mathbf{A}_j = (c_{j1}I, \ldots, c_{jn}I)$ for all j
(2) For any $\mathbf{X} \in (\mathcal{B}(\mathcal{H})^+)^n$, there exists $\alpha = (\alpha_1, \ldots, \alpha_n) \in \mathcal{C}_n \cap \mathbb{R}^n$ such that

$$\mathbf{w}(\sum_{j=1}^{j=n} \alpha_j \mathbf{A}_j + \mathbf{X}) = 1 + \mathbf{w}(\mathbf{X}).$$

To conclude this paper, we list several remarks.

(1) Note that Condition (1) in Theorem 2.5 does not imply that (2) holds for any $\mathbf{x} \in \mathfrak{A}^n$ in general. It suffices to take $n = 1$ and $a = e$.
(2) Observe that the proof uses the star-shapedness of the closure of the joint numerical range. Such a property is valid only for infinite-dimensional Hilbert spaces. It would be nice to investigate the finite-dimensional case.
(3) Assume first that $\Phi : \text{Her}(\mathfrak{A})^n \longrightarrow \text{Her}(\mathfrak{A})^n$. Let $\mathbf{e}_j$ be the n-tuple in $\text{Her}(\mathfrak{A})^n$ whose j entry is e and all other entries are 0. Write $\Phi(\mathbf{e}_j) = (a_{j1}, \ldots, a_{jn})$. By Theorem 2.5 the matrix $\Delta = (a_{jk})_{1 \leq j,k \leq n} \in \mathcal{M}_n(\mathfrak{A})$ is such that $\mathbf{a}_j \in \mathcal{Z}(\mathfrak{A})^n \cap \text{Her}(\mathfrak{A})^n$ and $\sum_{i=1}^{n} a_{ji} a_{ki} = \delta_{jk}\, e$. For any $\mathbf{x} = (x_1, \ldots, x_n) \in \text{Her}(\mathfrak{A})^n$ define $\Phi' = \Phi(.).\Delta^\top$. Straightforward computation shows that Φ' preserves the joint numerical radius and maps each $\mathbf{e}_j$ to itself. Hence the characterizations of Joint numerical radius isometry can be studied by using a similar reasoning as in [**11**].

References

[1] S. K. Berberian and G. H. Orland, *On the closure of the numerical range of an operator*, Proc. Amer. Math. Soc. **18** (1967), 499–503, DOI 10.2307/2035486. MR212588

[2] P. Binding, D. R. Farenick, and C.-K. Li, *A dilation and norm in several variable operator theory*, Canad. J. Math. **47** (1995), no. 3, 449–461, DOI 10.4153/CJM-1995-025-5. MR1346148

[3] B. Blackadar, *Operator algebras: Theory of C^*-algebras and von Neumann algebras*, Encyclopaedia of Mathematical Sciences, vol. 122, Springer-Verlag, Berlin, 2006. MR2188261

[4] F. F. Bonsall and J. Duncan, *Numerical ranges. II*, Cambridge University Press, New York-London, 1973. London Mathematical Society Lecture Notes Series, No. 10. MR0442682

[5] A. Bourhim and M. Mabrouk, *Numerical radius and product of elements in C^*-algebras*, Linear Multilinear Algebra **65** (2017), no. 6, 1108–1116, DOI 10.1080/03081087.2016.1228818. MR3615532

[6] A. Bourhim and M. Mabrouk, *Maps preserving the numerical radius distance between C^*-algebras*, Complex Anal. Oper. Theory **13** (2019), no. 5, 2371–2380, DOI 10.1007/s11785-019-00894-2. MR3979715

[7] G. Braatvedt, R. Brits, and H. Raubenheimer, *Spectral characterizations of scalars in a Banach algebra*, Bull. Lond. Math. Soc. **41** (2009), no. 6, 1095–1104, DOI 10.1112/blms/bdp094. MR2575340

[8] M. Brešar and Š. Špenko, *Determining elements in Banach algebras through spectral properties*, J. Math. Anal. Appl. **393** (2012), no. 1, 144–150, DOI 10.1016/j.jmaa.2012.03.058. MR2921656

[9] J.-T. Chan, *Numerical radius preserving operators on $B(H)$*, Proc. Amer. Math. Soc. **123** (1995), no. 5, 1437–1439, DOI 10.2307/2161132. MR1231293

[10] J.-T. Chan, *Numerical radius preserving operators on C^*-algebras*, Arch. Math. (Basel) **70** (1998), no. 6, 486–488, DOI 10.1007/s000130050223. MR1621998

[11] J.-T. Chan and K. Chan, *Two distance preservers of generalized numerical radii*, Linear Multilinear Algebra **62** (2014), no. 5, 674–682, DOI 10.1080/03081087.2013.784284. MR3195960

[12] A. T. Dash, *Joint numerical range* (English, with Serbo-Croatian summary), Glasnik Mat. Ser. III **7(27)** (1972), 75–81. MR324443

[13] J. Glimm, *Type I C^*-algebras*, Ann. of Math. (2) **73** (1961), 572–612, DOI 10.2307/1970319. MR124756

[14] E. Gutkin, E. A. Jonckheere, and M. Karow, *Convexity of the joint numerical range: topological and differential geometric viewpoints*, Linear Algebra Appl. **376** (2004), 143–171, DOI 10.1016/j.laa.2003.06.011. MR2014890

[15] P. R. Halmos, *A Hilbert space problem book*, 2nd ed., Graduate Texts in Mathematics, vol. 19, Springer-Verlag, New York-Berlin, 1982. Encyclopedia of Mathematics and its Applications, 17. MR675952

[16] R. V. Kadison and J. R. Ringrose, *Fundamentals of the theory of operator algebras. Vol. I: Elementary theory*, Graduate Studies in Mathematics, vol. 15, American Mathematical Society, Providence, RI, 1997. Reprint of the 1983 original. MR1468229

[17] C.-K. Li and Y.-T. Poon, *The joint essential numerical range of operators: convexity and related results*, Studia Math. **194** (2009), no. 1, 91–104, DOI 10.4064/sm194-1-6. MR2520042

[18] C.-K. Li and Y.-T. Poon, *Generalized numerical ranges and quantum error correction*, J. Operator Theory **66** (2011), no. 2, 335–351. MR2844468

[19] C.-K. Li, Y.-T. Poon, and N.-S. Sze, *Higher rank numerical ranges and low rank perturbations of quantum channels*, J. Math. Anal. Appl. **348** (2008), no. 2, 843–855, DOI 10.1016/j.jmaa.2008.08.016. MR2446039

[20] V. Müller, *The joint essential numerical range, compact perturbations, and the Olsen problem*, Studia Math. **197** (2010), no. 3, 275–290, DOI 10.4064/sm197-3-5. MR2607493

[21] G. J. Murphy, *C^*-algebras and operator theory*, Academic Press, Inc., Boston, MA, 1990. MR1074574

[22] B. T. Polyak, *Convexity of quadratic transformations and its use in control and optimization*, J. Optim. Theory Appl. **99** (1998), no. 3, 553–583, DOI 10.1023/A:1021798932766. MR1658026

Department of Mathematics, Faculty of Applied Sciences, Umm Al-Qura University 21955 Makkah, Saudi Arabia & Faculty of Sciences of Sfax, Department of Mathematics University of Sfax Tunisia

Email address: mbs_mabrouk@yahoo.fr

Weighted composition operators on non locally convex weighted spaces with operator-valued weights

Mohammed Klilou and Lahbib Oubbi

To the memory of Ali Klilou, Mohammed's father

ABSTRACT. In this paper, we propose a general framework for the study of the weighted spaces of vector-valued continuous functions. We consider operator-valued weights and assume no local convexity condition. Precisely, for a separated topological vector space A, a Hausdorff completely regular space X, and a family V of mappings (called weights) v from X into the algebra $\mathcal{L}(A)$ of all continuous operators on A, we consider the generalized weighted spaces

$$CV(X, A) = \{f : X \to A \text{ continuous} : vf \text{ is bounded on } X, v \in V\}$$

and

$$CV_0(X, A) := \{f \in CV(X, A) : vf \text{ vanishes at infinity}, v \in V\},$$

endowed with a linear topology τ_V associated with V. We then present some of their properties. We further investigate the weighted composition operators ψC_φ from a subspace E of $CV(X, A)$ into $CU(Y, A)$ or $CU_0(Y, A)$, where Y is a Hausdorff completely regular space, U is a family of weights on Y, $\psi : Y \to \mathcal{L}(A)$ and $\varphi : Y \to X$ are mappings, and $\psi C_\varphi(f)(y) := \psi(y)[f(\varphi(y))]$, $y \in Y$ and $f \in E$. In particular, we give necessary and sufficient conditions for such an operator to be continuous, bounded, or locally equicontinuous.

Introduction

In 1965, L. Nachbin [26] introduced the weighted spaces $CV(X)$ of scalar-valued continuous functions on X, where X is a Hausdorff completely regular space, V is a Nachbin family on X (i.e. a collection of non negative upper semicontinuous functions on X with appropriate conditions), and $CV(X) := \{f \in C(X) : \|f\|_v := \sup\{v(x)|f(x)|, x \in X\} < \infty, v \in V\}$. Such a space is endowed with the topology given by the seminorms $\| \|_v, v \in V$. Since then many authors have been interested in different aspects of the general theory of (Banach or) locally convex spaces (or algebras) in such spaces [3–5, 7, 12, 13, 27, 28, 30–32, 38–42] as well as in subspaces or generalizations of them [3–5, 7, 8, 16, 22–24, 29]. As first generalizations, many authors considered the weighted spaces $CV(X, E)$ of continuous functions with ranges in (a Banach or) a locally convex space E [1–5, 7, 8, 18, 29, 33–35, 41, 42].

2010 *Mathematics Subject Classification.* Primary 47B38 and 46E10 and 47A56, Secondary 46E40 .

Key words and phrases. Generalized Nachbin family ; Generalized weighted spaces of vector-valued continuous functions; Weighted composition operators ; Composition operators ; Multiplication operators .

Several issues were considered in such spaces such as those related to approximation properties [26, 27, 34, 41, 42], Tensor products [1, 2], inductive limits and their projective descriptions [3–5, 7, 8] and so on. Instead of considering the continuous functions, some authors considered holomorphic functions or harmonic ones on an open subset of $\mathbb{C}^n$, $n \geq 1$, satisfying the same weighted growth conditions either in the scalar case [9, 22–24] or in the vector-valued case [6, 10, 11].

Different types of operators between weighted spaces have been studied by several authors. J. S. Jeang and N. C. Wong [15] dealt with the weighted composition operators $\psi C_\varphi : f \mapsto \psi f \circ \varphi$ from $C_0(X)$ into $C_0(Y)$, where X and Y are Hausdorff locally compact spaces, $\psi \in C(Y)$ and φ a map from Y into X. For the Banach weighted spaces of analytic functions on the unit disk, the multiplication operators were the subject of [9]. In the vector-valued setting, J. E. Jamison and M. Rajagopalan [14] considered the weighted composition operators ψC_φ on the (special weighted) Banach space $C(K, A)$, where K is a compact space, A a Banach space, φ a self map on X, and ψ an $\mathcal{L}(A)$-valued function on K. Such weighted composition operators from the space $C_p(X, A)$ of all A-valued continuous functions f such that $f(X)$ is precompact were studied in [40] for an arbitrary completely regular space X and any locally convex space A.

For weighted spaces $CV(X, A)$ with A non-locally convex, the weighted composition operators were studied mainly in [17, 21, 25, 29, 37].

For all the aforementioned authors, the weights $v \in V$ determining the space and its topology are nonnegative upper semicontinuous real functions. Recently, C. Shekhar and B. S. Komal introduced in [36] systems of weights with ranges in the set of positive operators on a Hilbert space H and considered the associated generalized weighted spaces of H-valued continuous functions on X. This constitutes a nice generalization of the classical weighted spaces of Nachbin. At this point, notice that, in case of $CV(X, E)$ with real-valued weights, each $v(x) > 0$ defines actually an invertible, and bounded below operator on E by $T_{v(x)}(u) = v(x)u$. Such an operator is positive in case E is a Hilbert space. In [36] the authors investigate the multiplication operators on $CV_0(X, H)$. In [19], some issues in [36] have been fixed and the multiplication operators in such generalized weighted spaces have been studied.

The study of Weighted composition operators in Nachbin spaces of continuous functions with values in a normed space and operator-valued weights were investigated in [20].

In this paper, we consider Nachbin families on X, consisting of weights taking their values in the algebra $\mathcal{L}(A)$ of all continuous linear operators on an arbitrary separated topological vector space A. This yields a very general framework for the study of the weighted spaces. We specially give conditions under which such spaces are complete. However, our main purpose in this note is to investigate the weighted composition operators between a subspace E of a weighted space $CV(X, A)$ into a weighted space $CU(Y, A)$ or $CU_0(Y, A)$, where Y is a Hausdorff completely regular space and V and U are generalized Nachbin families on X and Y respectively. The consideration of subspaces of $CV(X, A)$ makes our results cover, at once, a large class of different spaces.

We produce in Section 1 some preliminaries, some notations and some first topological properties of the generalized weighted spaces $CV(X, A)$. In Section 2, we study the completeness of the spaces $CV(X, A)$ and $CV_0(X, A)$. The section 3 is devoted to the characterization of those weighted composition operators which map continuously some subspace E of $CV(X,A)$ into $CU(Y,A)$ or into $CU_0(Y,A)$. Section 4 focuses on the conditions under which ψC_φ is bounded, and Section 5 deals with locally equicontinuity of ψC_φ.

1. Preliminaries

Throughout this paper, unless the contrary is expressly stated, X and Y will stand for Hausdorff completely regular spaces and A for a non-trivial Hausdorff topological vector space over the field $\mathbb{K}$ ($= \mathbb{R}$ or $\mathbb{C}$). We will let $C(X,A)$ (resp. $C_b(X,A)$, $C_0(X,A)$, $\mathcal{K}(X,A)$) denote the vector space of all continuous (resp. continuous and bounded, continuous and vanishing at infinity, continuous with compact support) functions from X into A, while $\mathcal{F}(Z,A)$ will be the set of all A-valued functions on X. In case $A = \mathbb{K}$, we will write $C(X)$ (resp. $C_b(X)$, $C_0(X)$, $\mathcal{K}(X)$, $\mathcal{F}(X)$) respectively.

By $\mathcal{N}$ it will be meant the set of all closed, circled, and shrinkable 0-neighborhoods in A. Recall that a subset G of A is shrinkable if $r\overline{G} \subset \overset{\circ}{G}$ for every $0 \leq r < 1$, where $\overline{G}$ denotes the closure of G in A and $\overset{\circ}{G}$ its interior. Such sets constitute a base of 0-neighborhoods in any topological vector space and the gauge P_G of any such a $G \in \mathcal{N}$ is continuous and satisfies

$$G = \{a \in A : P_G(a) \leq 1\} \text{ and } \overset{\circ}{G} = \{a \in A : P_G(a) < 1\},$$

where

$$P_G(a) = \inf\{\alpha > 0 : a \in \alpha G\}, \quad a \in A.$$

We refer to [18] for details concerning such an issue. It is clear that $P_G(\lambda a) = |\lambda| P_G(a)$ for all $a \in A$ and $\lambda \in \mathbb{K}$, and that, whenever $H \in \mathcal{N}$ enjoys $H + H \subset G$, one has :

$$P_G(a+b) \leq P_H(a) + P_H(b), \ \forall a, b \in A.$$

Furthermore, a linear map $T : A \to B$ from A into another topological vector space B is continuous if, and only if,

$$\forall G \in \mathcal{N}', \exists H \in \mathcal{N} \ : \ P_G(T(a)) \leq P_H(a), \quad a \in A,$$

where $\mathcal{N}'$ is the set of all closed, circled, and shrinkable 0-neighborhoods in B. We will say that a linear mapping T from A into B is bounded below if,

$$\forall G \in \mathcal{N}, \exists H \in \mathcal{N}' \ : \ P_G(a) \leq P_H(T(a)), \quad a \in A.$$

A collection $\mathcal{T}$ of linear mappings from A into B is said to be equi-bounded below, if

$$\forall G \in \mathcal{N}, \exists H \in \mathcal{N}' \ : \ P_G(a) \leq P_H(T(a)), \quad a \in A, \ T \in \mathcal{T}.$$

By $\mathcal{L}_{bb}(A)$ we will mean the subset of $\mathcal{L}(A)$ consisting of bounded below operators. We will endow $\mathcal{L}(A)$ with the topology β (resp. σ) of uniform convergence on the finite (resp. bounded) subsets of A. The topology β is called the strong operator topology on $\mathcal{L}(A)$ and, some times, we will describe as strong, properties relative to this topology (e.g., strongly continuous, strongly bounded). A base of zero neighborhoods for β (resp. σ) consists of all sets of the form

$$N(B,G) := \{T \in \mathcal{L}(A) : T(B) \subset G\}, \quad G \in \mathcal{N}, B \in \mathcal{B},$$

where $\mathcal{B}$ stands for the collection of all finite (resp. bounded) subsets of A.

For every $a \in A$ and $G \in \mathcal{N}$, $\delta_{G,a}$ will denote the gauge evaluation at a. This is $\delta_{G,a}(T) := P_G(T(a))$, for every $T \in \mathcal{L}(A)$. An $\mathcal{L}(A)$-valued mapping v on X is said to be strongly upper semicontinuous if, for every $a \in A$ and $G \in \mathcal{N}$, the real-valued map $\delta_{G,a} \circ v$ is upper semicontinuous (u.s.c. in short) on X, i.e. the set $\{x \in X : P_G(v(x)a) < \alpha\}$ is open for every $\alpha \in \mathbb{R}$.

A mapping $\nu : X \to A$ is said to vanish at infinity on X if, for every $G \in \mathcal{N}$ and $\varepsilon > 0$, there exists a compact subset $K_{G,\varepsilon}$ of X such that $P_G(\nu(x)) < \varepsilon$ for all $x \notin K_{G,\varepsilon}$. Equivalently, for every $G \in \mathcal{N}$, there exists a compact subset K_G of X such that $\nu(x) \in \overset{\circ}{G}$ for all $x \notin K_G$, or equivalently, for every open neighborhood H of zero, there exists a compact subset K_H of X such that $\nu(x) \in H$ for all $x \notin K_H$. If the mapping $x \mapsto P_G(\nu(x))$ happens to be upper semicontinuous, for every $G \in \mathcal{N}$, then ν vanishes at infinity if and only if $\{x \in X : P_G(\nu(x)) \geq \varepsilon\}$ is compact for every $G \in \mathcal{N}$ and $\varepsilon > 0$. In [36] and subsequently in [19], it is introduced the notion of generalized Nachbin families in the framework of Hilbert spaces. Such families consist of positive operator-valued functions with some additional conditions. This notion was extended to normed spaces in [20]. Here, we introduce generalized Nachbin families in the general framework of arbitrary topological vector spaces as follows:

DEFINITION 1.1. An A-generalized Nachbin family on X is any collection V of $\mathcal{L}(A)$-valued functions on X such that:
i. Every $v \in V$ is strongly upper semicontinuous,
ii. $\forall x \in X$, $\bigcap\{\ker v(x), v \in V\} = \{0\}$,
iii. V is directed upward in the following sense : for all v_1, $v_2 \in V$, and $G \in \mathcal{N}$, there exist $v \in V$ such that $P_G(v_i(x)a) \leq P_G(v(x)a)$, $x \in X$, $a \in A$, and $i = 1, 2$.

There is no loss of generality in assuming that for every $v \in V$ and $\lambda > 0$, we also have $\lambda v \in V$.

Whenever V is an A-generalized Nachbin family on X, we will consider the so-called generalized weighted spaces of vector-valued continuous functions:

$$CV(X, A) := \{f \in C(X, A) : (vf)(X) \text{ is bounded in } A, \forall v \in V\}$$

and

$$CV_0(X, A) := \{f \in CV(X, A) : vf \text{ vanishes at infinity on } X, \forall v \in V\}.$$

These spaces will be endowed with the weighted linear topology τ_V for which a base of 0-neighborhoods consists of all the sets of the form

$$B_{G,v} := \{f \in CV(X, A), (vf)(X) \subset G\},$$

respectively

$$B_{G,v}^0 := \{f \in CV_0(X, A), (vf)(X) \subset G\},$$

where G and v run respectively over $\mathcal{N}$ and V. The gauge of such a set is denoted by $P_{G,v}$. This is

$$P_{G,v}(f) := \sup\{P_G(v(x)f(x)), x \in X\}, \quad f \in CV(X, A) \text{ (or } f \in CV_0(X, A)).$$

In all the following, we will drop the letter A from A-generalized Nachbin family.

Notice that the condition *ii.* in Definition 1.1 ensures that the topology τ_V is Hausdorff, while the condition *iii.*, ensures that the family $\{P_{G,v}; v \in V, G \in \mathcal{N}\}$ is directed upward. Consequently, the collection $(B_{G,v})_{G \in \mathcal{N}, v \in V}$ is a fundamental system of zero neighborhoods for τ_V.

If the mapping $x \mapsto P_G(v(x)g(x))$ happens to be u.s.c. on X for every $v \in V$, $G \in \mathcal{N}$, and every $g \in C(X, A)$, the following equality turns out to be true.

$$CV_0(X, A) := \{f \in C(X, A) : vf \text{ vanishes at infinity on } X, \forall v \in V\}.$$

Here are some examples of generalized Nachbin families, some of which are given in [20], whenever A is a normed space.

EXAMPLE 1.2. Let U be a usual Nachbin family (i.e. consisting of real-valued u.s.c. non negative functions) on X. Then, identifying $v(x)$ with the operator $a \mapsto v(x)a$, U is a generalized Nachbin family and the space $CU(X, A)$ and $CU_0(X, A)$ are exactly the classical weighted spaces.

EXAMPLE 1.3. Let U be a usual Nachbin family, $T \in \mathcal{L}(A)$ a non zero continuous operator on A, and $V := \{uT : u \in U\}$, with $uT(x) := u(x)T$. If T is injective, then V is a generalized Nachbin family on X and the inclusions $CU(X, A) \subset CV(X, A)$ and $CU_0(X, A) \subset CV_0(X, A)$ hold with continuous injections. In particular, if $T = I$, we are in the situation of Example 1.2.

EXAMPLE 1.4. Let U be a usual Nachbin family on X and let $R : X \to \mathcal{L}(A)$ be a continuous map, $\mathcal{L}(A)$ being endowed with the topology β. If $R(x)$ is injective for every $x \in X$, then $V := \{u(.)R(.) : u \in U\}$ is a generalized Nachbin family on X. Furthermore, setting $N_u := \{x \in X : u(x) > 0\}$, if $R(N_u)$ is an equicontinuous subset of $\mathcal{L}(A)$ for every $u \in U$, then $CU(X, A)$ and $CU_0(X, A)$ are subsets of $CV(X, A)$ and $CV_0(X, A)$ respectively and the injections are continuous.

EXAMPLE 1.5. To every compact subset K of X, assign a non zero operator $T_K \in \mathcal{L}(A)$ so that, whenever K_1 and K_2 are compact subsets of X with $K_1 \subset K_2$, for every $G \in \mathcal{N}$, $P_G(T_{K_1}(a)) \leq P_G(T_{K_2}(a))$, for all $a \in A$. Now, for every such a compact K, set $v_K := 1_K T_K$, where 1_K denotes the characteristic functional of K. Since K is compact, the mapping v_K is strongly upper semicontinuous. If, in addition, we assume that for every $x \in X$, the set $\bigcap \{\ker T_K, \ K \subset X \text{ compact and } x \in K\}$ is reduced to $\{0\}$, then $\mathcal{K} := \{\lambda v_K : K \subset X \text{ compact and } \lambda > 0\}$ is a generalized Nachbin family on X such that $C\mathcal{K}(X, A) = C\mathcal{K}_0(X, A) = C(X, A)$ algebraically. Furthermore, whenever $C(X, A)$ is endowed with the compact open topology, the inclusion $C(X, A) \subset C\mathcal{K}(X, A)$ is continuous.

EXAMPLE 1.6. Let $\mathcal{T} \subset \mathcal{L}(A)$ be a separating family such that for all $T_1, T_2 \in \mathcal{T}$, and all $G \in \mathcal{N}$, there exists $T \in \mathcal{T}$ such that:

$$P_G(T_i(a)) \leq P_G(T(a)), \quad a \in A, \quad i = 1, 2.$$

If we put for each $T \in \mathcal{T}$

$$\begin{aligned} v_T : X &\longrightarrow \mathcal{L}(A) \\ x &\longmapsto v_T(x) = T \end{aligned}$$

and $\mathcal{Z} := \{\lambda v_T : T \in \mathcal{T}, \lambda > 0\}$, then $\mathcal{Z}$ is a generalized Nachbin family on X so that $C_b(X, A) \subset C\mathcal{Z}(X, A)$ and $C_0(X, A) \subset C\mathcal{Z}_0(X, A)$ and the inclusions are continuous.

One may ask wether, for a generalized Nachbin family V, there is some (classical) Nachbin family U on X such that the equality $CV(X, A) = CU(X, A)$ holds algebraically and topologically. Actually the following example proves that this is not the case in general. It provides a generalized Nachbin family V such that there exists no Nachbin family U on X such that $CV(X, A)$ is a topological subspace of $CU(X, A)$.

EXAMPLE 1.7. Let $X = \mathbb{N}$ be the set of positive integers and A the algebra $\mathbb{C}^2$, equipped with the norm $\|(z_1, z_2)\| = \sqrt{|z_1|^2 + |z_2|^2}$. Define the mapping v from X into $\mathcal{L}(A) := M_2(\mathbb{C})$, the algebra of square matrices of order 2, by:

$$v(n) = \begin{pmatrix} n & 0 \\ 0 & n^2 \end{pmatrix}, \quad n \in \mathbb{N}.$$

As X is discrete, v is continuous. Moreover $v(n)$ is invertible for every n. Therefore $V := \{\lambda v : \lambda > 0\}$ is a generalized Nachbin family on X. We claim that there is no scalar Nachbin family U on X such that $CV(X, A)$ is a topological subspace of $CU(X, A)$. Indeed, assume that some such a Nachbin family U exists on X. Then there is $u \in U$ and a constant $d > 0$ such that:

$$\|f\|_v \leq d\|f\|_u, f \in CV(X, A).$$

But also, for every $u' \in U$, there exists some $c' = c(u') > 0$ such that:

$$c'\|f\|_{u'} \leq \|f\|_v \leq d\|f\|_u, f \in CV(X, A).$$

It follows that the relative topology τ_U on $CV(X, A)$ is defined by the single seminorm $\|\ \|_u$. Let $\delta_{m,n}$ denote the Kronecker delta function and let $c = c(u)$ satisfy:

(1.1) $$c\|f\|_u \leq \|f\|_v \leq d\|f\|_u, f \in CV(X, A),$$

Consider, for every $m \in \mathbb{N}$, the functions $g_m : n \mapsto (\frac{1}{m}\delta_{n,m}, 0)$ and $h_m : n \mapsto (0, \frac{1}{m^2}\delta_{n,m})$. Since g_m and h_m belong to $CV(X, A)$, and we have:

$$\|g_m\|_u = \sup\{u(n)\frac{1}{m}\delta_{n,m}, n \in \mathbb{N}\} = \frac{u(m)}{m},$$

$$\|g_m\|_v = \sup\{\frac{1}{m}\|v(n)(\delta_{n,m}, 0)\|, n \in \mathbb{N}\} = \frac{m}{m} = 1.$$

$$\|h_m\|_u = \sup\{u(n)\frac{1}{m^2}\delta_{n,m}, n \in \mathbb{N}\} = \frac{u(m)}{m^2},$$

$$\|h_m\|_v = \sup\{\frac{1}{m^2}\|v(n)(0, delta_{n,m})\|, n \in \mathbb{N}\} = \frac{m^2}{m^2} = 1.$$

Applying (1.1) once to g_m, we obtain $cu(m) \leq m \leq du(m)$, and another time to h_m, we get $cu(m) \leq m^2 \leq du(m)$. Therefore $m^2 \leq \frac{d}{c}m$ for every m, which is impossible since m is arbitrary.

If $f \in C(X)$ and $a \in A$ are given, we will denote by $f \otimes a$ the function defined on X by $f \otimes a(x) = f(x)a$, $x \in X$. Whenever z is an element of a topological space Z, we will designate by $\mathcal{V}_z$ the filter of neighborhoods of z.

2. Completeness of $CV(X, A)$ and $CV_0(X, A)$

In this section, we will investigate, for any generalized Nachbin family V on X, the completeness of $CV(X, A)$ and $CV_0(X, A)$. To this purpose, let us consider, for every $v \in V$, $G \in \mathcal{N}$, and $H \in \mathcal{N}$ the level set

$$N(v, G, H) := \{x \in X : P_G(a) \leq P_H(v(x)a), \forall a \in A\}.$$

Notice that, for $x \in N(v, G, H)$, $v(x)$ need not be bounded below, for G and H are fixed.

As in the scalar-valued weights case, $CV_0(X, A)$ is closed in $CV(X, A)$ as shows the following proposition.

PROPOSITION 2.1. *For every generalized Nachbin family V on X, $CV_0(X, A)$ is a closed subspace of $CV(X, A)$.*

PROOF. Let $f \in CV(X, A)$ be in the closure $\overline{CV_0(X, A)}^{CV(X,A)}$ of $CV_0(X, A)$. For arbitrary $G \in \mathcal{N}$, choose $H \in \mathcal{N}$ such that $H + H \subset \overset{\circ}{G}$. Then for all $v \in V$, there exists $g \in CV_0(X, A)$ such that $v(f - g)(X) \subset H$. Since g belongs to $CV_0(X, A)$, there exists a compact subset K of X such that $v(x)g(x) \in \overset{\circ}{H}$, for all $x \notin K$. Therefore, for such an x, we have: $v(x)f(x) = v(x)(f(x) - g(x)) + v(x)g(x) \in H + H \subset \overset{\circ}{G}$. This yields $f \in CV_0(X, A)$ since G and v are arbitrary. □

As in [20], we will say that X is a $V_{\mathbb{R}}$-space, if a mapping $f : X \to \mathbb{R}$ is continuous provided its restriction to every level set $N(v, G, H)$ is continuous, $v \in V$, $G \in \mathcal{N}$, and $H \in \mathcal{N}$.

Let us denote, for simplicity, $X_0 := \text{coz}(CV_0(X, A))$ and $X_1 := \text{coz}(CV(X, A))$.

THEOREM 2.2. *The space $CV(X, A)$ (resp. $CV_0(X, A)$) is complete whenever the following four conditions hold:*
1. $\forall x \in X_1$ *(resp. $\forall x \in X_0$), $\exists v \in V : v(x) \in \mathcal{L}_{bb}(A)$,*
2. *A is complete,*
3. *X is a $V_{\mathbb{R}}$-space,*
4. *For all $v \in V$, $G \in \mathcal{N}$, $H \in \mathcal{N}$, and $x \in N(v, G, H)$, there exists $\Omega \in \mathcal{V}_x$ such that $\{v(t) : t \in \Omega \cap N(v, G, H)\}$ is equi-bounded below.*

PROOF. Let $(f_i)_{i \in \mathfrak{I}}$ be a Cauchy net in $CV(X, A)$ (resp. $CV_0(X, A)$). By 1., for every $x \in X_1$ (resp. $x \in X_0$), the evaluation map $\delta_x : f \mapsto f(x)$ is continuous from $CV(X, A)$ (resp. $CV_0(X, A)$) into A. Therefore $(f_i(x))_{i \in \mathfrak{I}}$ is a Cauchy net in A. Since A is complete, $(f_i(x))_{i \in \mathfrak{I}}$ converges to some $f(x) \in A$. Extend the so-defined function f over X by putting $f = 0$ identically on $X \setminus X_1$ (resp. $X \setminus X_0$). We claim that f belongs to $CV(X, A)$ (resp. $CV_0(X, A)$) and that $(f_i)_{i \in \mathfrak{I}}$ converges to f in $CV(X, A)$ (resp. $CV_0(X, A)$).
Since X is a $V_{\mathbb{R}}$-space, in order to show that f is continuous on X, it suffices to show that its restriction to each $N(v, G, H)$ is. Let then $v \in V$, $G \in \mathcal{N}$, $H \in \mathcal{N}$, and $x \in N(v, G, H)$ be arbitrary. By 4., there exists $\Omega \in \mathcal{V}_x$ such that $\{v(t) : t \in \Omega \cap N(v, G, H)\}$ is equi-bounded below. Therefore, for every $I \in \mathcal{N}$, let I', $J \in \mathcal{N}$ satisfy $I' + I' + I' \subset I$ and $P_{I'}(a) \leq P_J(v(t)a)$, for every $a \in A$ and every $t \in \Omega \cap N(v, G, H)$. We then have in particular:

$$P_{I'}(f_i(t) - f_j(t)) \leq P_J(v(t)(f_i(t) - f_j(t))), t \in \Omega \cap N(v, G, H).$$

Whereby
$$P_{I'}(f_i(t) - f_j(t)) \leq P_{J,v}(f_i - f_j), t \in \Omega \cap N(v, G, H).$$

Since $(f_i)_{i \in \mathfrak{I}}$ is Cauchy, for J and v, there exists $i_0 \in \mathfrak{I}$ such that $f_i - f_j \in B_{J,v}$ whenever $i_0 \leq i$ and $i_0 \leq j$. Hence $f_{i_0}(t) - f_j(t) \in I'$, for all $t \in \Omega \cap N(v, G, H)$ and all j with $i_0 \leq j$. Passing to the limit on j, we get

(2.1) $\qquad f_{i_0}(t) - f(t) \in I', \ \forall t \in \Omega \cap N(v, G, H).$

By the continuity of f_{i_0}, there exists $\Omega' \in \mathcal{V}_x$ such that

(2.2) $\qquad f_{i_0}(t) - f_{i_0}(x) \in I', \ \forall t \in \Omega'.$

For $t \in \Omega' \cap \Omega \cap N(v, G, H)$, by (2.1) and (2.2), we have
$$f(t) - f(x) = (f(t) - f_{i_0}(t)) + (f_{i_0}(t) - f_{i_0}(x)) + (f_{i_0}(x) - f(x))$$
$$\in I' + I' + I' \subset I.$$

Then $f(t) - f(x) \in I$, for every $t \in \Omega' \cap \Omega \cap N(v, G, H)$. It follows that f, restricted to $N(v, G, H)$, is continuous. To achieve the proof, it is enough, since $CV_0(X, A)$ is closed in $CV(X, A)$, to show that $(f_i)_{i \in \mathfrak{I}}$ converges in $CV(X, A)$ to f.

Let then $u \in V$ and $K \in \mathcal{N}$ be arbitrary. Since $(f_i)_{i \in \mathfrak{I}}$ is Cauchy, there exists $i_0 \in \mathfrak{I}$ such that $f_i - f_j \in B_{K,u}$ whenever $i_0 \leq i$ and $i_0 \leq j$. i.e., $u(t)(f_i(t) - f_j(t)) \in K$. Since $u(t)$ is continuous and K is closed, passing to the limit on j, we get $u(t)(f_i(t) - f(t)) \in K$, $\forall t \in X$ and $i_0 \leq i$. Whereby
$$f_i - f \in B_{K,u}, i_0 \leq i.$$

Now, since $f = (f - f_i) + f_i$, f belongs to $CV(X, A)$. □

In case $(A, \| \ \|)$ is a Banach space, we may limit ourselves to the level sets of the form :
$$N(v, r) := \{x \in X : r\|a\| \leq \|v(x)a\|, \ \forall a \in A\}.$$

In this situation, Theorem 2.2 reduces to the following

COROLLARY 2.3 ([20]). *If $(A, \| \ \|)$ is a Banach space, then the space $CV(X, A)$ (resp. $CV_0(X, A)$) is complete whenever*
1. $\forall x \in X_1$ *(resp.* $\forall x \in X_0$*), $\exists v \in V : v(x) \in \mathcal{L}_{bb}(A)$,*
2. X *is a $V_\mathbb{R}$-space.*

REMARK 2.4. 1. A result, similar to Theorem 2.2, can be shown in case A is a locally bounded space. Actually, in this situation too, the level sets can also be restricted to those of the form $N(v, r) := N(v, G, \frac{1}{r}G)$, where G is a bounded 0-neighborhood from $\mathcal{N}$ and $r > 0$.
2. The condition 4. of Theorem 2.2 turns out to be equivalent to the fact that, for every $v \in V$, every $G, H \in \mathcal{N}$, and every $x \in N(v, G, H)$, there exists $\Omega \in \mathcal{V}_x$ such that:

(N) $\qquad \forall I \in \mathcal{N}, \exists J \in \mathcal{N} : \Omega \cap N(v, G, H) \subset \Omega \cap N(v, G \cap I, H \cap J).$

Indeed, let $v \in V$, $G \in \mathcal{N}$, and $H \in \mathcal{N}$ be given. For $x \in N(v, G, H)$, let $\Omega \in \mathcal{V}_x$ be such that $\{v(t) : t \in \Omega \cap N(v, G, H)\}$ is equi-bounded below. Then for every $I \in \mathcal{N}$, there exists $J \in \mathcal{N}$ such that
$$P_{G \cap I}(a) \leq P_J(v(t)a), \ \forall t \in \Omega, \ a \in A.$$
Whereby
$$P_{G \cap I}(a) \leq P_{H \cap J}(v(t)a), \ \forall t \in \Omega, \ a \in A$$
and then $t \in \Omega \cap N(v, G \cap I, H \cap J)$. Consequently $\Omega \cap N(v, G, H) \subset \Omega \cap N(v, G \cap I, H \cap J)$.

Assume now that the property (N) is satisfied and let $v \in V$, $G, H \in \mathcal{N}$ be given. By (N), there exists $\Omega \in \mathcal{V}_x$ such that for all $I \in \mathcal{N}$, there is $J \in \mathcal{N}$ with $\Omega \cap N(v, G, H) \subset \Omega \cap N(v, G \cap I, H \cap J)$. Therefore, every $t \in \Omega \cap N(v, G, H)$ belongs to $\Omega \cap N(v, G \cap I, H \cap J)$. This means $P_{G \cap I}(a) \leq P_{H \cap J}(v(t)a)$, $a \in A$. Thus
$$P_I(a) \leq P_{G \cap I}(a) \leq P_{H \cap J}(v(t)a), \quad t \in \Omega \cap N(v, G, H), \ a \in A.$$

Consequently
$$P_I(a) \leq P_{J'}(v(t)a), \quad a \in A, \ t \in \Omega \cap N(v, G, H) \text{ with } J' = H \cap J.$$
Thus $\{v(t) : t \in \Omega \cap N(v, G, H)\}$ is equi-bounded below.

3. Continuous weighted composition operators

Henceforth, U will stand for a generalized Nachbin family on Y. It is clear that a linear map $T : CV(X, A) \to CU(Y, A)$ is continuous if, and only if,
$$\forall u \in U, \forall G \in \mathcal{N}, \exists v \in V, \exists H \in \mathcal{N} \ : \ P_{G,u}(T(f)) \leq P_{H,v}(f), \ f \in CV(X, A).$$

Now, let $\varphi : Y \to X$ and $\psi : X \to \mathcal{L}(A)$ be arbitrary maps. The composition (resp. the multiplication) operator $C_\varphi : f \mapsto f \circ \varphi$ (resp. $M_\psi : f \mapsto \psi f$) associated with φ (resp. ψ) is defined from $CV(X, A)$ into $\mathcal{F}(Y, A)$ (resp. into $\mathcal{F}(X, A)$) by $C_\varphi(f)(y) = f(\varphi(y))$ (resp. $M_\psi(f)(x) := \psi(x)(f(x)), \ x \in X$).

The weighted composition operator associated with ψ and φ is the linear mapping ψC_φ defined from $CV(X, A)$ into $\mathcal{F}(Y, A)$ by $\psi C_\varphi(f)(y) = \psi_y(f(\varphi(y)))$. Notice that, whenever ψ is identically the identity of A, ψC_φ is nothing but C_φ and, whenever $X = Y$ and φ is the identity of X, ψC_φ is just M_ψ

Throughout all the remainder, unless stated otherwise, we will assume that E is both a linear subspace of $CV(X, A)$ and a $C_b(X)$-module such that ψC_φ applies E into $C(X, A)$. We will write $\text{coz}(E)$ to designate the cozero set $\{x \in X : f(x) \neq 0 \text{ for some } f \in E\}$ of E.

The following sets will come in force in the sequel:
$$Y_{E,\varphi} := \{y \in Y : \varphi(y) \in \text{coz}(E)\} = \text{coz}(C_\varphi(E)),$$
$$Y_{E,\varphi,\psi} := \text{coz}(\psi C_\varphi(E)).$$

The set $Y_{E,\varphi}$ (resp. $Y_{E,\varphi,\psi}$) is an open subset of Y, whenever $C_\varphi(E) \subset C(Y, A)$ (resp. $\psi C_\varphi(E) \subset C(Y, A)$). We refer to [29] (Proposition 1 and Proposition 2) for instances where, whenever $C_\varphi(f)$ is continuous for any function $f : X \to A$, $\psi C_\varphi(f)$ must also be continuous on Y. There, one can also find instances where the fact $\psi C_\varphi(E) \subset C(Y, A)$ forces φ and/or ψ to be continuous on $Y_{E,\varphi}$.

We will consider the following properties on the space E :

(M) $\qquad \forall a \in A, \ \forall G \in \mathcal{N}, \text{ and } \ \forall f \in E, \ P_G \circ f \otimes a \in E,$

(P) $\qquad \forall a \in A, \ \forall x \in \text{coz}(E), \ \exists f \in E : \ f(x) = a,$

(P') $\quad \forall a \in A, \ \forall x \in \text{coz}(E), \ \exists g \in C(X) : \ g(x) \neq 0 \text{ and } g \otimes a \in E,$

(S) $\quad \forall G \in \mathcal{N}, \ \forall v \in V, \text{ and } \forall f \in E, \text{ the mapping } \ x \mapsto P_G(v(x)f(x)) \text{ is u.s.c.}$

It is easily seen that, if E satisfies (M) or (P'), it also satisfies (P). This is also the case whenever X is locally compact and the inclusion $\mathcal{K}(X) \otimes A \subset E$ holds. Moreover, if E satisfies (P), then we have $Y_{E,\varphi,\psi} = Y_{E,\varphi} \cap \text{coz}(\psi)$. Finally, if E satisfies (P'), then for all $x \in \text{coz}(E)$ and all $a \in A$, there exists $f \in C(X)$ such that $0 \leq f \leq 1$, $f(x) = 1$, and $f \otimes a \in E$.

Now, we characterize the continuous operators ψC_φ from the subspace E into $CU(Y, A)$.

THEOREM 3.1. *Under each one of the conditions i., ii., or iii. below, the operator ψC_φ maps continuously E into $CU(Y, A)$ if, and only if, the following condition holds.*

$$\forall G \in \mathcal{N}, u \in U, \exists H \in \mathcal{N}, v \in V : P_G(u(y)\psi_y(a)) \leq P_H(v(\varphi(y))a),$$
(3.1)
$$\forall a \in A, \ y \in Y_{E,\varphi}.$$

i. *E satisfies (M),*
ii. *E satisfies both (P) and (S),*
iii. *X is locally compact and $\mathcal{K}(X) \otimes A \subset E$.*

PROOF. Necessity : Since $\psi C_\varphi : E \to CU(Y, A)$ is continuous, for every $G \in \mathcal{N}$ and $u \in U$, there exist $H \in \mathcal{N}$ and $v \in V$ such that

$$P_{G,u}(\psi C_\varphi(f)) \leq P_{H,v}(f), \ \forall f \in E.$$

Then for every $y \in Y$, one has

(3.2) $$P_G(u(y)\psi_y(f(\varphi(y)))) \leq \sup\{P_H(v(x)f(x)), \ x \in X\}.$$

Let y_0 in $Y_{E,\varphi}$ and a in A be given. Then $x_0 := \varphi(y_0)$ belongs to $\mathrm{coz}(E)$.
Under each one of the conditions i., ii., or iii., for every integer $n > 0$, we will exhibit an open neighborhood U_n of x_0 and a function $h_n \in E$ whose support is contained in U_n such that $h_n(x_0) = a$ and, whenever we apply (3.2) to h_n and let n tend to infinity, we get $P_G(u(y_0)\psi_{y_0}(a)) \leq P_H(v(\varphi(y_0))a)$ as desired.
Now, if E satisfies i., then there is some $f \in E$ and some $K \in \mathcal{N}$ such that $P_K(f(x_0)) = 1$. Set then

$$U_n := \{x \in X : P_K(f(x)) < 1 + \frac{1}{n} \ \text{and} \ P_H(v(x)a) < P_H(v(x_0)a) + \frac{1}{n}\}$$

and $h_n := g_n P_K \circ f \otimes a$, where $g_n \in C_b(X)$ is chosen so that $g_n(x_0) = 1$, $0 \leq g_n \leq 1$ and $\mathrm{supp}(g_n) \subset U_n$. Since v is strongly u.s.c. and P_K is continuous, U_n is an open neighborhood of x_0.
If E satisfies ii., then there is some $f \in E$ such that $f(x_0) = a$, then set

$$U_n := \{x \in X : \ P_H(v(x)f(x)) < P_H(v(x_0)f(x_0)) + \frac{1}{n}\}$$

and $h_n := g_n f$, where again $g_n \in C_b(X)$ is chosen so that $g_n(x_0) = 1$, $0 \leq g_n \leq 1$ and $\mathrm{supp}(g_n) \subset U_n$. Notice that, by (S), U_n is an open neighborhood of x_0.
Finally, if E satisfies iii., then there exists $f \in C(X, \mathbb{R}^+)$ such that $f(x_0) = 1$, $\mathrm{supp}(f)$ is compact, and $f \otimes a \in E$. As $P_H(v(x)f(x)a) = |f(x)|P_H(v(x)a)$, the mapping $x \mapsto P_H(v(x)f(x)a)$ is u.s.c. on X. Set then

$$U_n := \{x \in X : \ P_H(v(x)f(x)a) < P_H(v(x_0)a) + \frac{1}{n}\}$$

and $h_n = g_n f \otimes a$, where once again $g_n \in C_b(X)$ satisfies $g_n(x_0) = 1$, $0 \leq g_n \leq 1$ and $\mathrm{supp}(g_n) \subset U_n$.
Sufficiency : Let $f \in E$, $G \in \mathcal{N}$, and $u \in U$ be given. By (3.1), there exist $H \in \mathcal{N}$ and $v \in V$ such that

$$P_G(u(y)\psi_y(f(\varphi(y)))) \leq P_H(v(\varphi(y))f(\varphi(y))), \quad y \in Y.$$

Therefore,

$$P_{G,u}(\psi C_\varphi(f)) = \sup\{P_G(u(y)\psi_y(f(\varphi(y)))) : y \in Y\}$$
$$\leq \sup\{P_H(v(\varphi(y))f(\varphi(y))) : y \in Y\}$$
$$\leq \sup\{P_H(v(x)f(x)) : x \in \varphi(Y)\}$$
$$\leq P_{H,v}(f) < \infty.$$

This shows at once $\psi C_\varphi(f) \in CU(Y, A)$ and that ψC_φ is continuous. □

In case of multiplication operators, we get:

COROLLARY 3.2. *If $X = Y$, φ is the identity of X, and E satisfies one of the conditions i., ii, or iii. of Theorem 3.1, then M_ψ is continuous from E into $CU(X, A)$ if, and only if, the following condition holds.*

(3.3) $\quad \forall\, G \in \mathcal{N}, u \in U, \exists\, H \in \mathcal{N}, v \in V : P_G(u(x)\psi_x(a)) \leq P_H(v(x)a),$

(3.4) $\hfill \forall a \in A,\ \forall x \in coz(E).$

Similarly, in case of composition operators, we get:

COROLLARY 3.3. *If ψ is identically the identity of A and E enjoys one of the conditions i., ii, or iii. of Theorem 3.1, then C_φ is continuous from E into $CU(Y, A)$ if, and only if, the following condition holds.*

(3.5) $\quad \forall\, G \in \mathcal{N}, u \in U, \exists\, H \in \mathcal{N}, v \in V : P_G(u(y)a) \leq P_H(v(\varphi(y))a),$

(3.6) $\hfill \forall a \in A, y \in Y_{E,\varphi}.$

Whenever A happens to be a normed space with norm $\|\ \|$, the property (M) becomes: for all $f \in E$ and $a \in E$, the function $x \mapsto \|f(x)\|a$ belongs to E, and the property (S) becomes: $x \mapsto \|v(x)f(x)\|$ is upper semicontinuous, for all $v \in V$ and all $f \in E$. We obtain the following corollary to be compared with Theorem 4.1 of [20]:

COROLLARY 3.4. *If $(A, \|\ \|)$ is a normed space and E fulfils one of the conditions i., ii., or iii. of Theorem 3.1, then ψC_φ maps continuously E into $CU(Y, A)$ if, and only if, the following condition holds.*

(3.7) $\quad \forall u \in U, \exists v \in V : \|u(y)\psi_y(a)\| \leq \|v(\varphi(y))a\|,\ a \in A, y \in Y_{E,\varphi}.$

Next, we will investigate the continuity of ψC_φ from E into $CU_0(Y, A)$. To this aim, let us set as in [29]

$$\mathrm{Cst}(E) := \{K \subset X : \forall a \in A,\ \exists f \in E \text{ with } f = a \text{ identically on } K\}.$$

It is easily seen that every $v \in V$ is β-bounded on every $K \in \mathrm{Cst}(E)$. Indeed, for arbitrary $K \in \mathrm{Cst}(E)$, $a \in A$, and $G \in \mathcal{N}$, choose $f \in E$ such that $f = a$ identically on K. Then

$$\sup_{x \in K} P_G(v(x)a) = \sup_{x \in K} P_G(v(x)f(x)) \leq \sup_{x \in X} P_G(v(x)f(x)) = P_{G,v}(f) < \infty.$$

Therefore the set $\{v(x), x \in K\}$ is β-bounded since G and a are arbitrary.

For $v \in V$, $G \in \mathcal{N}$, and $f \in E$, set $N(G, v, f) := \{x \in X : v(x)f(x) \notin \overset{\circ}{G}\}$. Actually, we also have $N(G, v, f) = \{x \in X : P_G(v(x)f(x)) \geq 1\}$.

DEFINITION 3.5. The space E is said to satisfy the property (C) if $N(G, v, f)$ belongs to $\mathrm{Cst}(E)$, for every $v \in V$, $G \in \mathcal{N}$, and $f \in E$.

As we will see in Lemma 3.10, if E happens to be contained in $CV_0(X, A)$, then it satisfies the property (C) whenever it satisfies (S). In particular whenever all elements of V are strongly continuous.

THEOREM 3.6. *Assume that E satisfies both (C) and one of the conditions i., ii. or iii. of Theorem 3.1. If ψC_φ maps continuously E into $CU_0(Y, A)$, then (3.1) holds and*

$$\varphi^{-1}(K) \cap \{y \in Y : u(y)\psi_y(a) \notin \overset{\circ}{G}\}$$

is relatively compact, for all $K \in \mathrm{Cst}(E)$, $G \in \mathcal{N}$, $u \in U$, and $a \in A \setminus \{0\}$. The converse holds whenever, for all $G \in \mathcal{N}$, $v \in V$, and $f \in E$, $f(N(G, v, f))$ is precompact and $v(N(G, v, f))$ is equicontinuous on A.

PROOF. Assume that ψC_φ maps continuously E into $CU_0(Y, A)$. Then (3.1) follows from Theorem 3.1. Now, fix $K \in \mathrm{Cst}(E)$, $u \in U$, $G \in \mathcal{N}$, and $a \in A$, with $a \neq 0$. Choose $f \in E$ such that $f = a$ identically on K. As $\psi C_\varphi(f)$ belongs to $CU_0(Y, A)$, the set

$$S := \{y \in Y : u(y)\psi_y(f(\varphi(y))) \notin \overset{\circ}{G}\}$$

is relatively compact and contains $\varphi^{-1}(K) \cap \{y \in Y : u(y)\psi_y(a) \notin \overset{\circ}{G}\}$. Hence the latter is relatively compact. For the converse, by Theorem 3.1, the condition (3.1) implies that ψC_φ maps continuously E into $CU(Y, A)$. It remains to be shown the inclusion $\psi C_\varphi(E) \subset CU_0(Y, A)$. Let $f \in E$, $G \in \mathcal{N}$, and $u \in U$ be arbitrary and consider again the set S defined as above. We claim that S is relatively compact and therefore $\psi C_\varphi(f) \in CU_0(Y, A)$. Indeed, let $H \in \mathcal{N}$ satisfy $H + H \subset \overset{\circ}{G}$. By (3.1), there are $I \in \mathcal{N}$ and $v \in V$ such that:

(3.8) $$P_H(u(y)\psi_y(a)) \leq P_I(v(\varphi(y)a)), \ \forall a \in A, y \in Y_{E,\varphi}.$$

We will show that S is contained in some union of finitely many sets of the form $C_i := \{y \in Y : u(y)\psi_y(a_i) \notin \overset{\circ}{H}\}$, for some $a_i \in A \setminus \{0\}$. Since $K := N(I, v, f)$ belongs to $\mathrm{Cst}(E)$ and contains $\varphi(S)$, and since the set $v(K)$ is equicontinuous, there exists $J \in \mathcal{N}$ such that:

(3.9) $$v(x)J \subset I, \ \forall x \in N(I, v, f).$$

Moreover, since $f(K)$ is precompact, $f(\varphi(S))$ is also precompact in A. Therefore there are $y_1, \ldots, y_n \in S$ such that:

$$f(\varphi(S)) \subset \bigcup \{f(\varphi(y_i)) + J, i = 1, \ldots, n\}.$$

Thus, for $y \in S$, there is some $i \in \{1, \ldots, n\}$ such that $f(\varphi(y)) - f(\varphi(y_i)) \in J$. By (3.8) and (3.9), we obtain $u(y)\psi_y(f(\varphi(y)) - f(\varphi(y_i))) \in H$. Therefore $u(y)\psi_y(f(\varphi(y_i))) \notin \overset{\circ}{H}$. Otherwise,

$$u(y)\psi_y(f(\varphi(y))) = [u(y)\psi_y(f(\varphi(y)) - f(\varphi(y_i)))] + u(y)\psi_y(f(\varphi(y_i)))$$
$$\in H + \overset{\circ}{H} \subset \overset{\circ}{G},$$

which is a contradiction. Consequently

$$S \subset \bigcup \{\{y \in Y : u(y)\psi_y(a_i) \notin \overset{\circ}{H}\}, i = 1, \ldots, n\},$$

with $a_i = f(\varphi(y_i))$. □

It is worth mentioning that, whenever $E \subset CV_0(X, A)$, the set $N(G, v, f)$ is contained in a compact set. Therefore $f(N(G, v, f))$ is precompact. If, in addition, A happens to a barrelled locally convex space, then also $v(N(G, v, f))$ is equicontinuous, for it is strongly bounded. This gives plenty results as corollaries.

In general, in case of multiplication operators, we get:

COROLLARY 3.7. *Assume that $X = Y$, that φ is the identity of X, and that E satisfies both (C) and one of the conditions i., ii. or iii. of Theorem 3.1. If $f(N(G, v, f))$ is precompact and $v(N(G, v, f))$ is equicontinuous on A, for all $G \in \mathcal{N}$, $v \in V$, and $f \in E$, then M_ψ maps continuously E into $CU_0(X, A)$ if, and only if (3.3) holds and $K \cap \{y \in Y : u(y)\psi_y(a) \notin \overset{\circ}{G}\}$ is relatively compact, for all $K \in Cst(E)$, $G \in \mathcal{N}$, $u \in U$, and $a \in A \setminus \{0\}$.*

Similarly, in case of composition operators, we get:

COROLLARY 3.8. *Suppose that ψ is identically the identity of A and that E enjoys both (C) and one of the conditions i., ii. or iii. of Theorem 3.1. If $f(N(G, v, f))$ is precompact and $v(N(G, v, f))$ is equicontinuous on A, for all $G \in \mathcal{N}$, $v \in V$, and $f \in E$, then C_φ maps continuously E into $CU_0(X, A)$ if, and only if (3.5) holds and*

$$\varphi^{-1}(K) \cap \{y \in Y : u(y)a \notin \overset{\circ}{G}\}$$

is relatively compact, for all $K \in Cst(E)$, $G \in \mathcal{N}$, $u \in U$, and $a \in A \setminus \{0\}$.

In case $(A, \|\cdot\|)$ is a normed space, the sets $N(G, v, f)$ can be replaced by the sets $N(v, f) := \{x \in X : \|v(x)f(x)\| \geq 1\}$, $v \in V$ and $f \in E$. We then obtain:

COROLLARY 3.9. *Assume that A is a normed space and that E satisfies both (C) and one of the conditions i., ii. or iii. of Theorem 3.1. If ψC_φ maps continuously E into $CU_0(Y, A)$ then (3.7) holds and*

$$\varphi^{-1}(K) \cap \{y \in Y : \|u(y)\psi_y(a)\| \geq 1\}$$

is relatively compact for all $K \in Cst(E)$, all $u \in U$, and $a \in A \setminus \{0\}$.

The converse holds whenever $f(N(v, f))$ is precompact in A and $v(N(v, f))$ is equicontinuous on A, for all $v \in V$ and all $f \in E$.

The following lemma extends Lemma 6 of [29]:

LEMMA 3.10. *Assume that E satisfies one of the conditions (M), (P'), or that X is locally compact and $\mathcal{K}(X) \otimes A \subset E$. If $K \subset coz(E)$ is a nonempty compact set and $C \subset X$ a closed nonempty set such that $K \cap C = \emptyset$, then, for every $a \in A$, there exists $f \in E$ such that $f = a$ on K and $f = 0$ on C.*

PROOF. If E satisfies (M), the proof is given in [29] and a slight modification of the proof there yields the result in the two other situations. □

REMARK 3.11. In the conditions of Lemma 3.10, in case $E \subset CV_0(X, A)$, the property (C) is equivalent to the fact that $\overline{N(G, v, f)}^X \subset coz(E)$ for every $v \in V$, $G \in \mathcal{N}$, and $f \in E$, which holds if E satisfies (S).

THEOREM 3.12. *Assume that $E \subset CV_0(X,A)$ satisfies (C) and that, for every $v \in V$, $G \in \mathcal{N}$, and $f \in E$, the set $v(N(G,v,f))$ is equicontinuous. Then, under each one of the conditions i., ii'., or iii. below, ψC_φ is continuous from E into $CU_0(Y,A)$ if, and only if, (3.1) holds and $\varphi^{-1}(K) \cap \{y \in Y : u(y)\psi_y(a) \notin \overset{\circ}{H}\}$ is relatively compact, for every compact $K \subset coz(E)$, $H \in \mathcal{N}$, $u \in U$, and $a \in A\setminus\{0\}$.*
Where
i. E satisfies (M),
ii'. E satisfies (P') and (S),
iii. X is locally compact and $\mathcal{K}(X) \otimes A \subset E$.

PROOF. With Lemma 3.10 in mind and the fact that (P') implies (P) the necessity is the same as in the proof of Theorem 3.6.
Sufficiency : For every $v \in V$, $G \in \mathcal{N}$, $f \in E$, and $H \in \mathcal{N}$, the set $N(G,v,f)$ is relatively compact. By (C), there exists $g \in E$ such that $g = a$ on $N(G,v,f)$, with $a \in A \setminus \{0\}$. Then $N(G,v,f) \subset g^{-1}(\{a\})$ which is closed. Therefore the closure of $N(G,v,f)$ on X is subset of $coz(E)$. Then

$$\varphi^{-1}(N(G,v,f)) \cap \{y \in Y : u(y)\psi_y(a) \notin \overset{\circ}{H}\}$$

is relatively compact and $f(N(H,v,f))$ is precompact. We conclude as in the proof of Theorem 3.6. □

In case of multiplication operators, we get:

COROLLARY 3.13. *Assume that φ is the identity of $X = Y$, and that $E \subset CV_0(X,A)$ satisfies (C) and at least one of the conditions i., ii'. or iii. of Theorem 3.12. Assume also that, for every $v \in V$, $G \in \mathcal{N}$ and $f \in E$, the set $v(N(G,v,f))$ is equicontinuous. Then M_ψ is continuous from E into $CU_0(X,A)$ if, and only if, (3.1) holds.*

Similarly, in case of composition operators, we obtain :

COROLLARY 3.14. *If ψ is identically the identity of A, $E \subset CV_0(X,A)$ enjoys (C) and at least one of the conditions i., ii'. or iii. of Theorem 3.12, and if, for every $v \in V$, $G \in \mathcal{N}$ and $f \in E$, the set $v(N(G,v,f))$ is equicontinuous, then C_φ is continuous from E into $CU_0(X,A)$ if, and only if, (3.1) holds and*

$$\varphi^{-1}(K) \cap \{y \in Y : u(y)a \notin \overset{\circ}{G}\}$$

is relatively compact, for every compact $K \subset coz(E), G \in \mathcal{N}, u \in U$ and $a \in A\setminus\{0\}$.

In case A is a barrelled locally convex space, (in particular a Banach space), every β-bounded set in $\mathcal{L}(A)$ is equicontinuous, then also σ-bounded. By upper semi-continuity of the mapping $P_{G,v} : x \mapsto P_G(v(x)a)$, for every $a \in A$, $P_{G,v}$ is β-bounded on compact subsets of X. Since $N(G,v,f)$ is contained in some compact set, $v(N(G,v,f))$ is equicontinuous. We obtain

COROLLARY 3.15. *Assume that A is a barrelled normed space and that $E \subset CV_0(X,A)$ satisfies (C). Then, under each one of the conditions i., ii'. or iii. of Theorem 3.12, ψC_φ is continuous from E into $CU_0(Y,A)$ if, and only if, (3.7) holds and $\varphi^{-1}(K) \cap \{y \in Y : \|u(y)\psi_y(a)\| \geq 1\}$ is relatively compact, for every compact $K \subset coz(E)$, $u \in U$, and $a \in A \setminus \{0\}$.*

4. Bounded weighted composition operators

Recall that a linear map θ is said to be bounded if it maps some 0-neighborhood into a bounded set. Whenever the range of θ consists of continuous functions on some topological space Z, it is said to be locally Z_0-equicontinuous, if θ maps every bounded set into an equicontinuous set on $Z_0 \subset Z$.

THEOREM 4.1. *Assume that one of the conditions i., ii. or iii. of Theorem 3.1 is satisfied. Then ψC_φ is bounded from E into $CU(Y, A)$ if, and only if, there exist $H \in \mathcal{N}$ and $v \in V$ such that:*

$$\forall\, G \in \mathcal{N},\ u \in U;\ \exists\, \lambda > 0 : P_G(u(y)\psi_y(a)) \leq \lambda P_H(v(\varphi(y))a),$$
(4.1) $$\forall\, a \in A, y \in Y_{E,\varphi}.$$

PROOF. Necessity : Since ψC_φ is bounded from E into $CU(Y, A)$, there exist $H \in \mathcal{N}$ and $v \in V$ such that, for every $G \in \mathcal{N}$ and $u \in U$, there is some $\lambda > 0$ enjoying

$$P_{G,u}(\psi C_\varphi(f)) \leq \lambda,\ f \in B_{H,v}.$$

Therefore,

$$P_{G,u}(\psi C_\varphi(f)) \leq \lambda P_{H,v}(f),\ f \in E.$$

Then for every $y \in Y$, one has

(4.2) $$P_G(u(y)\psi_y(f(\varphi(y)))) \leq \lambda \sup\{P_H(v(x)f(x)), x \in X\}.$$

Let $y_0 \in Y_{E,\varphi}$ and $a \in A$ be given and put $x_0 := \varphi(y_0)$. Consider the function h_n constructed in the proof of Theorem 3.1 corresponding to each one of the three cases. Applying (4.2) to h_n and letting n tend to infinity, we get:

$$P_G(u(y_0)\psi_{y_0}(a)) \leq \lambda P_H(v(\varphi(y_0))a).$$

Sufficiency : Assume that, there exist $H \in \mathcal{N}$ and $v \in V$ so that (4.1) holds. Let $G \in \mathcal{N}$ and $u \in U$ be given. Then, for $f \in E$ and $y \in Y$, we have

$$P_G(u(y)\psi_y(f(\varphi(y)))) \leq \lambda P_H(v(\varphi(y)f(\varphi(y))),\ y \in Y.$$

In particular, for $f \in B_{H,v}$, we get

$$P_G(u(y)\psi_y(f(\varphi(y)))) \leq \lambda,\ y \in Y.$$

giving $P_{G,u}(\psi C_\varphi(f)) \leq \lambda,\ f \in B_{H,v}$. □

COROLLARY 4.2. *If $X = Y$, φ is the identity of X, and E satisfies one of the conditions i., ii. or iii. of Theorem 3.1, then M_ψ is bounded from E into $CU(X, A)$ if, and only if, there exist $H \in \mathcal{N}$ and $v \in V$ such that:*

$$\forall\, G \in \mathcal{N},\ u \in U;\ \exists\, \lambda > 0 : P_G(u(x)\psi_x(a)) \leq \lambda P_H(v(x)a), \forall\, a \in A, x \in coz(E).$$

Similarly

COROLLARY 4.3. *If $\psi(Y) = \{I_A\}$ and E satisfies one of the conditions i., ii. or iii. of Theorem 3.1, then C_φ is bounded from E into $CU(Y, A)$ if, and only if, there exist $H \in \mathcal{N}$ and $v \in V$ such that:*

$$\forall\, G \in \mathcal{N},\ u \in U;\ \exists\, \lambda > 0 : P_G(u(y)a) \leq \lambda P_H(v(\varphi(y))a), \forall\, a \in A, y \in Y_{E,\varphi}.$$

When $(A, \|\ \|)$ is a normed space, things become much easier, namely :

COROLLARY 4.4. *Assume $(A, \| \ \|)$ is a normed space and that E enjoys one of the conditions i., ii. or iii. of Theorem 3.1. Then ψC_φ is bounded from E into $CU(Y, A)$ if, and only if, the following condition holds.*

(4.3) $\quad \exists \, v \in V, \forall \, u \in U, \exists \, \lambda > 0 : \|u(y)\psi_y(a)\| \leq \lambda \|v(\varphi(y))a\|, a \in A, y \in Y_{E,\varphi}.$

Combining conveniently the proofs of the theorems 3.12 and 4.1, we obtain the following result.

THEOREM 4.5. *Assume that $E \subset CV_0(X, A)$ satisfies the property (C) and one of the conditions i., ii'. or iii. of Theorem 3.12. If, for every $v \in V$, $G \in \mathcal{N}$, and $f \in E$, $v(N(G, v, f))$ is equicontinuous, then ψC_φ is bounded from E into $CU_0(Y, A)$ if, and only if, (4.1) holds and $\varphi^{-1}(K) \cap \{y \in Y : u(y)\psi_y(a) \notin \overset{\circ}{H}\}$ is relatively compact, for every compact $K \subset coz(E)$, $H \in \mathcal{N}$, $u \in U$, and $a \in A \setminus \{0\}$.*

Whenever A is a normed space, one gets :

COROLLARY 4.6. *Assume $(A, \| \ \|)$ is a normed space and that $E \subset CV_0(X, A)$ satisfies (C) and one of the conditions i., ii'. or iii. of Theorem 3.12. Assume also that, for every $v \in V$, and $f \in E$, $v(N(v, f))$ is equicontinuous. Then ψC_φ is bounded from E into $CU_0(Y, A)$ if, and only if, (4.3) holds and $\varphi^{-1}(K) \cap \{y \in Y : \|u(y)\psi_y(a)\| \geq 1\}$ is relatively compact, for every compact $K \subset coz(E)$, $u \in U$, and $a \in A \setminus \{0\}$.*

5. Locally equicontinuous weighted composition operators

We now examine the local equicontinuity of ψC_φ. Notice that, in scalar-valued weights case, every $x \in X$ admits a neighborhood Ω such that every $v \in V$ is bounded on Ω.

DEFINITION 5.1. *X is said to be locally V-σ-bounding if for every $x \in X$, there exists $\Omega_x \in \mathcal{V}_x$ such that $\{v(t) : t \in \Omega_x\}$ is bounded in $\mathcal{L}_\sigma(A)$, for all $v \in V$.*

If A happens to be a barrelled locally convex space, X is locally V-σ-bounding if and only if it is locally V-β-bounding. Moreover, if X is locally compact, then every $x \in X$ admits a neighborhood Ω with $v(\Omega)$ β-bounded, for all $v \in V$. Therefore, if A is a barrelled locally convex space and X is locally compact, then X is locally V-σ-bounding. Moreover, if every $v \in V$ is σ-bounded, then X is, obviously, locally V-σ-bounding

THEOREM 5.2. *Assume that E fulfils one of the conditions i., ii'. or iii. of Theorem 3.12, that X is locally V-σ-bounding, and that, for all $x \in X$, $V(x) \cap \mathcal{L}_{bb}(A)$ is not empty. Then ψC_φ is locally $Y_{E,\varphi,\psi}$-equicontinuous, if, and only if, the following conditions hold:*
1. *φ is locally constant on $Y_{E,\varphi,\psi}$.*
2. *ψ is continuous from $Y_{E,\varphi,\psi}$ into $\mathcal{L}_\sigma(A)$.*

PROOF. Necessity : 1. Assume that φ is constant on no neighborhood of some $y_0 \in Y_{E,\varphi,\psi}$ and choose $f_0 \in E$ with $\psi_{y_0}(f_0(\varphi(y_0))) \neq 0$. Then every $\Omega \in \mathcal{V}_{y_0}$ contains some y_Ω with $\varphi(y_0) \neq \varphi(y_\Omega)$. Consider $f_\Omega \in C_b(X)$ such that $0 \leq f_\Omega \leq 1$, $f_\Omega(\varphi(y_\Omega)) = 0$ and $f_\Omega(\varphi(y_0)) = 1$. The set $C := \{g_\Omega := f_\Omega f_0, \Omega \in \mathcal{V}_{y_0}\}$ is bounded in E and then $\psi C_\varphi(C)$ is equicontinuous at y_0. Therefore, for every $G \in \mathcal{N}$, there exists $\Omega_0 \in \mathcal{V}_{y_0}$ such that:

$$\psi_y(g_\Omega(\varphi(y))) - \psi_{y_0}(g_\Omega(\varphi(y_0))) \in G, \ y \in \Omega_0, \ \Omega \in \mathcal{V}_{y_0}.$$

Hence, for every $\Omega \subset \Omega_0$ and $y = y_\Omega$, we get $\psi_{y_0}(f_0(\varphi(y_0))) \in G$. Since G is arbitrary, $\psi_{y_0}(f_0(\varphi(y_0))) = 0$, which is a contradiction.

2. Let y_0 be arbitrary in $Y_{E,\varphi,\psi}$ and let a bounded $B \subset A$ and $G \in \mathcal{N}$ be given. By 1., there exists a neighborhood Ω_0 of y_0 on which φ is constant with some value x_0. Since X is locally V-σ-bounding, there exists $\Omega_{x_0} \in \mathcal{V}_{x_0}$ such that, for every $v \in V$, the set $\{v(t) : t \in \Omega_{x_0}\}$ is σ-bounded. Therefore, for every $v \in V$ and every $I \in \mathcal{N}$, there exist $M_{I,v} > 0$ such that: $P_I(v(x)b) \leq M_{I,v}$, $\forall x \in \Omega_{x_0}$, $\forall b \in B$. In particular, there is $M_{G,v} > 0$ such that: $P_G(v(x)b) \leq M_{G,v}$, $\forall x \in \Omega_{x_0}$, $\forall b \in B$.

Under the condition *i*.: Choose $f_0 \in E$ so that $\psi_{y_0}(f_0(x_0)) \neq 0$ and $H \in \mathcal{N}$ with $P_H(f_0(x_0)) = 1$. Consider the open neighborhood $U := \{x \in X : \frac{1}{2} < P_H(f_0(x)) < \frac{3}{2}\}$ of x_0 and a function $g \in C_b(X)$ satisfying $g(x_0) = 1$, $0 \leq g \leq 1$, and $\operatorname{supp}(g) \subset U \cap \Omega_{x_0}$. Set then

$$K := \{gP_H \circ f_0 \otimes b, b \in B\}.$$

Since, for every $v \in V$ and every $I \in \mathcal{N}$,

$$P_{I,v}(gP_H \circ f_0 \otimes b) = \sup\{g(x)P_H(f_0(x))P_I(v(x)b) : x \in X\}$$
$$\leq \sup\{P_H(f_0(x))P_I(v(x)b) : x \in U \cap \Omega_{x_0}\}$$
$$\leq \frac{3}{2}M_{I,v},$$

K is bounded in E. Consequently $\psi C_\varphi(K)$ is equicontinuous at y_0. Therefore there is some y_0-neighborhood Ω contained in Ω_0 such that:

$$\psi_y\left(g(\varphi(y))P_H(f_0(\varphi(y))b)\right) - \psi_{y_0}\left(g(\varphi(y_0))P_H(f_0(\varphi(y_0)))b)\right) \in G, \ y \in \Omega, \ b \in B.$$

Equivalently,

$$\psi_y(b) - \psi_{y_0}(b) \in G, \ \forall y \in \Omega, \ \forall b \in B.$$

This yields $\psi_y - \psi_{y_0} \in N(B,G)$ for every $y \in \Omega$, showing that ψ is σ-continuous at y_0. Since y_0 is arbitrary in $Y_{E,\varphi,\psi}$, ψ is σ-continuous on $Y_{E,\varphi,\psi}$.

Under the condition *ii′*.: For $b \in B$, choose the function $f_b \in C(X)$ such that $f_b \otimes b \in E$, $f_b(x_0) = 1$, and $0 \leq f_b \leq 1$. Then the set $U := \{x \in X : \frac{1}{2} < f_b(x)\}$ is open and contains x_0. Choose $g_b \in C_b(X)$ such that $g_b(x_0) = 1$, $0 \leq g_b \leq 1$ and $\operatorname{supp} g_b \subset U \cap \Omega_{x_0}$ and set

$$K := \{g_b f_b \otimes b, b \in B\}.$$

It is easily seen that K is bounded in E and with a similar proof one gets the σ-continuity of ψ on $Y_{E,\varphi,\psi}$.

Under the condition *iii*.: There exists $f \in C(X)$ such that $f(x_0) = 1$ and $\operatorname{supp} f \subset \Omega_{x_0}$ is compact. Using the open set $U := \{x \in X : \frac{1}{2} < f(x) < \frac{3}{2}\}$ and arguing in the same way as above, we arrive again to the σ-continuity of ψ on $Y_{E,\varphi,\psi}$.

Sufficiency : Given a bounded set $\mathbb{B} \subset E$, $y_0 \in Y_{E,\varphi,\psi}$ and $G \in \mathcal{N}$. By assumption, there is some neighborhood Ω_0 of y_0 so that φ is constant on Ω_0 with some value x_0. Choose $v \in V$ with $v(x_0)$ bounded below. Then there exist $H \in \mathcal{N}$ such that $P_G(a) \leq P_H(v(x_0)a)$, $a \in A$. Then the set $B := \{f(x_0), f \in \mathbb{B}\}$ is bounded in A. Since ψ is σ-continuous at y_0, there is some other neighborhood Ω of y_0 such that $\Omega \subset \Omega_0$ and $\psi_y - \psi_{y_0} \in N(B,G)$, $y \in \Omega$. This is

$$\psi_y(f(x_0)) - \psi_{y_0}(f(x_0)) \in G, \quad y \in \Omega, \ f \in \mathbb{B}.$$

Therefore

$$\psi C_\varphi(f)(y) - \psi C_\varphi(f)(y_0) \in G, \quad y \in \Omega, \ f \in \mathbb{B}.$$

Hence $\psi C_\varphi(\mathbb{B})$ is equicontinuous at y_0 and then on $Y_{E,\varphi,\psi}$ since y_0 was arbitrary. □

A trivial consequence of Theorem 5.2 is

COROLLARY 5.3. *Assume that E satisfies the conditions of Theorem 5.2. If φ is not constant on any open set (in particular, if X has no isolated point and φ is one to one), then ψC_φ is locally $Y_{E,\varphi,\psi}$-equicontinuous from E into $C(Y, A)$ if, and only if, it is identically zero.*

References

[1] K.-D. Bierstedt, *Gewichtete Räume stetiger vektorwertiger Funktionen und das injektive Tensorprodukt. I* (German), J. Reine Angew. Math. **259** (1973), 186–210, DOI 10.1515/crll.1973.259.186. MR318871

[2] K.-D. Bierstedt, *Tensor products of weighted spaces*, Function spaces and dense approximation (Proc. Conf., Univ. Bonn, Bonn, 1974), Inst. Angew. Math., Univ. Bonn, Bonn, 1975, pp. 26–58. Bonn. Math. Schriften, No. 81. MR0493283

[3] K. D. Bierstedt and J. Bonet, *Some recent results on $\mathcal{V}C(X)$*, Advances in the theory of Fréchet spaces (Istanbul, 1988), NATO Adv. Sci. Inst. Ser. C Math. Phys. Sci., vol. 287, Kluwer Acad. Publ., Dordrecht, 1989, pp. 181–194. MR1083564

[4] K. D. Bierstedt and J. Bonet, *Completeness of the (LB)-spaces $\mathcal{V}C(X)$*, Arch. Math. (Basel) **56** (1991), no. 3, 281–285, DOI 10.1007/BF01190216. MR1091882

[5] K. D. Bierstedt and J. Bonet, *Weighted (LF)-spaces of continuous functions*, Math. Nachr. **165** (1994), 25–48, DOI 10.1002/mana.19941650104. MR1261361

[6] K. D. Bierstedt, J. Bonet, and A. Galbis, *Weighted spaces of holomorphic functions on balanced domains*, Michigan Math. J. **40** (1993), no. 2, 271–297, DOI 10.1307/mmj/1029004753. MR1226832

[7] K.-D. Bierstedt, R. Meise, and W. H. Summers, *A projective description of weighted inductive limits*, Trans. Amer. Math. Soc. **272** (1982), no. 1, 107–160, DOI 10.2307/1998953. MR656483

[8] J. Bonet, *On weighted inductive limits of spaces of continuous functions*, Math. Z. **192** (1986), no. 1, 9–20, DOI 10.1007/BF01162015. MR835386

[9] J. Bonet, P. Domański, and M. Lindström, *Pointwise multiplication operators on weighted Banach spaces of analytic functions*, Studia Math. **137** (1999), no. 2, 177–194. MR1734396

[10] J. Bonet, M. C. Gómez-Collado, D. Jornet, and E. Wolf, *Operator-weighted composition operators between weighted spaces of vector-valued analytic functions*, Ann. Acad. Sci. Fenn. Math. **37** (2012), no. 2, 319–338, DOI 10.5186/aasfm.2012.3723. MR2987071

[11] J. Bonet and E. Wolf, *A note on weighted Banach spaces of holomorphic functions*, Arch. Math. (Basel) **81** (2003), no. 6, 650–654, DOI 10.1007/s00013-003-0568-8. MR2029241

[12] A. Goullet de Rugy, *Espaces de fonctions pondérables* (French, with English summary), Israel J. Math. **12** (1972), 147–160, DOI 10.1007/BF02764659. MR324386

[13] W. Govaerts, *Homomorphisms of weighted algebras of continuous functions*, Ann. Mat. Pura Appl. (4) **116** (1978), 151–158, DOI 10.1007/BF02413872. MR506979

[14] J. E. Jamison and M. Rajagopalan, *Weighted composition operator on $C(X, E)$*, J. Operator Theory **19** (1988), no. 2, 307–317. MR960982

[15] J.-S. Jeang and N.-C. Wong, *Weighted composition operators of $C_0(X)$'s*, J. Math. Anal. Appl. **201** (1996), no. 3, 981–993, DOI 10.1006/jmaa.1996.0296. MR1400575

[16] L. A. Khan, *Weighted topology in the nonlocally convex setting* (English, with Serbo-Croatian summary), Mat. Vesnik **37** (1985), no. 2, 189–195. MR839746

[17] L. A. Khan and A. B. Thaheem, *Operator-valued multiplication operators on weighted function spaces*, Demonstratio Math. **35** (2002), no. 3, 599–605. MR1917102

[18] V. Klee, *Shrinkable neighborhoods in Hausdorff linear spaces*, Math. Ann. **141** (1960), 281–285, DOI 10.1007/BF01360762. MR131149

[19] M. Klilou and L. Oubbi, *Multiplication operators on generalized weighted spaces of continuous functions*, Mediterr. J. Math. **13** (2016), no. 5, 3265–3280, DOI 10.1007/s00009-016-0684-x. MR3554307

[20] M. Klilou and L. Oubbi, *Weighted composition operators on Nachbin spaces with operator-valued weights*, Commun. Korean Math. Soc. **33** (2018), no. 4, 1125–1140. MR3879821

[21] K. Kour and B. Singh, *WCOs on non-locally convex weighted spaces of continuous functions*, J. Indian Math. Soc. (N.S.) **66** (1999), no. 1-4, 17–25. MR1749625
[22] W. Lusky, *On weighted spaces of harmonic and holomorphic functions*, J. London Math. Soc. (2) **51** (1995), no. 2, 309–320, DOI 10.4064/sm175-1-2. MR1325574
[23] W. Lusky, *On the structure of $Hv_0(D)$ and $hv_0(D)$*, Math. Nachr. **159** (1992), 279–289, DOI 10.1002/mana.19921590119. MR1237115
[24] W. Lusky, *On the isomorphism classes of weighted spaces of harmonic and holomorphic functions*, Studia Math. **175** (2006), no. 1, 19–45, DOI 10.4064/sm175-1-2. MR2261698
[25] J. S. Manhas and R. K. Singh, *Weighted composition operators on nonlocally convex weighted spaces of continuous functions* (English, with English and Russian summaries), Anal. Math. **24** (1998), no. 4, 275–292, DOI 10.1007/BF02771088. MR1657631
[26] L. Nachbin, *Elements of approximation theory*, Van Nostrand Mathematical Studies, No. 14, D. Van Nostrand Co., Inc., Princeton, N.J.-Toronto, Ont.-London, 1967. MR0217483
[27] L. Nachbin, *Weighted approximation for algebras and modules of continuous functions: Real and self-adjoint complex cases*, Ann. of Math. (2) **81** (1965), 289–302, DOI 10.2307/1970617. MR176353
[28] L. Oubbi, *Weighted algebras of continuous functions*, Results Math. **24** (1993), no. 3-4, 298–307, DOI 10.1007/BF03322338. MR1244283
[29] L. Oubbi, *Weighted composition operators on non-locally convex weighted spaces*, Rocky Mountain J. Math. **35** (2005), no. 6, 2065–2087, DOI 10.1216/rmjm/1181069629. MR2210647
[30] L. Oubbi, *On different types of algebras contained in $CV(X)$*, Bull. Belg. Math. Soc. Simon Stevin **6** (1999), no. 1, 111–120. MR1674706
[31] L. Oubbi, *Multiplication operators on weighted spaces of continuous functions*, Port. Math. (N.S.) **59** (2002), no. 1, 111–124. MR1891597
[32] L. Oubbi, *Algèbres A-convexes à poids* (French, with English summary), Rev. Real Acad. Cienc. Exact. Fís. Natur. Madrid **89** (1995), no. 1-2, 99–110. MR1454351
[33] J. B. Prolla, *Weighted spaces of vector-valued continuous functions*, Ann. Mat. Pura Appl. (4) **89** (1971), 145–157, DOI 10.1007/BF02414945. MR308771
[34] J. B. Prolla, *Approximation of vector valued functions*, North-Holland Publishing Co., Amsterdam-New York-Oxford, 1977. North-Holland Mathematics Studies, Vol. 25; Notas de Matemática, No. 61. [Notes on Mathematics, No. 61]. MR0500122
[35] J. Prolla, *Topological algebras of vector-valued continuous functions*, Mathematical analysis and applications, Part B, Adv. in Math. Suppl. Stud., vol. 7, Academic Press, New York-London, 1981, pp. 727–740. MR634265
[36] C. Shekhar and B. S. Komal, *Multiplication operators on weighted spaces of continuous functions with operator-valued weights*, Int. J. Contemp. Math. Sci. **7** (2012), no. 37-40, 1889–1894. MR2959003
[37] B. Singh and K. Kour, *On weighted composition operators on non-locally convex function spaces*, Indian J. Pure Appl. Math. **28** (1997), no. 11, 1505–1512. MR1608589
[38] R. K. Singh and J. S. Manhas, *Multiplication operators on weighted spaces of vector-valued continuous functions*, J. Austral. Math. Soc. Ser. A **50** (1991), no. 1, 98–107. MR1094062
[39] R. K. Singh and J. S. Manhas, *Multiplication operators and dynamical systems*, J. Austral. Math. Soc. Ser. A **53** (1992), no. 1, 92–102. MR1164779
[40] R. K. Singh and W. H. Summers, *Composition operators on weighted spaces of continuous functions*, J. Austral. Math. Soc. Ser. A **45** (1988), no. 3, 303–319. MR957196
[41] W. H. Summers, *The bounded case of the weighted approximation problem*, Functional analysis and applications (Proc. Sympos. Analysis, Univ. Fed. Pernambuco, Recife, 1972), Springer, Berlin, 1974, pp. 177–183. Lecture Notes in Math., Vol 384. MR0385422
[42] W. H. Summers, *The general complex bounded case of the strict weighted approximation problem*, Math. Ann. **192** (1971), 90–98, DOI 10.1007/BF02052753. MR284800

Current address: M. Klilou, Faculty of Sciences, Mohammed V University in Rabat, Department of Mathematics, Center CeReMAR, Laboratory LMSA, Team GrAAF, 4. Avenue Ibn Batouta, Po.Box 1014 RP, Rabat, Morocco, E-mail m.klilou@hotmail.fr

Current address: L. Oubbi, Ecole Normale Supérieure, Mohammed V University in Rabat, Department of Mathematics, Center CeReMAR, Laboratory LMSA, Team GrAAF, PoBox 5118, Takaddoum, 10105, Rabat, Morocco, E-mail oubbi@daad-alumni.de

Zero product preserving maps on matrix rings over division rings

Matej Brešar and Peter Šemrl

ABSTRACT. Let $\mathbb{D}$ be a division ring and let $\phi : M_n(\mathbb{D}) \to M_n(\mathbb{D})$, $n \geq 2$, be a (not necessarily additive) map satisfying $\phi(A)\phi(B) = 0$ whenever $AB = 0$. We describe the form of ϕ under various assumptions on ϕ, n or $\mathbb{D}$, and provide examples showing that these assumptions are necessary.

1. Introduction

Let R be a ring. We say that a map $\phi : R \to R$ *preserves zero products* if for all $a, b \in R$, $ab = 0$ implies $\phi(a)\phi(b) = 0$. The goal is to describe the form of ϕ. This problem has been considered by many authors over many years, but mostly in the case where R is an algebra and ϕ is linear. In this setting, the usual conclusion is that ϕ is close to an algebra homomorphism – for example, it may be equal to an algebra homomorphism multiplied by a fixed central element. We refer to [**ABEV**] for a historic account on this topic.

The problem is, of course, much more difficult if we do not assume the linearity (or at least the additivity) of ϕ. We are aware of only one result in this direction [**S2**]. It concerns the case where $R = B(X)$, the algebra of all bounded linear operators on an infinite-dimensional Banach space X (see Theorem 2.1). We will consider the (more entangled!) case where X is finite-dimensional. In fact, in most of our results, we will consider a more general situation where $R = M_n(\mathbb{D})$, the ring of $n \times n$ matrices over a possibly noncommutative division ring $\mathbb{D}$. Optimal results will be obtained under the following assumptions:

(a) $n \geq 3$, ϕ preserves zero products in both directions, and either $\mathbb{D}$ is not isomorphic to any of its proper subrings or ϕ is bijective (Theorems 3.4 and 3.5).
(b) $n \geq 3$, $\mathbb{D} = \mathbb{R}$ or $\mathbb{D} = \mathbb{C}$, and ϕ is continuous and bijective (Theorem 4.1).
(c) ϕ is additive (Theorem 5.2).

We will provide several examples illustrating the theorems and justifying the assumptions that are imposed.

2010 *Mathematics Subject Classification.* 15A86, 16K40, 16S50.

Key words and phrases. Zero product preserving map, kernel-image preserving map, matrix ring, division ring, zero product determined ring.

Supported by ARRS Grants P1-0288, N1-0061, and J1-8133.

Different assumptions require different methods. In the case of (a) and (b), the proofs depend upon a version of the fundamental theorem of projective geometry. In the case of (b), we also make use of the invariance of domain theorem. The notion of a zero product determined ring is the concept behind the proof in the case of (c).

Section 2 is devoted to notation, terminology, and other preliminaries. The case (a) is considered in Section 3, (b) in Section 4, and (c) in Section 5.

2. Preliminaries

We begin by formulating the result from [S2]. To this end, we have to introduce some notation. Let X be a Banach space and $B(X)$ the algebra of all bounded linear operators on X. For a bounded linear operator $A : X \to X$ we denote by $\operatorname{Im} A$ and $\operatorname{Ker} A$ the image of A and the kernel of A, respectively. We say that $\theta : B(X) \to B(X)$ is a *kernel-image preserving map* if

$$\operatorname{Ker} \theta(A) = \operatorname{Ker} A \quad \text{and} \quad \overline{\operatorname{Im} \theta(A)} = \overline{\operatorname{Im} A}$$

for all $A \in B(X)$. Obviously, for any pair $A, B \in B(X)$ we have

$$AB = 0 \iff \operatorname{Im} B \subset \operatorname{Ker} A \iff \overline{\operatorname{Im} B} \subset \operatorname{Ker} A.$$

It is then clear that any kernel-image preserving map $\theta : B(X) \to B(X)$ preserves zero products in both directions, that is, for every pair $A, B \in B(X)$ we have

$$AB = 0 \iff \theta(A)\theta(B) = 0.$$

The following statement was proved in [S2].

THEOREM 2.1. *Let X be an infinite-dimensional real or complex Banach space and $\phi : B(X) \to B(X)$ a bijective map preserving zero products in both directions. Then there exist a bijective kernel-image preserving map $\theta : B(X) \to B(X)$ and a bounded bijective linear or (in the complex case) conjugate-linear operator $T : X \to X$ such that*

$$\phi(A) = T\theta(A)T^{-1}$$

for all $A \in B(X)$.

The structure of bijective kernel-image preserving maps can be easily described. We define an equivalence relation $\sim$ on $B(X)$ by

$$A \sim B \iff \operatorname{Ker} A = \operatorname{Ker} B \quad \text{and} \quad \overline{\operatorname{Im} A} = \overline{\operatorname{Im} B}.$$

It is straightforward to verify that $\theta : B(X) \to B(X)$ is a bijective kernel-image preserving map if and only if θ maps every equivalence class into itself and the restriction of θ to each equivalence class is a bijection of this class onto itself.

In this paper, we will be interested in the finite-dimensional case. In some sense, this case is more complicated than the infinite-dimensional one. Namely, the above theorem can be reformulated by saying that if X is an infinite-dimensional Banach space, then every bijective map on $B(X)$ preserving zero products in both directions is a product of a bijective kernel-image preserving map and a bounded linear or conjugate-linear spatial automorphism of $B(X)$. It is known that this is not true in the finite-dimensional case. Namely, if X is a finite-dimensional complex Banach space, then there exist discontinuous ring automorphisms of $B(X)$. By a ring automorphism we mean a bijective additive and multiplicative map. Clearly, such maps preserve zero products in both directions.

Nevertheless, we may ask whether an appropriate modification of Theorem 2.1 holds under the assumption that X is finite-dimensional. Moreover, we may ask if in the finite-dimensional case we have an analogue of this statement in the absence of the bijectivity assumption and/or under the weaker assumption that zero products are preserved in one direction only.

In the finite-dimensional case we can identify operators with matrices. As this paper is more abstract algebra than operator theory oriented, we will not be interested only in real or complex matrices, but in matrices over arbitrary division rings. Therefore some more notation is needed. When working with vector spaces over a division ring we have to distinguish between left and right vector spaces. In this regard, we will follow the conventions that are more standard in abstract algebra than in linear algebra.

Let $\mathbb{D}$ be a division ring. We denote by $M_n(\mathbb{D})$ the ring of all $n \times n$ matrices over $\mathbb{D}$. We will always consider $\mathbb{D}^n$, the set of all $1 \times n$ matrices, as a left vector space over $\mathbb{D}$. Correspondingly, we have the right vector space of all $n \times 1$ matrices ${}^t\mathbb{D}^n$. For $A \in M_n(\mathbb{D})$ we take the row space of A, that is the left vector subspace of $\mathbb{D}^n$ generated by the rows of A, and define the row rank of A to be the dimension of this subspace. Similarly, the column rank of A is the dimension of the right vector space generated by the columns of A. This space is called the column space of A. It turns out that these two ranks are equal for every matrix over $\mathbb{D}$ and this common value is called the *rank* of a matrix.

Let $a \in \mathbb{D}^n$ and ${}^tb \in {}^t\mathbb{D}^n$ be any nonzero vectors. Then ${}^tba = ({}^tb)a$ is a matrix of rank one. For a nonzero $x \in \mathbb{D}^n$ and a nonzero ${}^ty \in {}^t\mathbb{D}^n$ we denote by $R(x)$ and $L({}^ty)$ the subsets of $M_n(\mathbb{D})$ defined by

$$R(x) = \{ {}^tux : {}^tu \in {}^t\mathbb{D}^n \}$$

and

$$L({}^ty) = \{ {}^tyv : v \in \mathbb{D}^n \}.$$

Clearly, all the elements of these two sets are of rank at most one.

As always, we will identify $n \times n$ matrices with linear transformations mapping $\mathbb{D}^n$ into $\mathbb{D}^n$. Namely, each $n \times n$ matrix A gives rise to a linear operator defined by $x \mapsto xA$, $x \in \mathbb{D}^n$. The rank of the matrix A is equal to the dimension of the image $\operatorname{Im} A$ of the corresponding operator A. The kernel of an operator A is defined as $\operatorname{Ker} A = \{x \in \mathbb{D}^n : xA = 0\}$. It is the set of all vectors $x \in \mathbb{D}^n$ satisfying $x({}^ty) = 0$ for every ${}^ty$ from the column space of A.

By $\mathbb{P}(\mathbb{D}^n)$ and $\mathbb{P}({}^t\mathbb{D}^n)$ we denote the projective spaces over the left vector space $\mathbb{D}^n$ and the right vector space ${}^t\mathbb{D}^n$, respectively, $\mathbb{P}(\mathbb{D}^n) = \{[x] : x \in \mathbb{D}^n \setminus \{0\}\}$ and $\mathbb{P}({}^t\mathbb{D}^n) = \{[{}^ty] : {}^ty \in {}^t\mathbb{D}^n \setminus \{0\}\}$. Here, $[x]$ and $[{}^ty]$ denote the one-dimensional left vector subspace of $\mathbb{D}^n$ generated by x and the one-dimensional right vector subspace of ${}^t\mathbb{D}^n$ generated by ${}^ty$, respectively.

One of the main tools in our proofs will be a relatively recently obtained *non-surjective version of the fundamental theorem of projective geometry* [**F**]. We will present a slightly weaker version. We start with the projective space over the left vector space $\mathbb{D}^n$. A map $f : \mathbb{D}^n \to \mathbb{D}^n$ is called *semilinear* if we have

$$f(x + y) = f(x) + f(y)$$

for all $x, y \in \mathbb{D}^n$ and there exists a (not necessarily surjective) ring endomorphism $\tau : \mathbb{D} \to \mathbb{D}$ such that

$$f(\lambda x) = \tau(\lambda) f(x)$$

for every $\lambda \in \mathbb{D}$ and $x \in \mathbb{D}^n$. For a map $\alpha : \mathbb{P}(\mathbb{D}^n) \to \mathbb{P}(\mathbb{D}^n)$ we say that its image is *contained in a line* if there exist nonzero vectors $u, v \in \mathbb{D}^n$ such that $\alpha([x]) \subset [u] + [v]$ for every nonzero $x \in \mathbb{D}^n$. The non-surjective version of the fundamental theorem of projective geometry can be formulated in the following way. Let $\alpha : \mathbb{P}(\mathbb{D}^n) \to \mathbb{P}(\mathbb{D}^n)$ be an injective map whose image is not contained in a line. Assume that for all $x, y, z \in \mathbb{D}^n \setminus \{0\}$,

$$[x] \subset [y] + [z] \implies \alpha([x]) \subset \alpha([y]) + \alpha([z]).$$

Then there is an injective semilinear map $f : \mathbb{D}^n \to \mathbb{D}^n$ such that

$$\alpha([x]) = [f(x)], \quad x \in \mathbb{D}^n \setminus \{0\}.$$

Let $e_1, \ldots, e_n$ be the standard basis of $\mathbb{D}^n$. We define A to be the $n \times n$ matrix whose k-th row is $f(e_k)$, $k = 1, \ldots, n$. We claim that

$$f(x) = x_\tau A, \quad x \in \mathbb{D}^n,$$

where

$$x_\tau = \begin{bmatrix} x_1 & \ldots & x_n \end{bmatrix}_\tau = \begin{bmatrix} \tau(x_1) & \ldots & \tau(x_n) \end{bmatrix}.$$

Indeed, this is trivially true for $x \in \{e_1, \ldots, e_n\}$. For all other vectors $x \in \mathbb{D}^n$ the above equality is a straightforwad consequence of the semilinearity of f. It is important to observe that A is not invertible in general. However, the map $x \mapsto x_\tau A$, $x \in \mathbb{D}^n$, is injective, that is, $x_\tau A \neq 0$ for all nonzero vectors x.

One can now easily formulate an analogous statement for maps on the projective space over the right vector space ${}^t\mathbb{D}^n$.

Let n be a positive integer. In what follows we will always identify matrices $A \in M_n(\mathbb{D})$ with operators $A : \mathbb{D}^n \to \mathbb{D}^n$. Let τ be an endomorphism of the division ring $\mathbb{D}$. For a matrix $A \in M_n(\mathbb{D})$ we denote by A_τ the matrix obtained from $A = [a_{ij}]$ by applying τ entrywise,

$$A_\tau = [a_{ij}]_\tau = [\tau(a_{ij})].$$

Then we will say that $\xi : M_n(\mathbb{D}) \to M_n(\mathbb{D})$ is a τ-*kernel-image preserving map* if for every $A \in M_n(\mathbb{D})$ we have

$$\operatorname{Im} \xi(A) = \operatorname{Im} A_\tau \quad \text{and} \quad \operatorname{Ker} \xi(A) = \operatorname{Ker} A_\tau.$$

Note first that for every $A \in M_n(\mathbb{D})$ we have $\operatorname{rank} A = \operatorname{rank} A_\tau$, that is, $\dim \operatorname{Im} A = \dim \operatorname{Im} A_\tau$. The easiest way to verify this is to use the well-known fact that $\operatorname{rank} A = r$ if and only if there exist invertible matrices $T, S \in M_n(\mathbb{D})$ such that

$$A = T \begin{bmatrix} I_r & 0 \\ 0 & 0 \end{bmatrix} S.$$

Here, I_r is the $r \times r$ identity matrix and the zeros stand for zero matrices of the appropriate sizes. Then clearly,

$$A_\tau = T_\tau \begin{bmatrix} I_r & 0 \\ 0 & 0 \end{bmatrix} S_\tau$$

and T_τ and S_τ are invertible with $(T_\tau)^{-1} = (T^{-1})_\tau$ and $(S_\tau)^{-1} = (S^{-1})_\tau$. Moreover, since the equality $n = \dim \operatorname{Im} A + \dim \operatorname{Ker} A$ holds for matrices over an arbitrary division ring, we have also $\dim \operatorname{Ker} A = \dim \operatorname{Ker} A_\tau$.

We claim that ξ preserves zero products in both directions. Indeed, for $A, B \in M_n(\mathbb{D})$ we have

$$AB = 0 \iff A_\tau B_\tau = 0 \iff \operatorname{Im} A_\tau \subset \operatorname{Ker} B_\tau$$

$$\iff \operatorname{Im} \xi(A) \subset \operatorname{Ker} \xi(B) \iff \xi(A)\xi(B) = 0.$$

Let $\Gamma : \mathbb{D}^n \to \mathbb{D}^n$ be the map defined by

$$\Gamma\left(\begin{bmatrix} a_1 & a_2 & \ldots & a_n \end{bmatrix}\right) = \begin{bmatrix} \tau(a_1) & \tau(a_2) & \ldots & \tau(a_n) \end{bmatrix}.$$

For a subspace $U \subset \mathbb{D}^n$ we denote by U_τ the linear span of its Γ-image, that is,

$$U_\tau = \operatorname{span} \Gamma(U)$$

(note that we have $U_\tau = \Gamma(U)$ if τ is an automorphism, not just an endomorphism). For every pair of subspaces $U, V \subset \mathbb{D}^n$ with $\dim U + \dim V = n$ we set

$$\mathcal{S}(U, V) = \{ A \in M_n(\mathbb{D}) : \operatorname{Im} A = U \text{ and } \operatorname{Ker} A = V \}.$$

Clearly, $M_n(\mathbb{D})$ is a disjoint union of such subsets. A map $\xi : M_n(\mathbb{D}) \to M_n(\mathbb{D})$ is τ-kernel-image preserving if and only if for every pair of subspaces $U, V \subset \mathbb{D}^n$ with $\dim U + \dim V = n$ we have

$$\xi\left(\mathcal{S}(U, V)\right) \subset \mathcal{S}(U_\tau, V_\tau).$$

All the above observations will be used frequently without reference in what follows.

3. The general case

In this section, we consider the general situation where ϕ is an arbitrary map.

When starting to work on optimal finite-dimensional analogues of Theorem 2.1, our initial conjecture was that if $n \geq 3$ is an integer, $\mathbb{D}$ any division ring, and $\phi : M_n(\mathbb{D}) \to M_n(\mathbb{D})$ a map such that for every pair $A, B \in M_n(\mathbb{D})$ we have $AB = 0$ if and only if $\phi(A)\phi(B) = 0$, then there exist an endomorphism $\tau : \mathbb{D} \to \mathbb{D}$, a τ-kernel-image preserving map $\xi : M_n(\mathbb{D}) \to M_n(\mathbb{D})$, and an invertible matrix $T \in M_n(\mathbb{D})$ such that $\phi(A) = T\xi(A)T^{-1}$ for every $A \in M_n(\mathbb{D})$.

Clearly, the converse is true, that is, every map $\phi : M_n(\mathbb{D}) \to M_n(\mathbb{D})$ of the form $\phi(A) = T\xi(A)T^{-1}$, $A \in M_n(\mathbb{D})$, preserves zero products in both directions.

It should be explained why the assumption $n \geq 3$ is necessary in the above conjecture. Of course, we need it in order to apply the fundamental theorem of projective geometry. But it turns out that it is indispensable. Namely, in the following example we construct a bijective map $\phi : M_2(\mathbb{R}) \to M_2(\mathbb{R})$ preserving zero products in both directions that is not of the form as described in the conjecture.

EXAMPLE 3.1. Let $f : \mathbb{R} \to \mathbb{R}$ be any bijective function with $f(0) = 0$. Define $g : \mathbb{R} \to \mathbb{R}$ by $g(0) = 0$ and

$$g(x) = -\frac{1}{f\left(-\frac{1}{x}\right)}$$

whenever $x \neq 0$. We now define $\phi : M_2(\mathbb{R}) \to M_2(\mathbb{R})$ by $\phi(A) = A$ whenever A is invertible or $A = 0$. It remains to define $\phi(A)$ for all matrices

$$A = \begin{bmatrix} a_1 & a_2 \\ a_3 & a_4 \end{bmatrix}$$

of rank one. If A is such a matrix with $a_1 \neq 0$, then we define

$$\phi(A) = a_1 \begin{bmatrix} 1 & g\left(\frac{a_2}{a_1}\right) \\ f\left(\frac{a_3}{a_1}\right) & f\left(\frac{a_3}{a_1}\right) g\left(\frac{a_2}{a_1}\right) \end{bmatrix}.$$

In the case that A is of rank one, $a_1 = 0$, and $a_2 \neq 0$ (then, clearly, $a_3 = 0$) we set

$$\phi(A) = a_2 \begin{bmatrix} 0 & 1 \\ 0 & f\left(\frac{a_4}{a_2}\right) \end{bmatrix}.$$

Further, when A is of rank one, $a_1 = a_2 = 0$, and $a_3 \neq 0$ we define

$$\phi(A) = a_3 \begin{bmatrix} 0 & 0 \\ 1 & g\left(\frac{a_4}{a_3}\right) \end{bmatrix}.$$

And finally, for every $a_4 \neq 0$ we set

$$\phi\left(\begin{bmatrix} 0 & 0 \\ 0 & a_4 \end{bmatrix}\right) = \begin{bmatrix} 0 & 0 \\ 0 & a_4 \end{bmatrix}.$$

Since we can choose any bijective function fixing 0 for f, in general ϕ is not of the form $\phi(A) = T\xi(A)T^{-1}$ as in the above conjecture. To check that ϕ is bijective we first note that ϕ is a bijection of the set of all invertible matrices onto itself and it maps the zero matrix to itself. Thus we need to see that it maps the set of all rank one matrices bijectively onto itself. We write the set of all rank one matrices as the disjoint union of four sets:

- $\mathcal{S}_1 \subset M_2(\mathbb{R})$ is the set of all rank one matrices with a nonzero $(1,1)$-entry,
- $\mathcal{S}_2 \subset M_2(\mathbb{R})$ is the set of all rank one matrices with the zero $(1,1)$ entry and a nonzero $(1,2)$-entry,
- $\mathcal{S}_3 \subset M_2(\mathbb{R})$ is the set of all rank one matrices with the zero $(1,1)$ and $(1,2)$ entries and a nonzero $(2,1)$-entry,
- $\mathcal{S}_4 \subset M_2(\mathbb{R})$ is the set of all rank one matrices whose all entries are zero but the $(2,2)$-entry.

Clearly, if

$$A = \begin{bmatrix} a_1 & a_2 \\ a_3 & a_4 \end{bmatrix}$$

is a rank one matrix with $a_1 \neq 0$, then $a_4 = \frac{a_2 a_3}{a_1}$. It follows that $\phi(\mathcal{S}_j) \subset \mathcal{S}_j$, $j = 1, 2, 3, 4$, and we need to show that ϕ maps each $\mathcal{S}_j$ bijectively onto itself. Using the fact that both $f, g : \mathbb{R} \to \mathbb{R}$ are bijective functions one can easily complete the proof of the bijectivity of ϕ.

It remains to verify that for every pair $A, B \in M_2(\mathbb{R})$ we have

$$AB = 0 \iff \phi(A)\phi(B) = 0.$$

Assume first that A is invertible. Then $AB = 0$ if and only if $B = 0$ and $\phi(A)\phi(B) = 0 = A\phi(B)$ if and only if $\phi(B) = 0$ which is equivalent to $B = 0$. It is similarly easy to check that the above equivalence holds true when B is invertible or $A = 0$ or $B = 0$.

Hence, we can assume that both A and B are of rank one. We have to distinguish all possible cases, that is, $A \in \mathcal{S}_i$ and $B \in \mathcal{S}_j$, $i, j \in \{1, 2, 3, 4\}$. We will consider here only the first possibility and leave the others (which are easy exercises) to the reader. So, let A, B both belong to $\mathcal{S}_1$. Then we have

$$A = a \begin{bmatrix} 1 & x \\ y & xy \end{bmatrix} \quad \text{and} \quad B = b \begin{bmatrix} 1 & u \\ v & uv \end{bmatrix}$$

for some real numbers a, b, x, y, u, v with both a and b nonzero. Obviously, $AB = 0$ if and only if $xv = -1$. By the definition of ϕ we have

$$\phi(A) = a \begin{bmatrix} 1 & g(x) \\ f(y) & g(x)f(y) \end{bmatrix} \quad \text{and} \quad \phi(B) = b \begin{bmatrix} 1 & g(u) \\ f(v) & g(u)f(v) \end{bmatrix},$$

and thus, $\phi(A)\phi(B) = 0$ if and only if $1 + g(x)f(v) = 0$. Clearly this happens if and only if $1 + xv = 0$, that is, $AB = 0$.

The reader may have observed that instead of defining $\phi : M_2(\mathbb{R}) \to M_2(\mathbb{R})$ with formulas and verifying the required properties with somewhat lengthy (but trivial) computations, we could have used a more conceptual approach with two maps on projective spaces $\alpha : \mathbb{P}(\mathbb{R}^2) \to \mathbb{P}(\mathbb{R}^2)$ and $\beta : \mathbb{P}(^t\mathbb{R}^2) \to \mathbb{P}(^t\mathbb{R}^2)$ having a certain orthogonality property: if for $x \in \mathbb{R}^2 \setminus \{0\}$ and $^ty \in {}^t\mathbb{R}^2 \setminus \{0\}$ we denote $\alpha([x]) = [u]$ and $\beta([^ty]) = [^tv]$, then the requirement is that $x\,^ty = 0$ if and only if $u\,^tv = 0$. In the $n = 2$ case there are plenty of such pairs of maps α, β and they can be used to define the restriction of ϕ to the set of all rank one matrices (some special attention has to be paid to the bijectivity of a map ϕ defined in this way).

The following lemma seemingly supports the aforementioned conjecture. We denote by $M_n^{\leq 1}(\mathbb{D})$ the set of all $n \times n$ matrices with rank at most one.

LEMMA 3.2. *Let $n \geq 3$ be an integer and $\mathbb{D}$ any division ring. Let $\phi : M_n(\mathbb{D}) \to M_n(\mathbb{D})$ be a map such that for every pair $A, B \in M_n(\mathbb{D})$,*

$$AB = 0 \iff \phi(A)\phi(B) = 0.$$

Then there exist an endomorphism $\tau : \mathbb{D} \to \mathbb{D}$, a τ-kernel-image preserving map $\xi : M_n^{\leq 1}(\mathbb{D}) \to M_n^{\leq 1}(\mathbb{D})$, and an invertible matrix $T \in M_n(\mathbb{D})$ such that

$$\phi(A) = T\xi(A)T^{-1}$$

for every $A \in M_n^{\leq 1}(\mathbb{D})$.

PROOF. Let $^tx_1, \ldots, {}^tx_n \in {}^t\mathbb{D}^n$ and $y_1, \ldots, y_n \in \mathbb{D}^n$ be any two (linearly independent) n-tuples of vectors satisfying

$$y_i\,^tx_j = \delta_{ij}, \quad i, j = 1, \ldots, n.$$

If we denote $A_i = {}^tx_i y_i \in M_n(\mathbb{D})$, $i = 1, \ldots, n$, then clearly $A_i^2 = A_i$, $i = 1, \ldots, n$, and $A_i A_j = 0$ whenever $i \neq j$. Hence, we have

$$\phi(A_i)\phi(0) = 0$$

for all $i = 1, \ldots, n$,

$$\phi(A_i)\phi(A_i) \neq 0$$

for all $i = 1, \ldots, n$, and

$$\phi(A_i)\phi(A_j) = 0$$

for all $i, j \in \{1, \ldots, n\}$, $i \neq j$. In particular, each of the matrices (operators) A_i, $i = 1, \ldots, n$, is nonzero. We further conclude that $\{0\} \neq \operatorname{Im} \phi(A_i) \subset \operatorname{Ker} \phi(A_j)$ whenever $i \neq j$. Therefore

$$\sum_{i \neq j} \operatorname{Im} \phi(A_i) \subset \operatorname{Ker} \phi(A_j)$$

and since $\operatorname{Im} \phi(A_j) \not\subset \operatorname{Ker} \phi(A_j)$, we have

$$\operatorname{Im} \phi(A_j) \not\subset \sum_{i \neq j} \operatorname{Im} \phi(A_i)$$

for all $j = 1, \ldots, n$. It follows easily that for every $j \in \{1, \ldots, n\}$ the subspace $\sum_{i \neq j} \mathrm{Im}\, \phi(A_i)$ has dimension at least $(n-1)$, and since $\mathrm{Ker}\, \phi(A_j) \neq \mathbb{D}^n$, we infer that $\sum_{i \neq j} \mathrm{Im}\, \phi(A_i) = \mathrm{Ker}\, A_j$ is of dimension $n - 1$. We conclude that all $\phi(A_j)$'s are of rank one. Hence, all the images of operators $\phi(A_i)$ are one-dimensional subspaces and they are linearly independent. Thus, we have

$$\phi(A_i) = {}^t u_i v_i$$

for some nonzero vectors ${}^t u_1, \ldots, {}^t u_n \in {}^t\mathbb{D}^n$ and $v_1, \ldots, v_n \in \mathbb{D}^n$ and we further know that $v_1, \ldots, v_n \in \mathbb{D}^n$ are linearly independent. Similarly, ${}^t u_1, \ldots, {}^t u_n \in {}^t\mathbb{D}^n$ are linearly independent. It follows that $\phi(0) = 0$.

Let ${}^t x, {}^t y \in {}^t \mathbb{D}^n$ be any pair of linearly independent vectors. We choose vectors ${}^t x = {}^t x_1, {}^t y = {}^t x_2, {}^t x_3, \ldots, {}^t x_n \in {}^t \mathbb{D}^n$ and $y_1, \ldots, y_n \in \mathbb{D}^n$ as above. Further let matrices A_j, $j = 1, \ldots, n$, and vectors ${}^t u_1, \ldots, {}^t u_n \in {}^t \mathbb{D}^n$ and $v_1, \ldots, v_n \in \mathbb{D}^n$ be defined as in the previous paragraph.

Take any nonzero $w \in \mathbb{D}^n$ and observe that

$$A_j\, {}^t x w = 0, \quad j = 2, \ldots, n,$$

and

$$A_1\, {}^t x w \neq 0,$$

and therefore

$$ {}^t u_j v_j \phi({}^t x w) = 0$$

for every $j = 2, \ldots, n$. It follows that

$$v_j \phi({}^t x w) = 0$$

for every $j = 2, \ldots, n$, and

$$v_1 \phi({}^t x w) \neq 0.$$

Since $v_2, \ldots, v_n$ are linearly independent, we conclude that $\phi({}^t x w)$ is of rank one, that is, $\phi({}^t x w) = {}^t a b$ for some nonzero vectors ${}^t a$ and b (note that, in particular, we have verified that each rank one matrix is mapped into a matrix of rank one). Using

$$v_j\, {}^t u_1 = 0 \quad \text{and} \quad v_j\, {}^t a = 0, \quad j = 2, \ldots, n,$$

we see that u_1 and a are linearly dependent. In other words, for every $w \in \mathbb{D}^n$ there exists a vector $c \in \mathbb{D}^n$ such that

$$\phi({}^t x w) = {}^t u_1 c,$$

that is,

$$\phi(L({}^t x)) \subset L({}^t u_1).$$

As we have started with an arbitrary nonzero vector ${}^t x$ we see that for every nonzero ${}^t e \in {}^t \mathbb{D}^n$ there exists a nonzero ${}^t d \in {}^t \mathbb{D}^n$ such that

$$\phi(L({}^t e)) \subset L({}^t d).$$

Clearly, $L({}^t e) = L({}^t e')$ if and only if ${}^t e$ and ${}^t e'$ are linearly dependent. If, on the other hand, ${}^t e$ and ${}^t e'$ are linearly independent, then

$$L({}^t e) \cap L({}^t e') = \{0\}.$$

Thus, ϕ induces a map $\beta : \mathbb{P}({}^t \mathbb{D}^n) \to \mathbb{P}({}^t \mathbb{D}^n)$ such that

$$\beta([{}^t e]) = [{}^t d] \iff \phi(L({}^t e)) \subset L({}^t d).$$

Note that β is injective. Indeed, we have shown that for an arbitrary pair of linearly independent vectors ${}^tx$, ${}^ty$ we have $\beta([{}^tx]) \neq \beta([{}^ty])$.

Our next claim is that if for some nonzero ${}^tf \in {}^t\mathbb{D}^n$ we have $[{}^tf] \subset [{}^tx] + [{}^ty]$ then $\beta([{}^tf]) \subset \beta([{}^tx]) + \beta([{}^ty])$. Indeed, from $[{}^tf] \subset [{}^tx] + [{}^ty]$ we conclude that

$$A_j {}^tf = 0$$

for all $j = 3, \ldots, n$, implying that

$$\phi(A_j)\phi({}^tfg) = 0, \quad j = 3, \ldots, n,$$

where $g \in \mathbb{D}^n$ is any nonzero vector. Since $\phi(A_j) = {}^tu_j v_j$ and $\phi({}^tfg) = {}^tf'g'$ for some nonzero ${}^tf' \in \beta([{}^tf])$ and some nonzero $g' \in \mathbb{D}^n$, we see that

$$v_j {}^tf' = 0, \quad j = 3, \ldots, n.$$

We know that $v_1, \ldots, v_n$ are linearly independent, ${}^tu_1$ and ${}^tu_2$ are linearly independent, $v_j {}^tu_i = 0$, $i = 1, 2$, $j = 3, \ldots, n$, and consequently, ${}^tf'$ belongs to the linear span of ${}^tu_1$ and ${}^tu_2$. Since $\beta([{}^tf]) = [{}^tf']$, $\beta([{}^tx]) = [{}^tu_1]$ and $\beta([{}^ty]) = [{}^tu_2]$, we finally conclude that $\beta([{}^tf]) \subset \beta([{}^tx]) + \beta([{}^ty])$.

We have proved that for any pair of linearly independent vectors ${}^tx$ and ${}^ty$ we have

$$[{}^tf] \subset [{}^tx] + [{}^ty] \Rightarrow \beta([{}^tf]) \subset \beta([{}^tx]) + \beta([{}^ty]).$$

As the range of β is not contained in a line (the range of β contains elements $[{}^tu_1], \ldots, [{}^tu_n]$) we can apply the fundamental theorem of projective geometry to conclude that there exist a matrix $T \in M_n(\mathbb{D})$ and an endomorphism $\tau : \mathbb{D} \to \mathbb{D}$ such that

$$\beta([{}^tx]) = [T {}^tx_\tau], \quad {}^tx \in {}^tD^n \setminus \{0\}.$$

Since the range of β contains elements $[{}^tu_1], \ldots, [{}^tu_n]$, the matrix T is invertible. Using the definition of the map β we arrive at

$$\phi(L({}^tx)) \subset L(T {}^tx_\tau)$$

for every nonzero ${}^tx \in {}^t\mathbb{D}^n$. In the same way we see that there exist an invertible matrix S and an endomorphism $\sigma : \mathbb{D} \to \mathbb{D}$ such that

$$\phi(R(y)) \subset R(y_\sigma S)$$

for every nonzero $y \in \mathbb{D}^n$.

Hence, for every rank one matrix ${}^txy \in M_n(\mathbb{D})$ there exists a nonzero $\lambda \in \mathbb{D}$ (depending on ${}^txy$) such that

$$\phi({}^txy) = T {}^tx_\tau \lambda y_\sigma S.$$

The zero product preserving property then yields that for any pair of vectors $y \in \mathbb{D}^n$ and ${}^tu \in {}^t\mathbb{D}^n$ we have

$$y {}^tu = 0 \iff y_\sigma ST {}^tu_\tau = 0.$$

If we denote $C = ST$ then for any pair of n-tuples $y_1, \ldots, y_n \in \mathbb{D}$ and $u_1, \ldots, u_n \in \mathbb{D}$, the following is true:

$$\sum_{k=1}^n y_k u_k = 0 \iff \begin{bmatrix} \sigma(y_1) & \ldots & \sigma(y_n) \end{bmatrix} C \begin{bmatrix} \tau(u_1) \\ \vdots \\ \tau(u_n) \end{bmatrix} = 0.$$

Choosing $y_1 = u_2 = 1$ and $y_2 = \ldots = y_n = u_1 = u_3 = \ldots = u_n = 0$ we see that the $(1, 2)$-entry of C is zero. In the same way we show that all off-diagonal entries of

C are zero. Thus, C is a diagonal matrix with diagonal entries $c_1, \ldots, c_n$ and then the above equivalence can be rewritten as

$$\sum_{k=1}^{n} y_k u_k = 0 \iff \sum_{k=1}^{n} \sigma(y_k) c_k \tau(u_k) = 0.$$

Choosing $y_1 = u_1 = 1$, $y_2 = \lambda$, $u_2 = -\frac{1}{\lambda}$, and $y_3 = \ldots = y_n = u_3 = \ldots = u_n = 0$, where $\lambda \in \mathbb{D}$ is any nonzero element, we arrive at

$$c_1 - \sigma(\lambda) c_2 \tau(\lambda)^{-1} = 0$$

for every nonzero $\lambda \in \mathbb{D}$. If we set $\lambda = 1$ we conclude that $c_1 = c_2$, and then similarly, $c_1 = \ldots = c_n = c$. In particular, $ST = cI$ and

$$\sigma(\lambda) = c\tau(\lambda) c^{-1}, \quad \lambda \in \mathbb{D}.$$

It follows that for every rank one matrix ${}^t xy \in M_n(\mathbb{D})$ there exists a nonzero $\lambda \in \mathbb{D}$ (depending on ${}^t xy$) such that

$$\phi({}^t xy) = T\, {}^t x_\tau \lambda (c y_\tau c^{-1})(cT^{-1}) = T\, {}^t x_\tau \mu y_\tau T^{-1},$$

where we have denoted $\mu = \lambda c$. Clearly,

$$\operatorname{Im} {}^t x_\tau \mu y_\tau = \operatorname{Im}({}^t xy)_\tau \quad \text{and} \quad \operatorname{Ker} {}^t x_\tau \mu y_\tau = \operatorname{Ker}({}^t xy)_\tau.$$

Thus, the map $\xi : M_n^{\leq 1}(\mathbb{D}) \to M_n^{\leq 1}(\mathbb{D})$ given by $\xi({}^t xy) = {}^t x_\tau \mu y_\tau$ has the desired properties. $\square$

In spite of the fact that the above statement shows that the behavior of ϕ on the set of all rank one matrices is as expected, the conjecture turns out to be false.

EXAMPLE 3.3. Recall (see [**K**]) that there exists an endomorphism $\tau : \mathbb{C} \to \mathbb{C}$ such that $\tau(\mathbb{C})$ is a proper subfield of $\mathbb{C}$. Even more, it is possible to find an endomorphism τ such that $\mathbb{C}$ is an infinite-dimensional vector space over $\tau(\mathbb{C})$. In particular, it is possible to find complex numbers $c_1, \ldots, c_n$ that are linearly independent in the vector space $\mathbb{C}$ over the field $\tau(\mathbb{C})$. We define $\phi : M_n(\mathbb{C}) \to M_n(\mathbb{C})$ by

$$\phi(A) = A_\tau$$

for all $A \in M_n(\mathbb{C}) \setminus \{I\}$ and

$$\phi(I) = {}^t cc,$$

where $c = \begin{bmatrix} c_1 & \ldots & c_n \end{bmatrix}$. Using the fact ${}^t ccB \neq 0$ and $B\,{}^t cc \neq 0$ for every nonzero complex $n \times n$ matrix B with all entries in $\tau(\mathbb{C})$ we can easily verify that ϕ preserves zero products in both directions, but is not of the form as in the above conjecture.

Nevertheless, the conjecture is true for a large class of division rings. We call a division ring $\mathbb{D}$ an EAS division ring if every ring endomorphism $\tau : \mathbb{D} \to \mathbb{D}$ is automatically surjective. In other words, $\mathbb{D}$ is EAS if it is not isomorphic to any of its proper subrings. The field of real numbers and the field of rational numbers are well-known to be EAS. Obviously, every finite field is EAS. Another example is the division ring of quaternions (see, for example, [**S1**]). In the above counterexample we have used the fact that the field of complex numbers is not EAS.

THEOREM 3.4. *Let $n \geq 3$ be an integer and $\mathbb{D}$ any EAS division ring. Let $\phi : M_n(\mathbb{D}) \to M_n(\mathbb{D})$ be a map such that for every pair $A, B \in M_n(\mathbb{D})$,*

$$AB = 0 \iff \phi(A)\phi(B) = 0.$$

Then there exist an automorphism $\tau : \mathbb{D} \to \mathbb{D}$, *a τ-kernel-image preserving map* $\xi : M_n(\mathbb{D}) \to M_n(\mathbb{D})$, *and an invertible matrix* $T \in M_n(\mathbb{D})$ *such that*

$$\phi(A) = T\xi(A)T^{-1}$$

for every $A \in M_n(\mathbb{D})$.

PROOF. We apply Lemma 3.2. Because of the EAS assumption τ must be an automorphism. After composing ϕ first by a similarity transformation $A \mapsto T^{-1}AT$ and then by the map $A \mapsto A_{\tau^{-1}}$ we can assume with no loss of generality that every matrix ${}^txy$ of rank one is mapped by ϕ to ${}^tx\lambda y$ for some nonzero $\lambda \in \mathbb{D}$ depending on ${}^txy$. From

$${}^txyA = 0 \iff \phi({}^txy)\phi(A) = 0 \quad \text{and} \quad A\,{}^txy = 0 \iff \phi(A)\phi({}^txy) = 0$$

we conclude that for every $A \in M_n(\mathbb{D})$, ${}^tx \in {}^t\mathbb{D}^n$, and every $y \in \mathbb{D}^n$ we have

$$yA = 0 \iff y\phi(A) = 0 \quad \text{and} \quad A\,{}^tx = 0 \iff \phi(A)\,{}^tx = 0$$

yielding that

$$\operatorname{Im} A = \operatorname{Im} \phi(A) \quad \text{and} \quad \operatorname{Ker} A = \operatorname{Ker} \phi(A).$$

□

For general division rings we need to add the bijectivity assumption to get the desired conclusion.

THEOREM 3.5. *Let $n \geq 3$ be an integer and $\mathbb{D}$ any division ring. Let $\phi : M_n(\mathbb{D}) \to M_n(\mathbb{D})$ be a bijective map such that for every pair $A, B \in M_n(\mathbb{D})$ we have*

$$AB = 0 \iff \phi(A)\phi(B) = 0.$$

Then there exist an automorphism $\tau : \mathbb{D} \to \mathbb{D}$, a bijective τ-kernel-image preserving map $\xi : M_n(\mathbb{D}) \to M_n(\mathbb{D})$, and an invertible matrix $T \in M_n(\mathbb{D})$ such that

$$\phi(A) = T\xi(A)T^{-1}$$

for every $A \in M_n(\mathbb{D})$.

PROOF. Again we apply Lemma 3.2. As before there is no loss of generality in assuming that T is the identity matrix. We need to show that τ is bijective. Once we will prove this, the proof can be completed in exactly the same way as the proof of the previous theorem.

We observe that ϕ^{-1} has exactly the same properties as ϕ. Thus, ϕ^{-1} maps rank one matrices to rank one matrices. Let $c \in \mathbb{D}$ be any nonzero element. By bijectivity, there exists a rank one matrix ${}^txy$ that is mapped by ϕ into ${}^te_1(e_1 + ce_2)$, that is,

$${}^tx_\tau \lambda y_\tau = {}^te_1(e_1 + ce_2)$$

for some nonzero element λ which further implies that

$$y_\tau = \begin{bmatrix} \tau(y_1) & \tau(y_2) & \ldots & \tau(y_n) \end{bmatrix}$$

and $e_1 + ce_2$ are linearly dependent. Consequently, c belongs to the range of τ. Hence, τ is surjective, as desired. □

One may ask whether we can prove the above theorem under the weaker assumption that zero products are preserved in one direction only. The next example shows that the answer is, in general, negative.

EXAMPLE 3.6. We need to construct a bijective map $\phi : M_n(\mathbb{D}) \to M_n(\mathbb{D})$ satisfying $\phi(A)\phi(B) = 0$ whenever $AB = 0$ that is not of the form as in the conclusion of the above theorem. We will consider the special case when $\mathbb{D} = \mathbb{F} \in \{\mathbb{R}, \mathbb{C}\}$. As usual, we write E_{11} for the matrix unit all of whose entries are zero but the $(1,1)$ entry that is equal to 1, and I for the identity matrix. Define $\phi : M_n(\mathbb{F}) \to M_n(\mathbb{F})$ by $\phi(A) = A$ whenever

$$A \notin \{E_{11}, I, 2E_{11}, 2I, 3E_{11}, 3I, \ldots\},$$

$$\phi(I) = E_{11},$$

$$\phi(nI) = (n-1)I, \quad n = 2, 3, \ldots,$$

and

$$\phi(nE_{11}) = (n+1)E_{11}, \quad n = 1, 2, \ldots$$

It is easy to check that ϕ is bijective. So, assume that $AB = 0$ for some $A, B \in M_n(\mathbb{F})$ and we need to show that then $\phi(A)\phi(B) = 0$. This is obviously true when $A \neq I$ and $B \neq I$ since in this case we have $\phi(A) = \lambda A$ and $\phi(B) = \mu B$ for some nonzero scalars λ, μ. If $A = I$ or $B = I$ then $AB = 0$ implies that the other one must be the zero matrix and thus, $\phi(A)\phi(B) = 0$ in this case as well.

Of course, using similar ideas one can construct more complicated examples. Moreover, the composition of two bijective maps preserving zero products in one direction is again a bijective map with the same preserving property, which yields a variety of further examples.

4. The case where ϕ is continuous

The map ϕ in Example 3.6 is not continuous. It is perhaps somewhat surprising that in the presence of the continuity assumption we can get a nice result on maps preserving zero products in one direction only. Let $\mathbb{F}$ be either the field of real numbers or the field of complex numbers. As before we say that a map $\theta : M_n(\mathbb{F}) \to M_n(\mathbb{F})$ is a kernel-image preserving map if

$$\operatorname{Ker} \theta(A) = \operatorname{Ker} A \quad \text{and} \quad \operatorname{Im} \theta(A) = \operatorname{Im} A$$

for all $A \in M_n(\mathbb{F})$. In the complex case we say that a map $\theta : M_n(\mathbb{C}) \to M_n(\mathbb{C})$ is a conjugate-kernel-image preserving map if

$$\operatorname{Ker} \theta(A) = \operatorname{Ker} \overline{A} \quad \text{and} \quad \operatorname{Im} \theta(A) = \operatorname{Im} \overline{A}$$

for all $A \in M_n(\mathbb{C})$. Here, $\overline{A} = \overline{[a_{ij}]} = [\overline{a_{ij}}]$ is the matrix obtained from A by applying the complex conjugation entrywise.

The goal of this section is to prove the following theorem.

THEOREM 4.1. *Let $\mathbb{F}$ be either the field of real numbers or the field of complex numbers and $n \geq 3$ a positive integer. Assume that $\phi : M_n(\mathbb{F}) \to M_n(\mathbb{F})$ is a bijective continuous map such that for every pair $A, B \in M_n(\mathbb{F})$,*

$$AB = 0 \implies \phi(A)\phi(B) = 0.$$

Then either there exist a continuous bijective kernel-image preserving map $\theta : M_n(\mathbb{F}) \to M_n(\mathbb{F})$ and an invertible matrix $T \in M_n(\mathbb{F})$ such that

$$\phi(A) = T\theta(A)T^{-1}$$

for every $A \in M_n(\mathbb{F})$, or $\mathbb{F} = \mathbb{C}$ and there exist a continuous bijective conjugate-kernel-image preserving map $\theta : M_n(\mathbb{C}) \to M_n(\mathbb{C})$ and an invertible matrix $T \in M_n(\mathbb{C})$ such that
$$\phi(A) = T\theta(A)T^{-1}$$
for every $A \in M_n(\mathbb{C})$.

PROOF. Obviously, $\phi(0) = 0$. Indeed, we have $\phi(A)\phi(0) = 0$ for every $A \in M_n(\mathbb{F})$ and therefore, by the surjectivity of ϕ we see that $B\phi(0) = 0$ for every $B \in M_n(\mathbb{F})$, and thus, $\phi(0) = 0$.

Our next claim is that for every nonzero $x \in \mathbb{F}^n$ there exists a nonzero $y \in \mathbb{F}^n$ such that
$$\phi(R(x)) \subset R(y).$$
Without loss of generality we can assume that $x = e_1$, that is, $R(x) = R(e_1)$ is the set of all matrices of the form
$$\begin{bmatrix} * & 0 & \ldots & 0 \\ * & 0 & \ldots & 0 \\ \vdots & \vdots & \ddots & \vdots \\ * & 0 & \ldots & 0 \end{bmatrix}.$$
We denote by $\mathcal{V}$ the linear subspace of $M_n(\mathbb{F})$ consisting of all matrices of the form
$$\begin{bmatrix} 0 & 0 & \ldots & 0 \\ * & * & \ldots & * \\ \vdots & \vdots & \ddots & \vdots \\ * & * & \ldots & * \end{bmatrix}.$$
Clearly, $AB = 0$ for every $A \in R(e_1)$ and every $B \in \mathcal{V}$.

Let $A_0 \in R(e_1)$ be any nonzero matrix. Since ϕ is injective we have $\phi(A_0) \neq 0$. We can find invertible matrices $P, Q \in M_n(\mathbb{F})$ such that
$$P\phi(A_0)Q = \begin{bmatrix} I_r & 0 \\ 0 & 0 \end{bmatrix},$$
where $r > 0$ is the rank of A_0. Since $\phi(A_0)\phi(B) = 0$ for every $B \in \mathcal{V}$ we have
$$Q^{-1}\phi(B) = \begin{bmatrix} 0_r & 0 \\ * & * \end{bmatrix}, \quad B \in \mathcal{V},$$
where 0_r stands for the $r \times r$ zero matrix. Hence, the $n(n-1)$-dimensional space $\mathcal{V}$ is mapped by ϕ injectively and continuously into some $n(n-r)$-dimensional subspace of $M_n(\mathbb{F})$. By the invariance of domain theorem this is possible only if $r = 1$.

Hence, the map $X \mapsto Q^{-1}\phi(X)$ is an injective continuous map from $\mathcal{V}$ into itself, and since the zero matrix is mapped into itself and this map is open by the invariance of domain theorem, we can find a positive real number ε and $B_0 \in \mathcal{V}$ such that
$$Q^{-1}\phi(B_0) = \begin{bmatrix} 0 & 0 & \ldots & 0 & 0 \\ \varepsilon & 0 & \ldots & 0 & 0 \\ 0 & \varepsilon & \ldots & 0 & 0 \\ \vdots & \vdots & \ddots & \vdots & \vdots \\ 0 & 0 & \ldots & \varepsilon & 0 \end{bmatrix}.$$

Since for every $A \in R(e_1)$ we have $\phi(A)QQ^{-1}\phi(B_0) = 0$ we conclude that $\phi(A)Q \in R(e_1)$. In other words, $\phi(R(e_1)) \subset R(e_1 Q^{-1})$.

Thus, we have shown that for every nonzero $x \in \mathbb{F}^n$ there exists a nonzero $y \in \mathbb{F}^n$ such that $\phi(R(x)) \subset R(y)$, and similarly, for every nonzero ${}^t u \in {}^t\mathbb{F}^n$ there exists a nonzero ${}^t v \in {}^t\mathbb{F}^n$ such that $\phi(L({}^t u)) \subset L({}^t v)$.

Let $x, u \in \mathbb{F}^n$ be any linearly independent vectors. Assume that we have $\phi(R(x)) \subset R(y)$ and $\phi(R(u)) \subset R(y)$ for some nonzero $y \in \mathbb{F}^n$. All three subspaces $R(x)$, $R(y)$, and $R(u)$ are homeomorphic to $\mathbb{F}^n$ and thus, by the invariance of domain theorem, both $\phi(R(x))$ and $\phi(R(u))$ are open subsets in $R(y)$ and both of these open subsets contain the zero matrix. But then $\phi(R(x)) \cap \phi(R(u))$ is a nonempty open set contradicting the bijectivity of ϕ and the fact that
$$R(x) \cap R(u) = \{0\}.$$

Hence, ϕ induces an injective map $\alpha : \mathbb{P}(\mathbb{F}^n) \to \mathbb{P}(\mathbb{F}^n)$ such that for any nonzero $x, y \in \mathbb{F}^n$ we have
$$\alpha([x]) = ([y]) \iff \phi(R(x)) \subset R(y).$$
Similarly, there is an injective map $\beta : \mathbb{P}({}^t\mathbb{F}^n) \to \mathbb{P}({}^t\mathbb{F}^n)$ such that for any nonzero ${}^t u, {}^t v \in {}^t\mathbb{F}^n$ we have
$$\beta([{}^t u]) = ([{}^t v]) \iff \phi(L({}^t u)) \subset L({}^t v).$$

Let $x, y \in \mathbb{F}^n$ be nonzero vectors such that $\phi(R(x)) \subset R(y)$, and let k be an integer, $1 \le k \le n$. Assume that $U \subset R(x)$ is a linear subspace of dimension k. Then we can find vectors ${}^t u_1, \ldots, {}^t u_k \in {}^t\mathbb{F}^n$ such that
$${}^t u_1 x, \ldots, {}^t u_k x \in U,$$
and if we denote
$$\phi({}^t u_j x) = {}^t v_j y, \quad j = 1, \ldots, k,$$
then ${}^t v_1, \ldots, {}^t v_k$ are linearly independent. Indeed, if this was not the case then $\phi(U)$ would be contained in some $(k-1)$-dimensional subspace of $R(y)$ contradicting the invariance of domain theorem.

In the next step we will prove that for any nonzero $x, y, z \in \mathbb{F}^n$ we have
$$[x] \subset [y] + [z] \implies \alpha([x]) \subset \alpha([y]) + \alpha([z]).$$
The case when y and z are linearly dependent is trivial. So, assume they are linearly independent. Denote $\alpha([x]) = [x']$, $\alpha([y]) = [y']$, and $\alpha([z]) = [z']$. Choose nonzero vectors $w, t \in \mathbb{F}^n$ satisfying $\phi(R(w)) \subset R(t)$. Then the linear subspace $W \subset R(w)$ of all matrices A satisfying $yA = zA = 0$ is of dimension $n-2$, and by the previous paragraph we can find linearly independent vectors ${}^t a_1, \ldots, {}^t a_{n-2}$ such that
$${}^t a_1 t, \ldots, {}^t a_{n-2} t \in \phi(W).$$
From $yA = zA = 0$, $A \in W$, we conclude that $xA = 0$ for every $A \in W$, and consequently,
$$x'\, {}^t a_j = y'\, {}^t a_j = z'\, {}^t a_j = 0, \quad j = 1, \ldots, n-2.$$
Since ${}^t a_1, \ldots, {}^t a_{n-2}$ are linearly independent and y' and z' are linearly independent we conclude that x' belongs to the linear span of y' and z', as desired.

Thus, we can apply the fundamental theorem of projective geometry to find a field endomorphism $\tau : \mathbb{F} \to \mathbb{F}$ and an invertible $n \times n$ matrix T (the invertibility follows from the fact described in the paragraph before the previous one) such that
$$\alpha([x]) = [x_\tau T], \quad x \in \mathbb{F}^n \setminus \{0\}.$$

We repeat the same arguments for β and then conclude that there exists another field endomorphism $\sigma : \mathbb{F} \to \mathbb{F}$ and an invertible $n \times n$ matrix S such that for every rank one matrix ${}^t yx \in M_n(\mathbb{F})$ we have
$$\phi({}^t yx) = \lambda S\, {}^t y_\sigma x_\tau T$$
for some nonzero scalar λ (note that rank one matrix cannot be mapped to the zero matrix because ϕ is bijective).

Take any sequence (μ_n) of complex numbers converging to 0. Then
$${}^t e_1(e_1 + \mu_n e_2) \to {}^t e_1 e_1,$$
and if we denote
$$\phi({}^t e_1 e_1) = \lambda_0 S\, {}^t e_1 e_1 T,$$
then
$$\phi({}^t e_1(e_1 + \mu_n e_2)) = \lambda_n (S\, {}^t e_1)(e_1 + \tau(\mu_n)e_2) T \to \lambda_0 S\, {}^t e_1 e_1 T,$$
yielding that $\lambda_n \to \lambda_0$, which together with
$$\lambda_n \tau(\mu_n) \to 0$$
implies that $\tau(\mu_n)$ converges to 0. This yields that τ is continuous at 0, and hence continuous everywhere. It follows that τ is either the identity or the complex conjugation (of course, in the real case the only endomorphism is the identity and therefore this last paragraph could be omitted if we were interested in the real case only). Similarly, σ is either the identity or the complex conjugation.

Using the same ideas as in the proof of Lemma 3.2 we show that either both τ and σ are the identity, or they are both the complex conjugation, and $S = cT^{-1}$ for some nonzero scalar c. After composing ϕ with a similarity transformation $X \mapsto TXT^{-1}$, and the map $X \mapsto \overline{X}$, if necessary, we may assume with no loss of generality that for every rank one matrix R there exists a nonzero scalar λ (depending on R) such that
$$\phi(R) = \lambda R.$$

It remains to show that for any pair of subspaces $U, V \in \mathbb{F}^n$ satisfying $\dim U + \dim V = n$ we have
$$\phi(\mathcal{M}(U, V)) = \mathcal{M}(U, V),$$
where
$$\mathcal{M}(U, V) = \{A \in M_n(\mathbb{D}) : \operatorname{Im} A \subset U \text{ and } V \subset \operatorname{Ker} A\}.$$
Once we know this, an easy inductive argument shows that then actually
$$\phi(\mathcal{S}(U, V)) = \mathcal{S}(U, V),$$
which completes the proof.

Thus, let $U, V \in \mathbb{F}^n$ be subspaces satisfying $\dim U + \dim V = n$. Set $\dim U = r$ and choose invertible matrices $P, Q \in M_n(\mathbb{F})$ such that $UQ = \operatorname{span}\{e_1, \ldots, e_r\}$ and $VP^{-1} = \operatorname{span}\{e_{r+1}, \ldots, e_n\}$. Then $A \in \mathcal{M}(U, V)$ if and only if
$$\operatorname{Im}(PAQ) = \operatorname{Im}(AQ) \subset UQ = \operatorname{span}\{e_1, \ldots, e_r\}$$
and
$$\operatorname{Ker}(PAQ) = \operatorname{Ker}(PA) \supset VP^{-1} = \operatorname{span}\{e_{r+1}, \ldots, e_n\},$$
that is, we have $A \in \mathcal{M}(U, V)$ if and only if the matrix PAQ has the block form
$$PAQ = \begin{bmatrix} * & 0 \\ 0 & 0 \end{bmatrix},$$

where $*$ stands for some $r \times r$ matrix. If $A \in \mathcal{M}(U,V)$ and R is any rank one matrix of the form
$$R = \begin{bmatrix} 0 & 0 \\ * & * \end{bmatrix},$$
where the blocks are of the same size as in the block matrix representation of PAQ, then $PAQR = 0$, and therefore $AQR = 0$ which further yields that $0 = \phi(A)\phi(QR) = \phi(A)\lambda(QR)$ for some nonzero scalar λ. Thus, $(P\phi(A)Q)R = 0$ for every such rank one matrix R yielding that $P\phi(A)Q$ is of the form
$$P\phi(A)Q = \begin{bmatrix} * & 0 \\ * & 0 \end{bmatrix}.$$
We repeat the same trick with rank one matrices but this time with multiplication on the left side to conclude that
$$P\phi(A)Q = \begin{bmatrix} * & 0 \\ 0 & 0 \end{bmatrix}.$$
In other words, we have
$$\phi(\mathcal{M}(U,V)) \subset \mathcal{M}(U,V).$$

Clearly, $\mathcal{M}(U,V)$ is homeomorphic to $\mathbb{F}^{r^2}$. By the invariance of domain theorem, $\phi(\mathcal{M}(U,V))$ is open in $\mathcal{M}(U,V)$. We need to prove that it is also closed. Once we will do this we will know that we have the equality
$$\phi(\mathcal{M}(U,V)) = \mathcal{M}(U,V),$$
as desired.

Thus, the last step in the proof is to show that $\phi(\mathcal{M}(U,V))$ is closed in $\mathcal{M}(U,V)$. By the invariance of domain theorem the bijective map ϕ maps open sets into open sets, and therefore closed sets into closed sets, and thus $\phi(\mathcal{M}(U,V))$ is closed in $M_n(\mathbb{F})$. It follows that $\phi(\mathcal{M}(U,V)) \subset \mathcal{M}(U,V)$ is closed in $\mathcal{M}(U,V)$. □

5. The case where ϕ is additive

In this final section, we consider the case where $\phi : M_n(\mathbb{D}) \to M_n(\mathbb{D})$, with $n \geq 2$ and $\mathbb{D}$ a division ring, is an *additive* zero product preserving map. There are two obvious types of such maps. The first type consists of additive maps ϕ satisfying
$$\phi(A)\phi(B) = 0$$
for all $A, B \in M_n(\mathbb{D})$. The second type consists of maps of the form
$$\phi(A) = C\phi_0(A)$$
where ϕ_0 is a ring endomorphism of $M_n(\mathbb{D})$ (i.e., a multiplicative additive map from $M_n(\mathbb{D})$ to itself) and C is a matrix commuting with each $\phi_0(A)$. Our goal is to show that there are no other types than these two.

Some comments concerning the second type are in order. First, every ring endomorphism ϕ_0 of $M_n(\mathbb{D})$ is of the form
$$\phi_0(A) = TA_\tau T^{-1},$$
where T is an invertible matrix in $M_n(\mathbb{D})$ and τ is a ring endomorphism of $\mathbb{D}$ (and, as always, $A_\tau = [a_{ij}]_\tau = [\tau(a_{ij})]$). We believe this is a folklore result; in any case, it can be proved by standard methods. From this it follows easily that C is a scalar matrix, $C = cI$, where $c \in D$ satisfies $c\tau(a) = \tau(a)c$ for all $a \in \mathbb{D}$. As the next

example shows, this does not necessarily mean that c lies in the center of $\mathbb{D}$ (and hence C does not necessarily lie in the center of $M_n(\mathbb{D})$).

EXAMPLE 5.1. Let W be the Weyl algebra over a characteristic 0 field F in countably infinitely many variables. That is, W is the algebra generated by x_i, y_i, $i = 1, 2, \ldots$, and relations
$$x_i y_j - y_j x_i = \delta_{ij}, \quad x_i x_j = x_j x_i, \quad y_i y_j = y_j y_i$$
for all i, j. It is well-known that a Weyl algebra in finitely many variables is a (right and left) Ore domain [**B2**, Example 7.21]; since every pair of elements in W lies in such a (sub)algebra, W is an Ore domain as well. The classical (right or left) ring of quotients of W is therefore a division ring, which we denote by $\mathbb{D}$ (see [**B2**, Section 7.3] for details). Let $\tau_0 : W \mapsto W$ be the algebra endomorphism determined by
$$\tau_0(x_i) = x_{i+1} \quad \text{and} \quad \tau_0(y_i) = y_{i+1}.$$
We can extend τ_0 to an endomorphism τ of $\mathbb{D}$ [**B2**, Proposition 7.14]. It is clear that $c = x_1$ commutes with every element in $\tau(\mathbb{D})$. However, c does not commute with y_1 and so does not belong to the center of $\mathbb{D}$.

Our approach is based on the concept of a *zero product determined ring* [**BGS**]. This is a ring, let us call it R, with the following property: if G is an additive Abelian group and $\beta : R \times R \to G$ is an additive map satisfying $\beta(x, y) = 0$ whenever $xy = 0$, then there exists an additive map $\alpha : R \to G$ such that $\beta(x, y) = \alpha(xy)$ for all $x, y \in R$. If S is any unital ring and $n \geq 2$, then $R = M_n(S)$ is a zero product determined ring [**BGS**, Theorem 2.1] (see also [**B1**, Section 4] for a more conceptual proof). This is clearly applicable to our problem. Indeed, if $\phi : M_n(\mathbb{D}) \to M_n(\mathbb{D})$, $n \geq 2$, is an additive zero product preserving map, then the map $\beta : M_n(\mathbb{D}) \times M_n(\mathbb{D}) \to M_n(\mathbb{D})$ given by $\beta(A, B) = \phi(A)\phi(B)$ satisfies $\beta(A, B) = 0$ whenever $AB = 0$, and hence there exists an additive map $\alpha : M_n(\mathbb{D}) \to M_n(\mathbb{D})$ such that
$$\phi(A)\phi(B) = \beta(A, B) = \alpha(AB)$$
for all $A, B \in M_n(\mathbb{D})$. (It should be mentioned that the existence of a map α satisfying the last formula can be also extracted from the arguments in [**CKLW**, Section 2].)

Our last theorem reads as follows.

THEOREM 5.2. *Let $n \geq 2$ be an integer and $\mathbb{D}$ any division ring. Let $\phi : M_n(\mathbb{D}) \to M_n(\mathbb{D})$ be an additive map such that for every pair $A, B \in M_n(\mathbb{D})$,*
$$AB = 0 \implies \phi(A)\phi(B) = 0.$$
Then either
$$\phi(A)\phi(B) = 0$$
for all $A, B \in M_n(\mathbb{D})$ or there exist $C \in M_n(\mathbb{D})$ and a ring endomorphism ϕ_0 of $M_n(\mathbb{D})$ such that
$$\phi(A) = C\phi_0(A) = \phi_0(A)C$$
for all $A \in M_n(\mathbb{D})$.

PROOF. As we have just explained, there exists an additive map $\alpha : M_n(\mathbb{D}) \to M_n(\mathbb{D})$ such that
$$\alpha(AB) = \phi(A)\phi(B)$$

for all $A, B \in M_n(\mathbb{D})$. Setting
$$C = \phi(I)$$
we clearly have
$$\alpha(A) = \phi(A)C = C\phi(A)$$
for all $A \in M_n(\mathbb{D})$, and hence
$$C\phi(AB) = \phi(A)\phi(B)$$
for all $A, B \in M_n(\mathbb{D})$.

If C is invertible then
$$\phi_0(A) = C^{-1}\phi(A)$$
defines a ring endomorphism of $M_n(\mathbb{D})$ that clearly satisfies the conditions from the statement of the theorem. Assume, therefore, that C is not invertible. Then C has a nontrivial kernel. From $C\phi(AB) = \phi(A)\phi(B)$ we thus see that the intersection of the kernels of all $\phi(A)$ is nontrivial. Choosing an appropriate basis we may therefore assume that the n-th row of each matrix $\phi(A)$ is zero. That is, writing
$$\phi(A) = [\phi_{ij}(A)]$$
where ϕ_{ij} are additive maps from $M_n(\mathbb{D})$ to $\mathbb{D}$, we have $\phi_{nj} = 0$ for every j. Similarly, let us write
$$\alpha(A) = [\alpha_{ij}(A)]$$
where $\alpha_{ij} : M_n(\mathbb{D}) \to \mathbb{D}$ are additive maps. Suppose $\alpha \neq 0$. Then $\alpha_{rs} \neq 0$ for some r and s. From $\phi(A)\phi(B) = \alpha(AB)$ we infer that
$$\sum_{k=1}^{n-1} \phi_{rk}(A)\phi_{ks}(B) = \alpha_{rs}(AB).$$
Since α_{rs} is additive, there exists a matrix unit E_{pq} and $d \in \mathbb{D}$ such that $\alpha_{rs}(dE_{pq}) \neq 0$. Writing dE_{pi} for A and E_{jq} for B in the above identity we thus obtain
$$\sum_{k=1}^{n-1} \phi_{rk}(dE_{pi})\phi_{ks}(E_{jq}) = \delta_{ij}\alpha_{rs}(dE_{pq}).$$
That is, the vectors
$$x_i = \begin{bmatrix} \phi_{r1}(dE_{pi}) & \phi_{r2}(dE_{pi}) & \cdots & \phi_{r,n-1}(dE_{pi}) \end{bmatrix} \in \mathbb{D}^{n-1}, \ i = 1, \ldots, n,$$
and
$$y_j = \begin{bmatrix} \phi_{1s}(E_{jq}) & \phi_{2s}(E_{jq}) & \cdots & \phi_{n-1,s}(E_{jq}) \end{bmatrix} \in \mathbb{D}^{n-1}, \ j = 1, \ldots, n,$$
satisfy
$$x_i \, {}^t y_j = 0 \iff i \neq j$$
for all $i, j = 1, \ldots, n$. However, this is impossible since $x_1, \ldots, x_n$, being vectors in an $(n-1)$-dimensional vector space, are linearly dependent. This contradiction shows that $\alpha = 0$, yielding that $\phi(A)\phi(B) = 0$ for all $A, B \in M_n(\mathbb{D})$. $\square$

Theorem 5.2 is a generalization of [**CKLW**, Corollary 2.4] and [**B1**, Corollary 5.2].

References

[ABEV]   J. Alaminos, M. Brešar, J. Extremera, and A. R. Villena, *Maps preserving zero products*, Studia Math. **193** (2009), no. 2, 131–159, DOI 10.4064/sm193-2-3. MR2515516

[B1]   M. Brešar, *Multiplication algebra and maps determined by zero products*, Linear Multilinear Algebra **60** (2012), no. 7, 763–768, DOI 10.1080/03081087.2011.564580. MR2929643

[B2]   M. Brešar, *Introduction to noncommutative algebra*, Universitext, Springer, Cham, 2014. MR3308118

[BGS]   M. Brešar, M. Grašič, and J. S. Ortega, *Zero product determined matrix algebras*, Linear Algebra Appl. **430** (2009), no. 5-6, 1486–1498, DOI 10.1016/j.laa.2007.11.018. MR2490691

[CKLW]   M. A. Chebotar, W.-F. Ke, P.-H. Lee, and N.-C. Wong, *Mappings preserving zero products*, Studia Math. **155** (2003), no. 1, 77–94, DOI 10.4064/sm155-1-6. MR1961162

[F]   C.-A. Faure, *An elementary proof of the fundamental theorem of projective geometry*, Geom. Dedicata **90** (2002), 145–151, DOI 10.1023/A:1014933313332. MR1898158

[K]   H. Kestelman, *Automorphisms of the field of complex numbers*, Proc. London Math. Soc. (2) **53** (1951), 1–12, DOI 10.1112/plms/s2-53.1.1. MR0041206

[S1]   P. Šemrl, *Generalized symmetry transformations on quaternionic indefinite inner product spaces: an extension of quaternionic version of Wigner's theorem*, Comm. Math. Phys. **242** (2003), no. 3, 579–584, DOI 10.1007/s00220-003-0957-7. MR2020281

[S2]   P. Šemrl, *Maps on idempotents*, Studia Math. **169** (2005), no. 1, 21–44, DOI 10.4064/sm169-1-2. MR2139640

Faculty of Mathematics and Physics, University of Ljubljana, and Faculty of Natural Sciences and Mathematics, University of Maribor, Slovenia
Email address: matej.bresar@fmf.uni-lj.si

Faculty of Mathematics and Physics, University of Ljubljana, Slovenia
Email address: peter.semrl@fmf.uni-lj.si

Published Titles in This Subseries

750 **A. Bourhim, J. Mashreghi, L. Oubbi, and Z. Abdelali, Editors,** Linear and Multilinear Algebra and Function Spaces, 2020

743 **H. Garth Dales, Dmitry Khavinson, and Javad Mashreghi, Editors,** Complex Analysis and Spectral Theory, 2020

680 **Sergei Gukov, Mikhail Khovanov, and Johannes Walcher, Editors,** Physics and Mathematics of Link Homology, 2016

655 **A. C. Cojocaru, C. David, and F. Pappalardi, Editors,** SCHOLAR—a Scientific Celebration Highlighting Open Lines of Arithmetic Research, 2015

654 **Carlo Gasbarri, Steven Lu, Mike Roth, and Yuri Tschinkel, Editors,** Rational Points, Rational Curves, and Entire Holomorphic Curves on Projective Varieties, 2015

638 **Javad Mashreghi, Emmanuel Fricain, and William Ross, Editors,** Invariant Subspaces of the Shift Operator, 2015

630 **Pierre Albin, Dmitry Jakobson, and Frédéric Rochon, Editors,** Geometric and Spectral Analysis, 2014

622 **S. Ejaz Ahmed, Editor,** Perspectives on Big Data Analysis, 2014

606 **Chantal David, Matilde Lalín, and Michelle Manes, Editors,** Women in Numbers 2, 2013

605 **Omid Amini, Matthew Baker, and Xander Faber, Editors,** Tropical and Non-Archimedean Geometry, 2013